Techniques and Mechanisms in Electrochemistry

Techniques and Mechanisms in Electrochemistry

P.A. Christensen

and

A. Hamnett

Department of Chemistry
University of Newcastle upon Tyne

BLACKIE ACADEMIC & PROFESSIONAL
An Imprint of Chapman & Hall
London · Glasgow · New York · Tokyo · Melbourne · Madras

Published by
Blackie Academic & Professional, an imprint of Chapman & Hall,
Wester Cleddens Road, Bishopbriggs, Glasgow G64 2NZ

Chapman & Hall, 2–6 Boundary Row, London SE1 8HN, UK

Blackie Academic & Professional, Wester Cleddens Road, Bishopbriggs, Glasgow G64 2NZ, UK

Chapman & Hall Inc., One Penn Plaza, 41st Floor, New York NY 10119, USA

Chapman & Hall Japan, Thomson Publishing Japan, Hirakawacho Nemoto Building, 6F, 1-7-11 Hirakawa-cho, Chiyoda-ku, Tokyo 102, Japan

DA Book (Aust.) Pty Ltd., 648 Whitehorse Road, Mitcham 3132, Victoria, Australia

Chapman & Hall India, R. Seshadri, 32 Second Main Road, CIT East, Madras 600 035, India

First edition 1994

© 1994 Chapman & Hall

Typeset in 10/12 pt Times New Roman by
Thomson Press (India) Ltd., New Delhi

Printed in Great Britain at the Alden Press, Oxford

ISBN   0 7514 0129 3

A catalogue record for this book is available from the British Library

Library of Congress Cataloging-in-Publication data
Christensen, P.A., 1960–
    Techniques and mechanisms in electrochemistry / P.A. Christensen and A. Hamnett. -- 1st ed.
        p.   cm.
    Includes bibliographical references and index.
    ISBN 0-7514-0129-3 (PB : acid-free)
    1. Electrochemistry. 2. Surface chemistry.   I. Hamnett, A.
    II. Title.
    QD553.C59   1993
    541.3'7--dc20
                                                93-11366
                                                CIP

Preface

It is hard to overstate the importance of electrochemistry in the modern world: the ramifications of the subject extend into areas as diverse as batteries, fuel cells, effluent remediation and re-cycling, clean technology, electro-synthesis of organic and inorganic compounds, conversion and storage of solar energy, semiconductor processing, material corrosion, biological electron transfer processes and a wide range of highly specific analytical techniques. The impact of electrochemistry on the lives of all of us has increased immeasurably, even in recent years, but this increase has not been reflected in the level or content of courses taught at universities, many of which portray the subject as a collection of arcane recipes and poorly understood formulae of marginal importance to the mainstream of chemistry.

This approach reached its nadir with the recent extraordinary furore surrounding the purported discovery of cold fusion, where two electrochemists claimed to have shown that the fusion of deuterium nuclei could be effected under ambient conditions by the electrochemically induced intercalation of deuterium atoms into palladium. Whatever the truth behind such claims, their discussion revealed a lamentable lack of knowledge of modern electro-chemistry, not only among science writers for the popular press, but among many professional chemists and physicists whose acquaintance with the subject seems, for the most part, to have stopped somewhere about the time of Nernst. In a year in which Professor R. Marcus has been awarded the Nobel Prize for his work in electron transfer, the status of the subject seems, at last, to be on the rise, and it is our intention in this book to present the subject as an integral part of modern physical chemistry, shorn of both excessive mathematical complexity and over-reliance on techniques solely employing electrical measurement.

Electrochemistry is, first and foremost, a branch of surface science, and is specifically concerned with the electrified interface, that is the behaviour of surfaces in contact with a liquid and in the presence of an electric field, and includes the effect of the field on the structural and chemical properties of the interface. As with all surface science, the *structure* of the surface is expected to play an important role in determining its properties and modern studies have borne this out. However, it is the chemical reactivity of the surface as a function of electric field that is the core subject matter of electrochemistry and the one of most concern to us in this text.

In the most straightforward of cases, the existence of an electric field simply leads to a charge separation across the interface. This charge separation is particularly significant for the physico-chemical properties of small charged particles, and its quantitative understanding forms the basis of modern colloid science. For macroscopic surfaces, the charge and field distributions are likely to have a major impact on the rate of adsorption and resultant orientation of molecular species and on the rate of electron transfer to and from such species across the interface. It is this latter quantity that is central to modern electrochemistry and this text is written primarily to illustrate how the understanding of the electrified interface and its connection to other aspects of surface chemistry has been enormously enhanced in recent years. We will concentrate on the electrified liquid–solid interface but it should be emphasised that this is *not* the only type of electrified interface known; it is, however, overwhelmingly the most important from the standpoint of the technology referred to above.

Our treatment consists of three chapters, the first of which sets the scene and also acts as a summary of many of the techniques and aims set out in detail later in the book. The second covers the techniques that are of most importance in modern electrochemistry, the third covers some more complex specific reactions where a multiplicity of techniques have been used.

This book is written primarily for mid to advanced level undergraduates and postgraduates of chemistry but is also intended for electrochemists working in any of the vast range of industries exploiting electrochemical technology.

The authors would like to thank all those who have kindly contributed material for this book, particularly Professor A.A. Gewirth, Dr H.A.O. Hill, F.R.S. and Professor A.R. Hillman. We would also like to thank Professor R. Parsons, F.R.S. for his invaluable help in elucidating the history of the platinum cyclic voltammogram, and Daniel Read for doing a really special job of reading the entire manuscript. Finally, we would like to thank Dean and Margot for keeping the wheels well oiled.

P.A.C.
A.H.

Contents

3   Examples of the application of electrochemical methods   228

Index   375

List of symbols

Symbols most commonly used in this text. Where a symbol has been employed more than once the minor usage is given in brackets.

α asymmetry parameter for a multi-electron process (polarisability, Azimuthal angle)

Å   Ångstroms, 10^{-10} m

B   magnetic field strength, Tesla, T

β asymmetry parameter for a one-electron process

C   coulomb, the unit of electric charge

$\left.\begin{matrix} c_i \\ [i] \end{matrix}\right\}$ concentration of species i in mole dm^{-3}

c^0 standard concentration $= 1$ mole dm^{-3}

E   electric field strength, volts m^{-1}

E electromotive force, emf (in an electrochemical cell); equilibrium electrode potential in volts

E^0 Standard emf; standard equilibrium electrode potential

E_r equilibrium electrode rest potential

E   Non-equilibrium electrode potential

$\left.\begin{matrix} E_{p,a} \\ E_{p,c} \end{matrix}\right\}$ potential at which a peak in the anodic or cathodic current occurs

ε relative permittivity or dielectric constant

ε_0 permittivity of free space, $8.854 \times 10^{-12}\,F\,m^{-1}$

e_0 charge on the electron, $1.6022 \times 10^{-10}\,C$

eV   electron volts, units of energy $= 8056\,cm^{-1}$

η activation overpotential

η_1 solvent viscosity, $kg\,m^{-1}\,s^{-1}$

F Faraday, 96 487 coulombs

f F/RT (the frequency of rotation in hertz)

Γ surface excess molar concentration of any species in the interfacial region (Lorentzian broadening parameter)

γ surface tension

Hz   units of rotation, hertz $=$ cycles per second

h Planck's constant, $6.6261 \times 10^{-34}\,J\,s$

$\hbar$ $h/2\pi$

I electric current, amperes, A (light intensity, nuclear spin)

$\left.\begin{matrix} I_{p,a} \\ I_{p,c,} \end{matrix}\right\}$ peak anodic and cathodic current

i tunnelling current; the square root of -1

J ionic flux, mole $cm^{-2}\,s^{-1}$

J current density, $A\,cm^{-2}$

κ screening constant, m^{-1}

$1/\kappa$ screening length, m

K equilibrium constant

k_B	Boltzmann constant, $1.3807 \times 10^{-23} \, J\,K^{-1}$
k	rate constant
k^0	standard rate constant
$\log$	logarithm to base 10
$\log_e$	logarithm to base e
λ	wavelength, nanometres (10^{-9} m)
m_0	electron mass, $9.1094 \times 10^{-31} \, kg$
m^*	electron effective mass
μ	chemical potential (dipole moment, ionic mobility)
μ_B	the Bohr magneton, $9.2740 \times 10^{-24} \, J\,T^{-1}$
N	number of species
v	kinematic viscosity $= \eta_1/\rho$ (where $\rho =$ density) (frequency of light)
$\bar{v}$	wavenumber, $1/\lambda$, cm^{-1}
ω	angular velocity ($2\pi f$, where $f =$ frequency of rotation in hertz)
Ω	ohms, units of resistance
P	polarisation, nett dipole moment densities per unit volume
ϕ	electrode potential, volts
$\left.\begin{array}{l} p_{\Delta E} \\ p_\eta \end{array}\right\}$	reaction order
p	pressure, $N\,m^{-2}$
p^0	The pressure corresponding to 1 atmosphere, $1.01 \times 10^5 \, N\,m^{-2}$
Q	electric charge, coulombs
R	gas constant, $8.3145 \, J\,mol^{-1}\,K^{-1}$
r_+, r_-	ionic radius
S	magnitude and direction of the spin angular momentum
ρ	charge density, $C\,m^{-3}$ (density, $g\,cm^{-3}$)
σ	surface charge density, $C\,m^{-2}$ (conductivity, $\Omega^{-1}\,cm^{-1}$)
T	temperature, K
t	time, s
θ	fractional coverage of an adsorbed species
v	rate of reaction or scan rate (velocity, volume of a single molecule)
V	voltage applied to a cell
ξ	asymmetry parameter for Temkin adsorption isotherm kinetics
z	numerical charge on an ion
ze_0	charge on an ion in coulombs

Potentials of reference electrodes in aqueous solutions at 25°C (from *Reference Electrodes*, eds. D.J.G. Ives and G.J. Janz, Academic Press, New York, 1961)

	Potential of reference vs. NHE/volts
H_2/H^+ ($[H^+] = 1$ M): normal hydrogen electrode (NHE)	0
Hg/HgO 0.1 M NaOH: mercury/mercury oxide (MMO)	0.926
Hg/Hg_2SO_4 0.5 M H_2SO_4: mercury/mercury sulphate (MMS)	0.68
Hg/Hg_2SO_4 sat. K_2SO_4: mercury/mercury sulphate	0.64
Hg/Hg_2Cl_2 0.1 M KCl: calomel electrode	0.3337
Hg/Hg_2Cl_2 1 M KCl: normal calomel electrode (NCE)	0.2801
Hg/Hg_2Cl_2 sat. KCl: saturated calomel electrode (SCE)	0.2412
Hg/Hg_2Cl_2 sat. NaCl: saturated sodium calomel electrode (SSCE)	0.2360
$Ag/AgCl$ sat. KCl: silver/silver chloride electrode	0.197

The reversible hydrogen electrode (RHE) consists of Pt in contact with hydrogen at 1 atmosphere *in the same solution as that employed in the electrochemical cell*. On the RHE scale, therefore, hydrogen evolution always occurs at 0 V.

Introduction to modern electrochemistry

1

A solid–liquid interface will have three aspects to its structure; the atomic structure of the solid electrode, the structure of any adsorbed layer and the structure of the liquid layer above the electrode. All three of these are of fundamental importance in the understanding of the electron transfer processes at the core of electrochemistry and we must consider all three if we are to arrive at a fundamental understanding of the subject.

1.1
Structure of surfaces

1.1.1 Structure of the solid electrode

The surface of a solid will usually have a strong structural relationship to that of the underlying bulk. Even a polycrystalline material actually consists of crystalline grains separated by boundaries across which the lattice structure is discontinuous, and the surface will, in turn, show small areas of well-defined structure, again separated by boundaries. There are materials that exhibit extreme properties at either end of the polycrystalline range. The simplest example is that of a substance that is *monocrystalline* on a macroscopic scale. For such materials, the axes of the unit cell are in the same direction throughout the sample and the surface structure is, at least in principle, determined entirely by the angles made between the surface perpendicular and the three unit cell axes. At the other extreme, there may be no long-range structure at all, with only local coordination at all well defined. Such materials are termed amorphous and in recent years amorphous or 'glassy' metals have been extensively studied since they are unexpectedly resistant to corrosion.

Mono- or single-crystal materials are undoubtedly the most straightforward to handle conceptually, however, and we start our consideration of electrochemistry by examining some simple substances to show how the surface structure follows immediately from the bulk structure; we will need this information in chapter 2, since modern single-crystal studies have shed considerable light on the mechanism of many prototypical electrochemical reactions. The great majority of electrode materials are either elemental metals or metal alloys, most of which have a face-centred or body-centred cubic structure, or one based on a hexagonal close-packed array of atoms.

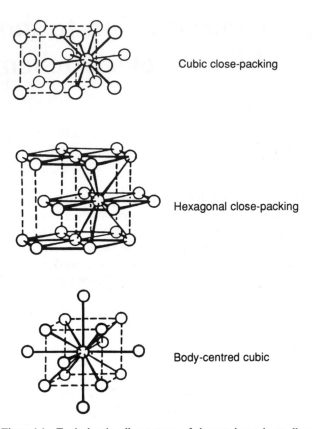

Cubic close-packing

Hexagonal close-packing

Body-centred cubic

Figure 1.1   Typical unit cell structures of elemental metals or alloys.

The corresponding unit cells are shown in Figure 1.1 and an examination of simple ball-and-stick models (which the reader is *strongly* urged to carry out) shows that the face-centred cubic (fcc) and hexagonal close-packed (hcp) structures correspond to the only two possible ways of close-packing spheres, in which each sphere has twelve nearest neighbours.

The direction of *planes* in a lattice is described in a manner that, at first sight, seems rather strange but which is, in fact, derived directly from standard techniques in 3-D geometry. In essence, the unit cell is drawn and the plane of interest translated until it intercepts all three axes *within* the unit cell but as far away from the origin as possible. The point of intersection of the plane with the axes then determines the label given to the plane; if the intersection takes place at a fraction $(1/h)$ of the a-axis, $(1/k)$ of the b-axis and $(1/l)$ of the c-axis, then the plane is referred to as the '(hkl) plane'. As indicated, this apparently rather strange method arises because if the axes a, b, c of the unit cell are mutually perpendicular, and of equal length, then the equation of any point x, y, z in the (hkl) plane can always be written:

$$hx + ky + lz = \text{constant} \tag{1.1}$$

Essentially, our method corresponds to choosing *hkl* values that are as small as possible but it is also conventional to choose values that have no common factor. The only problem with this method may arise if the plane is parallel to one or two unit cell axes. Under these conditions, the intersection point clearly has no meaning and we put the corresponding value of *h*, *k* or *l* equal to zero.

Once the values of *hkl* are found, then the arrangement of atoms on these surfaces is easily obtained, and Figure 1.2 shows the commonest *low-index* form of these surfaces. If the common surfaces of the fcc structure are examined, it will be seen that the surface structure changes quite remarkably. The (111) surface is clearly a close-packed structure but the (100) surface has a square arrangement of metal atoms and the fcc (110) surface, which shows grooves running parallel to the *c*-axis, is even more remarkable. The coordination of the surface atoms clearly is also very different, with the coordination evidently 9 in the (111) surface, 8 in the (100) surface and a remarkable 6 in the (110) surface, as compared to 12 in the bulk.

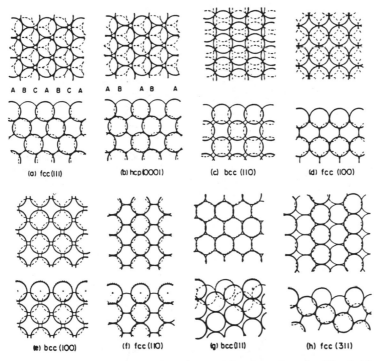

Figure 1.2   Atomic arrangement on various clean metal surfaces. In each of the sketches (a) to (h) the upper and lower diagrams represent top and side views, respectively. Atoms drawn with dashed lines lie behind the plane of those drawn with thick lines. Atoms in unrelaxed positions (i.e. in the positions they occupy in the bulk) are shown as dotted lines. From G.A. Somorjai, *Chemistry in Two Dimensions*, Cornell University Press, London, 1981, p. 133. For the Miller index convention in hexagonal close-packed structures, see also G.A. Somorjai loc. cit. Used by permission of Cornell University Press.

The rather low coordination in the (100) and (110) surfaces will clearly lead to some instability and it is perhaps not surprising that the ideal surface structures shown in Figure 1.2 are frequently found in a rather modified form in which the structure changes to increase the coordination number. Thus, the (100) surfaces of Ir, Pt and Au all show a topmost layer that is close-packed and buckled, as shown in Figure 1.3, and the (110) surfaces of these metals show a remarkable reconstruction in which one or more alternate rows in the $\langle 001 \rangle$ direction are removed and the atoms used to build up small facets of the more stable (111) surface, as shown in Figure 1.4. These reconstructions have primarily been characterised on bare surfaces under high-vacuum conditions and it is of considerable interest and importance to note that chemisorption on such reconstructed surfaces can cause them to snap back to the unreconstructed form even at room temperature. Recently, it has also been shown that reconstructions at the liquid–solid interface also

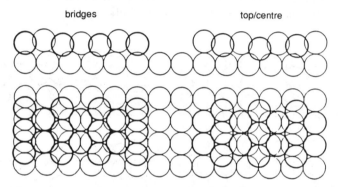

Figure 1.3 The arrangement of atoms on the reconstructed (100) crystal faces of gold, iridium and platinum. Side and top views are illustrated. From G.A. Somorjai, *Chemistry in Two Dimensions*, Cornell University Press, London, 1981, p. 145. Used by permission of Cornell University Press.

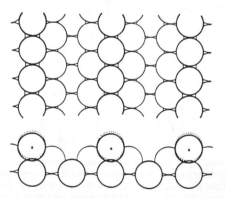

Figure 1.4 The atomic arrangement in the 'missing row' model of the reconstructed iridium (110) crystal surface. From G.A. Somorjai, *Chemistry in Two Dimensions*, Cornell University Press, London, 1981, p. 146. Used by permission of Cornell University Press.

take place, with gold providing some spectacular examples. Again, the presence of adsorbed electrolyte does seem to stabilise the unreconstructed surface under certain circumstances and this suppression of reconstruction can be understood in terms of the stabilisation of low-coordinate metal atoms through coordination to the solvent.

For compound materials the situation is both more complex and less well explored. Even the elemental semiconductors, such as silicon, show complex surface reconstructions on low-index faces under high-vacuum conditions, and compound semiconductors, such as GaAs, show even more intriguing behaviour. For the zinc-blende structure of GaAs, the As and Ga atoms occupy alternate layers in the $\langle 111 \rangle$ direction, so that (111) faces may either be all Ga or all As, the latter surface being denoted $(\bar{1}\bar{1}\bar{1})$ for clarity. These differences have macroscopic consequences, one of which is that dissolution in a suitable etchant can give a smooth surface for (111) but shows an 'orange peel' texture for $(\bar{1}\bar{1}\bar{1})$. For more polar materials, such as oxides, stable surfaces must be electrostatically neutral, since otherwise very large internal electrostatic fields will develop. Thus, the (001) surface of rutile oxides, such as TiO_2, is stable, whereas the (100) surface shows some complex reconstruction in part also facilitated by the ease of reduction of Ti^{4+} to Ti^{3+}. Stoichiometrically driven reconstructions are only now being explored and the whole area is one of considerable contemporary interest.

Also of great interest is the behaviour of higher index faces. Some examples of such surfaces are given in Figure 1.5 and a careful examination shows that such surfaces usually consist of terraces of close-packed atoms separated by either simple steps or by a combination of steps and kinks. Study of these surfaces has shown that such step and kink sites are highly active in promoting dissociative chemisorption and catalytic activity, and constitute 'active sites' on the surface. Furthermore, although the monatomic step-terrace structure is the most stable for the bare metal, chemisorption can lead to multiple steps and wider terraces, giving rise to a hill-and-valley appearance to the surface. The kinetics of formation of these types of surface have been an area of extensive study in the last few years; it is clear that these reconstructions must involve motion of atoms across the terraces to form kink sites, new steps and finally new terraces.

The existence of active sites on surfaces has long been postulated, but confidence in the geometric models of kink and step sites has only been attained in recent years by work on high index surfaces. However, even a lattice structure that is unreconstructed will show a number of random defects, such as vacancies and isolated adatoms, purely as a result of statistical considerations. What has been revealed by the modern techniques described in chapter 2 is the extraordinary *mobility* of surfaces, particularly at the liquid–solid interface. If the metal atoms can be stabilised by coordination, very remarkable atom mobilities across the terraces are found, with reconstruction on Au(100), for example, taking only minutes to complete at room temperature in chloride-containing electrolytes. It is now clear that the

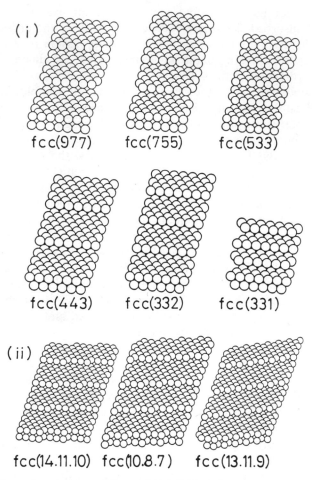

Figure 1.5   The surface structures of (i) several high-Miller index stepped surfaces with different terrace widths and step orientations; (ii) several high Miller-index surfaces with differing kink concentrations in the steps. From G.A. Somorjai, *Chemistry in Two Dimensions*, Cornell University Press, London, 1981, pp. 160 and 161. Used by permission of Cornell University Press.

approximation of considering the metal surface as a disinterested spectator in electrochemical reactions must be replaced by a model in which surface reconstructions are driven by the presence or absence of strongly adsorbed species in solution, and the surface can snap from one form to another with remarkable ease.

1.1.2   *The structure of the electrified interface*

The distribution of charge at an electrified interface is a central factor in electrochemical activity and has been carefully studied throughout the last

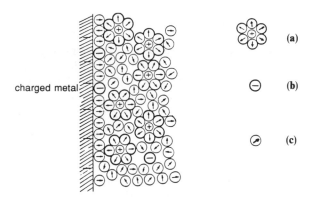

Figure 1.6 A schematic representation of the 'double layer' at the interface between an electrode and the electrolyte. (a) Solvated positive ion, (b) unsolvated negative ion and (c) water molecules showing dipole direction.

100 years. This is an area in which both physicists and chemists have made major contributions and the modern consensus for the structure of the electrified interface is shown in Figure 1.6. The evidence for the picture shown in Figure 1.6 is reviewed in detail in chapter 2 but the following points need to be made:

1. The total charge integrated over the entire interface comprising electrode and associated electrolyte is zero.
2. The charge within the metal is confined to an extremely thin layer (< 1 Å) at the surface. There is no charge separation within the metal since if there were there would be an associated potential difference and this would cause a current to flow. Potential V and charge density ρ (coulomb [C] m^{-3}) are related by the Poisson equation:

$$d^2V/dx^2 = -\rho/\varepsilon\varepsilon_0 \qquad (1.2)$$

in one dimension, where x is the distance, ε the relative permittivity of the material, and ε_0 is a fundamental physical constant termed the permittivity of free space. Its value is 8.854×10^{-12} C V^{-1} m^{-1} and the smallness of this number essentially prevents substantial uncompensated charge density appearing at the interface.
3. The metal is conceived, in the model of Figure 1.6, as being terminated by a mathematical plane; this is not a very good approximation, as the previous section has indicated, though it is sufficient to derive at least a semi-quantitative theory.
4. The anions and cations in solution are normally hydrated and although hydration is a dynamic process, it is usually possible to identify a small number of water molecules in the hydration shell immediately surrounding the ion whose exchange rate is slow in comparison with other processes that might take place as the ions approach and recede from the interface. The picture of Figure 1.6 is, in any case, intended to represent a dynamic

situation, the ions shown being continually replaced by others from solution.

5. Most cations are strongly solvated, since their radii are small, and the free energy of solvation is approximately proportional to z^2/r_+, where ze_0 is the cation charge in coulombs and r_+ its ionic radius. The result of this is that even if the charge on the electrode is *negative*, there is usually little tendency for these cations to shed their water molecules and adsorb directly on the metal surface. Thus, the distance of closest approach of *cations* is determined by the radius of the inner solvent coordination sphere, and if the metal surface itself constitutes a plane, then the cation nuclei, at the distance of closest approach, will also constitute a plane termed the *outer Helmholtz plane (OHP)*.

6. By contrast, anions have larger radii and tend to be more weakly hydrated. In addition, they are able to form relatively strong ionic/covalent bonds to the surface of the metal electrode and, as a result, frequently find it energetically feasible to shed their inner hydration sphere, or at least part of it, and adsorb directly on the surface. The plane formed by the nuclei of anions directly adsorbed on the metal surface is termed the *inner Helmholtz plane (IHP)*.

7. In the absence of specific adsorption, for example if the anion is strongly hydrated, such as the fluoride anion, or very weakly coordinating, such as ClO_4^-, there will clearly be a well-defined potential at which the metal electrode has *no nett charge*, with the result that equal numbers of cations and anions will be present in the OHP. This potential is termed the potential of zero charge (pzc) and its determination is an important step forward in the understanding of the distribution of charge at the interface. It is, of course, possible, at least in principle, to define a pzc even when specific adsorption takes place, but the meaning is now less clear cut, particularly if some covalency is associated with the bond. As an example, OH^- on Au(111) is adsorbed essentially as the anion but on the Au(100) surface the OH^- ion is strongly covalently bonded and is present essentially as the neutral covalently bonded OH unit. Clearly, these effects will strongly influence the measured pzc.

8. One important area of ionosorption is found in metallic and semiconducting *oxides*. There are many such oxides and their importance in electrocatalysis has steadily increased in recent years. One example of a *metallic* oxide is RuO_2: although this is formally an ionic oxide, extensive Ru–O–Ru interactions allow the electrons on the formally Ru^{4+} ions to move freely from one to another. In the language of solid-state physics this covalency leads to an electronic bandwidth sufficient to sustain itinerant electronic motion, and in fact the material has an electronic conductivity comparable to that found in many elemental metals. When this oxide is immersed in aqueous solution, the surface becomes hydrated and this neutral surface may then gain protons in acidic solution or lose protons in base, becoming positively or negatively charged respectively. The extent of protonation will also depend, in a complex way, on the potential across the interface

and it is evident that we can define one pH as the point at which the surface has no nett charge: this is termed the pH of zero zeta potential (pzzp), a term borrowed from colloid chemistry, where the zeta potential is a measure of the electric field induced by the charge on the colloidal particle surface.

9. Finally, particularly when the total ionic concentration in the electrolyte is below ~ 0.1 M, the total charge on metal, IHP and OHP will not cancel in general and a diffuse charged layer develops in the electrolyte, extending out into the solution. This diffuse layer is characterised by an exponential decay of the potential at points well away from the surface and the potential is given by $V \sim e^{-\kappa x}$, where κ depends on the square root of the electrolyte concentration and is termed the screening constant. For concentrations higher than ~ 0.5 M, the screening length is shorter than the distance from the electrode surface to the OHP and the interface can then be thought of as a parallel plate capacitance but as the concentration falls and κ *decreases*, the width of the diffuse layer increases. This has a very significant effect on electron transfer kinetics, as discussed below.

1.1.3 Adsorption on the electrode surface

In addition to the specific adsorption of ions, it is also possible to adsorb neutral organic molecules at the interface and both Butler and Frumkin independently arrived at equivalent expressions for the free energy of adsorption of a neutral dipolar species, A, replacing the solvent, S. The free energy change takes the form:

$$\Delta G = [\tfrac{1}{2}(\alpha_S - \alpha_A)\mathbf{E}^2 + (\mathbf{P}_S - \mathbf{P}_A)\cdot\mathbf{E}]v_A \tag{1.3}$$

where α_S and α_A are the polarisabilities of solvent and A respectively, $\mathbf{P}_S$ and $\mathbf{P}_A$ the corresponding polarisations (i.e. the nett dipole moment densities normal to the surface per unit volume), v_A the volume of a molecule of A, and $\mathbf{E}$ the electric field at the surface. The surface excess molar concentration at the interface is then given by:

$$\Gamma_E = \Gamma_0 \exp(-\Delta G/k_B T) \tag{1.4}$$

where Γ_0 is the concentration of organic adsorbate in the *absence* of the electric field (i.e. at the pzc), and k_B is Boltzmann's constant.

In general, $\mathbf{P}_S \gg \mathbf{P}_A$, so that $\Delta G > 0$, and the coverage by organic adsorbates is *maximal* at or near the pzc.

At a given potential or, more exactly, at a given potential *difference* across the interface, the variation of coverage of A with concentration of A constitutes the *isotherm*. The simplest commonly encountered isotherm is that due to Langmuir and arises if:

(1) The surface is uniform and all adsorption sites are equivalent for all values of the fractional coverage, θ, of the surface by A.

(2) There is no lateral interaction between adsorbed molecules regardless of coverage θ.

(3) The equilibrium adsorption coverage is attained rapidly and reversibly.

The relationship of coverage to concentration of A in the bulk of the electrolyte is then given by:

$$\frac{\theta_A}{(1 - \theta_A)^N} = \exp\left\{\frac{-\Delta\mu^O}{k_B T}\right\} \cdot c_A/c_A^O \tag{1.5}$$

where c_A^O is the standard concentration of 1 mole dm^{-3}, N is the number of water molecules displaced, θ_A is the *equivalent* coverage ($\equiv N \cdot N_A/N_{total}$), and $\Delta\mu^O$ the change in free energy on adsorption of a molecule of A at unit concentration (more accurately activity) of A and at a coverage of θ_A^* such that $\theta_A^*/(1 - \theta_A^*)^N = 1$; if $N = 1$, then $\theta_A^* = 0.5$.

The idea that the surface will adsorb A with essentially the same enthalpy for all values of θ is clearly very simplified and attempts have been made to establish more realistic isotherms. Perhaps the best known of these is the Temkin isotherm which is based on the notion that the magnitude of the enthalpy of adsorption would *decrease* as the coverage *increases*. If the free energy of adsorption at $\theta = 0$ is ΔG_0^O, then the simplest assumption is that

$$\Delta G_\theta^O = \Delta G_0^O + X\theta RT$$

where X is the so-called heterogeneity parameter. Incorporating this into the basic theory of adsorption leads to:

$$\theta \approx (1/X)\log_e\left\{\frac{1 + K_0 c}{1 + K_0 c \cdot \exp(-X)}\right\} \tag{1.6}$$

where K_0 is the adsorption equilibrium constant appropriate to $X = 0$. As might be expected, an increase in X leads to a decrease in θ for a given concentration c. Temkin's treatment is valid provided $0.1 < \theta < 0.9$; furthermore, if X is large enough such that $K_0 c \gg 1$ but $K_0 \exp(-X) \ll 1$, then:

$$\theta \sim (1/X) \cdot \log_e(K_0 c) \tag{1.7}$$

a well-known isotherm in surface chemistry. Temkin's isotherm can actually arise for a number of microscopic conditions:

(1) If the surface is heterogeneous, consisting of minute patches whose free energy of adsorption differs incrementally.

(2) If the surface potential alters as a result of adsorption as $N\mu/\varepsilon\varepsilon_0$, where μ is the dipole moment of the adsorbate and N the number of adsorbed molecules.

(3) If the surface of the metal can be described as a two-dimensional electron gas, where the calculated adsorption enthalpy will change linearly with θ.

Temkin's isotherm can describe the effects of surface heterogeneity or of surface modification on adsorption but we should also take into account the lateral interactions between adsorbed molecules. For the adsorption of simple ions, for example, the interaction energy, neglecting image charges (equal and opposite charges induced in the metal as a result of the adsorbate), has the form:

$$U \sim \theta^{1/2} \tag{1.8}$$

Incorporation of image charges alters this relationship to $U \sim \theta^{3/2}$, and both forms have been claimed in the literature, giving rise to isotherms that have the general form:

$$\log_e[\theta/(1-\theta)] \sim K + \log_e c + AQ_M - B\theta^y \tag{1.9}$$

where Q_M is the charge on the metal, K is a constant, y is a constant taking values of $1/2, 3/2$, etc., B is a constant that depends on the adsorbate/substrate dipole and A is a constant that depends on the gain in electrostatic energy on specific adsorption of an ionic species.

Finally, combining (1.9) with the Temkin isotherm leads to the most general isotherm commonly used to describe the coverage of neutral or ionic species at the electrode surface:

$$\log_e[\theta/(1-\theta)] \sim K + \log_e c + X\theta + AQ_M - B\theta^y \tag{1.10}$$

For neutral dipolar molecules, the value of B depends on the dipole moment itself, and the value of y is $3/2$ or, in the presence of image forces, $5/2$. The former of these has been verified experimentally for the adsorption of such molecules as 2-chloropyridine on mercury.

It is, of course, the case that the form of (1.10) does not give us any information about the *orientation* of the organic molecule on the surface. Some techniques for establishing this *in situ* are discussed in chapter 2 and it is now recognised that this orientation may itself be a function of the concentration of A in solution. A well-known example is the molecule hydroquinone on platinum: at low concentrations *ex-situ* electron spectroscopic investigations have shown that the molecule lies flat with the two O–H bonds ionised. However, at concentrations in excess of 1 mM, the adsorbed molecules begin to lie vertically, with two C–H bonds of the benzene ring now pointing at the surface. This change in orientation has profound effects on the electrochemical activity of adsorbed species: when the platinum electrode is swept to positive potentials, the hydroquinone oriented *flat* is completely oxidised to CO_2 and water but that oriented vertically is only partially oxidised to maleic acid.

1.2.1 Thermodynamic considerations

**1.2
Electron transfer**

Without any doubt, it is the ability of the electrified interface to facilitate electron transfer that makes its study so important. All species that are electroactive will undergo some form of electron transfer or exchange near the interface. Experimentally, the rate of such electron transfer processes varies enormously, being extremely rapid for the process:

$$2H^+ + 2e^- \rightleftharpoons H_2(g): \text{platinised platinum} \qquad (1.11)$$

and for the process:

$$2Cl^- \rightleftharpoons Cl_2(g) + 2e^-: RuO_2/TiO_2 \text{ mixtures} \qquad (1.12)$$

but very much slower for the process:

$$CH_3OH + H_2O \longrightarrow CO_2 + 6H^+ + 6e^-: \text{platinum} \qquad (1.13)$$

Provided the reaction is, in some sense, reversible, so that equilibrium can be attained, and provided the reactants and products are all gas-phase, solution or solid-state species with well-defined free energies, it is possible to define the free energies for *all* such reactions under any defined reaction conditions with respect to a standard process; this is conventionally chosen to be the hydrogen evolution/oxidation process shown in (1.11). The relationship between the relative free energy of a process and the *emf* of a hypothetical cell with the reaction (1.11) as the *cathode* process is given by the expression $\Delta G = -nFE$, or, for the free energy and potential under standard conditions, $\Delta G^0 = -nFE^0$, where n is the number of electrons involved in the process, F is Faraday's constant and E is the emf.

We can exemplify this by considering reactions (1.11) and (1.12) above:

$$\Delta G_1^0 = -2FE^0_{H^+/H_2} \qquad (1.14)$$

$$\Delta G_2^0 = -2FE^0_{Cl_2/Cl^-} \qquad (1.15)$$

Hence:

$$E^0_{Cl_2/Cl^-} - E^0_{H^+/H_2} = -(\Delta G_2^0 - \Delta G_1^0)/2F = -\Delta G_3^0/2F \qquad (1.16)$$

But ΔG_3^0 is the free energy for the process:

$$2H^+ + 2Cl^- \longrightarrow H_2(g) + Cl_2(g) \qquad (1.17)$$

which is known to be $262.5\,kJ\,mol^{-1}$. Since F has the value $96\,450\,C\,mol^{-1}$, we find that the emf for the cell:

$$Cl_2 | RuO_2, TiO_2 | H^+, Cl^-, aq | Pt | H_2$$

is $-1.36\,V$ at $298\,K$ and $1\,atm$ pressure.

Conventionally, $E^0_{H^+/H_2} = 0$ provided the concentration, or, more exactly, the activity of H^+ is unity and the pressure is one atm, so $E^0_{Cl_2/Cl^-} = 1.36\,V$.

Two points emerge from this discussion. Simple application of the expression for ΔG in terms of the equilibrium constant K shows that for the chlorine reaction:

$$E_{Cl_2/Cl^-} \approx E^0_{Cl_2/Cl^-} + (RT/2F)\ln\{p_{Cl_2}/[Cl^-]\} \qquad (1.18)$$

Thus, if the potential of the cell were raised above $1.36\,V$, we would expect a rapid evolution of Cl_2 to take place and if the potential applied to the cell were lower than $1.36\,V$, then the Cl_2 would be rapidly reduced to Cl^-, in both cases to ensure that equation (1.18) is satisfied by adjusting the ratio $p_{Cl_2}/[Cl^-]$ locally at the electrode surface.

For the cell above, this thermodynamic treatment does, in fact, give the correct predictions provided that the value of E_{Cl_2/Cl^-} remains close to $E^0_{Cl_2/Cl^-}$. However, there are a number of problems with the treatment that must be addressed in a more thorough-going account:

1. For very many systems for which E^0 is well defined, the expected current behaviour is not described in any way by the Nernst equation (1.18). The current shows no marked rise as the potential is changed from E^0 and when the potential is raised or lowered sufficiently for electron transfer to take place the resulting electrochemical reactions are often complex.

2. Even for the Cl^- oxidation process on the RuO_2/TiO_2 electrode, for which the individual electron transfer processes are sufficiently fast for thermodynamic equilibrium to be maintained, at least close to E^0, it is highly *unlikely* that two electrons are transferred simultaneously from two Cl^- ions that are exactly the right distance apart for Cl_2 bond formation to take place. It is far more likely that some kind of radical intermediate is involved which is stabilised by complexation on the surface.

3. As the potential is increased, the ratio of p_{Cl_2} to $[Cl^-]$ should increase without limit according to (1.18). In fact, given that p_{Cl_2} cannot substantially exceed one atm under ambient conditions, this implies that $[Cl^-]$ must become vanishingly small at the electrode surface. The current will then be determined entirely *by the rate of transport of* Cl^- *to the electrode surface* and indeed quite generally the current will be determined by two factors: (a) the rate of electron transfer at the electrode surface; (b) the rate of transport of material to the electrode surface.

1.2.2   Rate of electron transfer

For simplicity, we consider first the oxidation of a simple hydrated cation $[Fe(H_2O)_6]^{2+}$ to $[Fe(H_2O)_6]^{3+}$. This is essentially the simplest electron-transfer process that we can envisage and modern theoretical studies have suggested the following sequence of events:

1. The $[Fe(H_2O)_6]^{2+}$ ion approaches the interface to within a critical distance of $\lesssim 15\,\text{Å}$.

2. Electron transfer takes place from a t_{2g} orbital that is doubly occupied on the Fe^{2+} ion. The total electronic energy of the $[Fe(H_2O)_6]^{2+}$ species can be plotted vs. the Fe–O bond distance in the hydrate on the right-hand part of Figure 1.7. On the left-hand side of this diagram is plotted the electronic energy of the $[Fe(H_2O)_6]^{3+}$ ion together with the energy of the electron at the Fermi level of the metal. The latter scales linearly with potential, so that we can imagine the left-hand side of the diagram moving

vertically up and down with potential, moving *down* as the potential of the electrode is made more *positive*.

3. The electronic energy of the complex is also affected by the orientation of water molecules outside the primary hydration sphere. These water molecules are more free to re-orientate than those held firmly within the inner coordination sheath, and the electronic energy is affected mainly through the water dipoles interacting with the charge on the ion. In principle, this effect could be represented by a multi-dimensional energy diagram, but for the present we will retain the simplistic view of Figure 1.7 without attempting, in this case, to include explicitly the effects of water molecule re-orientation (or 'librational' modes).

4. For a given potential it is clear that there will be a single intersection point between the curves representing $[Fe(H_2O)_6]^{2+}$ and $[Fe(H_2O)_6]^{3+}$. At this point, the electronic energies of the two systems $[Fe(H_2O)_6]^{2+}$. and $([Fe(H_2O)_6]^{3+} + e_M^-)$ are equal, and the electron can tunnel adiabatically between the two states (adiabatically means that the electron can move without any energy being lost or gained in the process).

5. The transfer of the electron takes place very rapidly compared to nuclear motion, and will only take place when the combination of internal and librational coordinates is such that the curves interact. Thus, the $[Fe(H_2O)_6]^{2+}$ species must first distort and/or experience a dipole moment field from the instantaneous positions of the water molecules such that it attains the cross-over point. At this point, the electron *may* tunnel from the $[Fe(H_2O)_6]^{2+}$ ion to the metal, leaving behind an $[Fe(H_2O)_6]^{3+}$ ion with a non-equilibrium geometry. This then relaxes by heat transfer to the solvent to the equilibrium point, q_0.

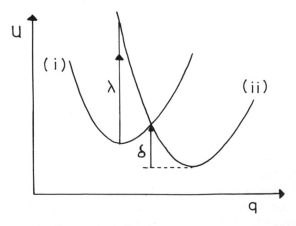

Figure 1.7   A schematic representation of electron transfer between $Fe(H_2O)_6^{2+}$ and a metal electrode. The figure represents (i) the total electronic energy of the $Fe(H_2O)^{3+}$ ion together with the energy of the electron at the Fermi level of the metal; (ii) the total electronic energy of the $Fe(H_2O)_6^{2+}$ ion, plotted vs. the Fe–O bond distance in the hydrates.

6. Plainly, the activation energy for the oxidation of $[Fe(H_2O)_6]^{2+}$ is given by δ in Figure 1.7. At some electrode potential, the electronic energies of $[Fe(H_2O)_6]^{2+}$ and $e_M^- + [Fe(H_2O)_6]^{3+}$ will be the same. At this potential, examination of Figure 1.7 shows that the activation energy for the electron transfer process is simply related to the optical transition energy, λ (i.e. the vertical transition energy from the bottom of the Fe^{3+} curve to the point on the Fe^{2+} curve vertically above this). In fact, provided the force constants for the vibrational modes are the same for both the oxidised (O) and reduced (R) components of the couple, and the modes are harmonic, then it is easy to show that $\delta = \lambda/4$, and in many texts, λ is referred to as the 'reorganisation energy'.

The potential difference of significance here is that between the metal and the ion in the OHP, which is denoted $\Delta\phi$; this can be systematically varied by varying the electrode potential. If $\Delta\phi = \Delta\phi_0$ when the two curves for $O + e_M^-$ and R are symmetrical, then:

$$\delta_R = (\lambda - e_0[\Delta\phi - \Delta\phi_0])^2/4\lambda \qquad (1.19)$$

from simple geometric considerations, again assuming harmonic vibrational modes.

7. In a similar way, the activation energy for the reduction of $[Fe(H_2O)_6]^{3+}$, O, is given by:

$$\delta_O = (\lambda + e_0[\Delta\phi - \Delta\phi_0])^2/4\lambda \qquad (1.20)$$

and, assuming a simple Arrhenius-type expression, we can write the rate of the forward reaction (oxidation of R) as:

$$v_+ = [R]_0 \cdot \kappa \cdot Z \exp(-\delta_R/k_B T) \qquad (1.21)$$

and the reduction rate as:

$$v_- = [O]_0 \cdot \kappa \cdot Z \exp(-\delta_O/k_B T) \qquad (1.22)$$

where κ is the tunnelling probability, Z is the thermal velocity of the ions ($\approx (k_B T/2\pi m)^{1/2}$, m is the mass of the ion) and k_B is Boltzmann's constant. The tunnelling probability will depend critically on the distance x from the electrode surface; empirically, it has the form:

$$\kappa = \kappa_0 e^{-\xi x} \qquad (1.23)$$

at larger x values, where ξ normally has the value $1-2 \text{Å}^{-1}$. Thus, the tunnelling probability will increase rapidly as the ion approaches the OHP and it is sufficient normally just to consider electron transfer at that point.

Also, if, as is usual, $\lambda \gg e_0[\Delta\phi - \Delta\phi_0]$, then equation (1.19) can be expanded as:

$$\delta_R \approx (\lambda/4) \cdot (1 - 2e_0[\Delta\phi - \Delta\phi_0]/\lambda) = \lambda/4 - e_0[\Delta\phi - \Delta\phi_0]/2 \qquad (1.24)$$

Similarly:

$$\delta_O \approx (\lambda/4) \cdot (1 + 2e_0[\Delta\phi - \Delta\phi_0]/\lambda) = \lambda/4 + e_0[\Delta\phi - \Delta\phi_0]/2 \qquad (1.25)$$

so finally the current takes the form:

$$I \sim v_+ - v_- \sim \kappa Z \cdot e^{-\lambda/4k_BT} \cdot \{[R]_0 e^{e_0(\Delta\phi - \Delta\phi_0)/2k_BT} - [O]_0 e^{-e_0(\Delta\phi - \Delta\phi_0)/2k_BT}\}$$

(1.26)

which is the familiar Butler–Volmer equation, albeit slightly disguised. This extremely simplified derivation does allow us to see that, in general, the rate of electron transfer will decrease very rapidly as λ increases, and that the current will, for $\Delta\phi \gg \Delta\phi_0$, increase exponentially with $\Delta\phi$, and hence with E.

The assumptions underlying our derivation are rather drastic: we have assumed that there is no energy involved in transporting the reactant from the bulk of the solution to the OHP, an assumption only justified if there is a high concentration of electrolyte in the solvent. We have also assumed that the force constants for oxidised and reduced forms are identical, that the potential energy curves are parabolic, that activation is dominated by a single mode, and that tunnelling takes place from or to a single energy level in the electrode. In fact, the corrections that need to be made to incorporate more rigorous theories affect the form of (1.26) only relatively slightly, the most important being a change from a factor of 1/2 in the exponent by a more general factor, leading to the familiar form:

$$I \sim \kappa Z \cdot e^{-\lambda/4k_BT} \cdot \{[R]_0 e^{(1-\beta)e_0(\Delta\phi - \Delta\phi_0)/k_BT} - [O]_0 e^{-\beta e_0(\Delta\phi - \Delta\phi_0)/k_BT}\}$$

(1.27)

where β is termed the asymmetry parameter, which will only have the value 1/2 if the force constants for the oxidised and reduced forms of the couple are equal. A further simplification is that for high electrolyte concentrations, we can replace $[\Delta\phi - \Delta\phi_0]$ by $[E - E^0]$, where E is the electrode potential with respect to a standard reference electrode in solution, and E^0 is the Nernstian potential of the redox couple with respect to the same reference. The current I in equation (1.27) can be more simply expressed by combining $\kappa Z \cdot e^{-\lambda/4k_BT}$ into a single rate constant, k^0. Since Z is the thermal velocity and $[R]_0$ has units of concentration, it follows that the current is expressed per unit area. We must also include the Faraday, since every mole of R oxidised required one mole of electrons or F coulombs. Finally we obtain:

$$I = I_a - I_c = FAk^0\{[R]_0 \exp[f(1-\beta)(E-E^0) - [O]_0 \exp[-f\beta(E-E^0)]\}$$

where $f = F/RT$. At the rest potential, E_r, the rates of the cathodic, I_c, and anodic I_a, currents are equal, and no net current flows:

$$|I_a| = |I_c| = I_0.$$

Further, since at the rest potential the concentrations of [O] and [R] at the surface of the electrode must equal the bulk concentrations [O*] and

[R*], the exchange current can be expressed as:

$$I_0/FA = [O^*]k^0 \exp[-\beta f(E_r - E^0)] \tag{1.28}$$

$$I_0/FA = [R^*]k^0 \exp[(1-\beta)f(E_r - E^0)] \tag{1.29}$$

Equating these two, we have:

$$E_r = E^0 + (RT/F)\cdot\log_e([O^*]/[R^*])$$

which is, of course, the Nernst equation. Substituting back, we find:

$$I_0/FA = k^0[O^*]^{1-\beta}[R^*]^{\beta}$$

an equation we will need below.

8. Clearly, the *intrinsic* rate of electron transfer is dominated by λ and modern theories have shown that λ can be expressed as the sum of two terms: λ_i, the contribution from vibrational modes within the charged ionic unit, and λ_0, the contribution from the solvent dipole re-orientation effects. Expressions for both of these terms have been given by Marcus:

$$\lambda_0 \approx (n^2 e_0^2/8\pi a\varepsilon_0)\cdot[(1/\varepsilon_{opt}) - (1/\varepsilon_s)] \tag{1.30}$$

where ε_{opt} is the relative permittivity of the solvent in the visible region of the spectrum, ε_s is the static relative permittivity, a is the ionic radius and n is the number of electrons transferred.

$$\lambda_i \approx \Sigma_j[k_j k_j^*/(k_j + k_j^*)]\cdot(\Delta q_j)^2 \tag{1.31}$$

where k_j, k_j^* are the force constants associated with normal coordinates in the product and reactant respectively, and Δq_j is the change in the equilibrium normal coordinate, j, accompanying the reaction. Clearly, if only one coordinate dominates and the force constants are the same for both oxidised and reduced forms, (1.31) reduces to the expression already derived above.

The expressions (1.30) and (1.31) have been extensively tested and in many simple cases agreement is reasonably good. Thus, for $[V(H_2O)]^{3+}$, the value of $\lambda/4$ from experiment is 0.379 eV, and the calculated value is 0.389 eV. Similarly, for reduction of naphthalene in DMF, the observed value of $\lambda/4$ is 0.237 eV and the calculated value 0.247 eV. There are, unfortunately, a number of exceptions. For example, the value of $\lambda/4$ for $[Ce(H_2O)_6]^{4+}$ is substantially *under*estimated theoretically, possibly as a consequence of the neglect of hydrolysis of the ion, and for others, such as $[Cr(H_2O)]^{2+}$, the complexity of vibronic coupling is thought to invalidate the simple treatment given above.

Thus far, we have only considered the state of the system in an equilibrium in which no net current can flow. Consider now the case where the potential is stepped to E, where $E < E_r$, such that the equilibrium is perturbed. Current now flows as O is reduced to R, producing new concentrations of O and R at the surface, giving rise to anodic and cathodic currents:

$$I_a/FA = [R]_o k^O \exp[(1-\beta)f(E-E^O)] \tag{1.32}$$

$$I_c/FA = [O]_o k^O \exp[-\beta f(E-E^O)] \tag{1.33}$$

The net current is now given by:

$$I/FA = [R]_o k^O \exp[(1-\beta)f(E-E^O)] - [O]_o k^O \exp[-\beta f[E-E^O)].$$

We can express I in terms of E_r straightforwardly. Using the expressions for the Nernst equation and I_O above:

$$\begin{aligned} I/FA &= k^O\{[R]_o([O^*]^{1-\beta}/[R^*]^{1-\beta})\exp[(1-\beta)f\eta] \\ &\quad - [O]_o([O^*]^{-\beta}/[R^*]^{-\beta})\exp[-\beta f\eta]\} \\ &= I_o/FA\{([R]_o/[R^*])\exp[(1-\beta)f\eta] - ([O]_o/[O^*])\exp[-\beta f\eta]\} \end{aligned} \tag{1.34}$$

where $\eta = (E - E_r)$ is the activation overpotential, the deviation of the potential away from its equilibrium value. If the kinetics at the electrode are slow, then the surface concentrations of O and R can again be taken to be approximately equal to their bulk values and equation (1.34) becomes:

$$I = I_O[(\exp([1-\beta]f\eta) - \exp(-\beta f\eta)] \tag{1.35}$$

This is the most commonly employed form of the Butler–Volmer equation as it does not involve the unmeasurable surface concentration terms. It must be remembered, however, that equation (1.35) is only applicable under the conditions where $[O]_O \approx [O^*]$ and $[R]_O \approx [R^*]$. We must now examine this equation in some detail, as its form dictates the nature of a number of electrochemical techniques for exploring reaction mechanisms.

From equation (1.34) it can be seen that the net current depends on three factors:

1. The exchange current density, I_O. From equations (1.28) and (1.29) it is clear that I_O is a direct measure of the kinetics of the reaction at the equilibrium potential E_r. A high exchange current density indicates a facile electrochemical reaction that will rapidly become limited by transport at potentials away from E^O.
2. The overpotential, η, can be regarded as an activation energy; the smaller I_O, and hence the more sluggish the kinetics, the higher η has to be to provide a given current.
3. The concentration terms in equation (1.34) are a manifestation of the reactant supply to the electrode and hence the relative importance of mass transport. The maximum current in a given direction will be obtained when the bulk and surface concentrations of the primary reactant electroactive species are equal.

The Butler–Volmer equation can be employed only when O and R are chemically stable on the timescale of the experiment. Within this requirement, we can envisage two limiting cases:

1. Large I_O, facile kinetics. From equation (1.34):

$$I/I_O = [([R]_o/[R^*])\exp([1-\beta]f\eta) - ([O]_o/[O^*])\exp(-\beta f\eta)]$$

if I_O is very large, then $I/I_O \rightarrow 0$, and:

$$([R]_o[R^*])\exp([1-\beta]f\eta) = ([O]_o/[O^*])\exp(-\beta f\eta)$$

and since $\eta = (E - E_r)$:

$$[O^*]/[R^*] = ([O]_o/[R]_o)\exp(-f[E - E_r])$$

From the Nernst equation above we have:

$$E_r = E^0 + (1/f)\log_e([O]^*/[R^*])$$

so:

$$[O^*]/[R^*] = \exp[f(E_r - E^0)]$$

and

$$[O]_o/[R]_o = \exp[f(E - E^0)] \qquad (1.36)$$

This is an extremely important equation since it tells us that, in the limit of very facile kinetics, the surface concentrations of O and R are constrained to satisfy the local Nernst equation. Under these conditions, the net current is always dictated by the diffusion of the electroactive species to the electrode; i.e. the flux of O.

2. Low I_O, quasi-reversible kinetics. Under these circumstances, we can use the approximation represented by equation (1.35) in which we assume that $[O]_0 \approx [O^*]$. The equation is generally applied under two limiting conditions:

(i) Relatively large positive, or negative, overpotentials; i.e. $|\eta| > 0.12\,\text{V}$. In this case, one of the terms in equation (1.35) becomes negligible. Thus, at large positive overpotentials, such that the cathodic process becomes very slow due to the large activation barrier, the net current is given by:

$$I = I_O \exp([1-\beta]f\eta)$$

which can be re-written in the form:

$$\eta = (1/[1-\beta]f)\cdot\log_e I - (1/[1-\beta]f)\cdot\log_e I_O \qquad (1.37)$$

Equation (1.37) is of the form $\eta = a + b\log I$; an empirical observation first reported by Tafel. Thus, a Tafel plot of η vs $\log_e I$ giving a straight line at high overpotentials is indicative of quasi-reversible kinetics. The slope gives β and the intercept (obtained via the extrapolation back to $\eta = 0$) gives I_O; see Figure 1.8.

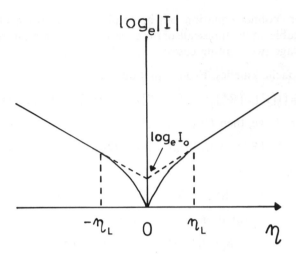

Figure 1.8   A schematic representation of a Tafel plot of $\log_e |I|$ vs. η, showing linearity at high overpotentials. At values of the overpotential $< |\eta_L|$, the current shows a linear dependence on overpotential.

(ii) At small overpotentials, $|\eta| \leqslant 0.02\,\text{V}$, where the system is very close to equilibrium, we can use the approximation that $e^x \approx 1 + x$, thus:

$$I \approx I_0[(1 + [1 - \beta]f\eta) - (1 + [-\beta f\eta])]$$
$$I \approx I_0 f\eta \qquad\qquad (1.38)$$

i.e. the interface behaves like an Ohmic conductor, with $I \propto$ potential. Hence, at potentials near E_r, the current is linearly dependent on the overpotential. However, if I_0 is too low then the current may be difficult to measure; this method is thus only practicable for large I_0.

In practice, the Butler–Volmer equation is only obeyed in any case for potentials close to E_r, or, more generally, for small currents, not through any neglect of factors such as anharmonicity but rather because the rate of transport of the ions to the electrode becomes rate-limiting, a problem we turn to next.

1.2.3   *Rate of transport of solution species*

There are, in principle, three ways in which material may be transported to the electrode surface: diffusion, convection and migration. Of these, perhaps the most straightforward is migration, which simply consists of the movement of a charged particle under the influence of an electric field. Experimentally, it is well established that after an extremely short time an ion in solution in an electric field will behave as if it had acquired a steady velocity in the direction of the field. The reason why a *steady* velocity is established rather than the continuous acceleration apparently predicted by Newton's laws of motion lies in the fact that the actual motion of the particle is very complex, consisting of innumerable collisions with other ions and solvent molecules. However, when a macroscopic averaging is carried out, this incessant buffeting is found to constitute a frictional drag (κ_R) and phenomenologically this drag increases with velocity v according to Stokes' formula:

$$\kappa_R = 6\pi\eta r v \qquad (1.39)$$

where r is the ionic radius, and η_1 is the solvent viscosity.

The force on the ion exerted by the electric field is $ze_0\mathbf{E}$, where ze_0 is the charge on the ion (in coulombs) and $\mathbf{E}$ is the field (in $V\,m^{-1}$). Once steady state conditions are established:

$$ze_0\mathbf{E} = 6\pi r v_{ss}\eta_1 \qquad (1.40a)$$

and the steady-state velocity, v_{ss}, is given by:

$$v_{ss} = ze_0\mathbf{E}/6\pi\eta_1 r \equiv \pm \mu\mathbf{E} \qquad (1.40b)$$

where μ is the ionic mobility and the $\pm$ sign arises from the fact that z can be positive ($+$) or negative ($-$).

Diffusion is a more subtle phenomenon and depends on the fact that if a concentration gradient exists between two points in a solution, there will be a flux of material from the region of higher to lower concentration. If we consider Figure 1.9, this shows the presence of a stream of material through two planes, at x and $x + \delta x$. If the concentration *gradient* at x is dc/dx, where c is the concentration of the species at x, then the flux of material per unit area, $\mathbf{J}$, across the plane at x in the *positive direction of x*, is given by Fick's first law:

$$\mathbf{J} = -D\,dc/dx \qquad (1.41)$$

where D is termed the diffusion coefficient. Unlike the migration of ions considered above, diffusion may take place for any species, charged or uncharged. However, if we are considering the movement of *ions* and an

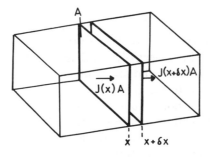

Figure 1.9   A schematic representation of a stream of flux, $\mathbf{J}$, of material through two planes of area A at distance x and $(x + \partial x)$.

electric field $\mathbf{E}$ is also present, we must consider the combined flux due to both migration and diffusion:

$$J_{com} = \pm \mu c\mathbf{E} - D\,dc/dx \tag{1.42}$$

where $\mathbf{E}$ must be in the *positive* x-direction. Since $\mathbf{E} = -dV/dx$, where V is the potential, this implies that for $\mathbf{E} > 0$, $dV/dx < 0$ and V *decreases* with x.

Equation (1.42) is fundamental in electrochemistry and lies at the basis of all classical electrochemical techniques. As stated, it is not easy to use, since both flux and concentration will be time dependent in general and we need an explicit relationship between c and t. If we return to Figure 1.9, however, it can be seen that the rate of *accumulation* of ions in the thin layer between x and $x + \delta x$ must be given by:

rate of accumulation =

$$\frac{\text{flux of ions } into \text{ layer at } x - \text{flux of ions } out \text{ of layer at } x + \delta x}{\text{volume of layer}}$$

Hence:

$$\partial c/\partial t = \frac{(J)_x - (J)_{x+\delta x}}{A\delta x} = -\partial/\partial x \cdot \{\pm \mu c\mathbf{E} - D\partial c/\partial x\} \tag{1.43}$$

where we have assumed that the area of the plane, A, is unity and we must now use *partial* derivatives since c is a function both of position x and time t. Assuming that μ is *not* a function of x, we have, finally,

$$\partial c/\partial t = D\partial^2 c/\partial x^2 \mp \mu\partial(c\mathbf{E})/\partial x \tag{1.44}$$

Although this may look very unpleasant, techniques for solving equations of this sort both analytically and numerically are very well established; in fact it is usually possible to solve the equation analytically, at least for the simpler geometries often encountered in cell design. Moreover, the structure of the electrode–electrolyte interface allows us to make a substantial simplification under normal circumstances in the region beyond the diffuse-layer boundary. We can take $\mathbf{E}$ as *zero* in this region, as indicated above, since the potential in the electrolyte bulk must be constant, and so:

$$\partial c/\partial t = D\partial^2 c/\partial x^2 \tag{1.45}$$

save in the region very close to the electrode surface. In fact, if a larger concentration of some inert electrolyte such as $NaClO_4$ is used in the solvent, then equation (1.45) will be valid for all other ionic species present at lower concentrations right up to the OHP.

The third form of transport is *convection*, which involves the bulk movement of the solvent. Again, with reference to Figure 1.9, if the solvent as a whole is moving with velocity v_f, then clearly the flux across the plane is $v_f c$, and by analogy with the arguments above:

$$(\partial c/\partial t)_{con} = -v_f\partial c/\partial x \tag{1.46}$$

There are very many situations in which well-defined *patterns* of convection can be established, and analytical expressions for v_f derived. Such situations usually involve forced convection, in which the movement of the liquid is determined by rotation, agitation, forced flow over a flat surface, etc. Once the functional form of v_f is known, solutions for c as a function of x are sought so that values of the current can be found and compared with those obtained experimentally.

A simple example is the rotating disc electrode described in detail in chapter 2. The horizontal spinning disc draws liquid up and then flings it out sideways, creating a continuous but steady-state convection pattern. If the distance down from the disc is denoted by z and the distance across the disc surface by the radial distance, r, then it is not difficult to show that:

$$v_{fz} = -Cz^2, \text{ where } C = 0.510\omega^{3/2}v^{-1/2} \tag{1.47}$$

and ω is the angular velocity of the disc and v is the kinematic viscosity ($\equiv \eta_1/\rho$, where ρ is the density). Note that this is independent of r.

The transport of c is combined convection and diffusion so:

$$\partial c/\partial t = D\partial^2 c/\partial z^2 - v_f \partial c/\partial z = D\partial^2 c/\partial z^2 + Cz^2 \cdot \partial c/\partial z \tag{1.48}$$

At steady state, $\partial c/\partial t = 0$, so an expression for c can be obtained immediately. Obviously, close to the surface, transport of c is dominated by diffusion but further away from the surface convection dominates.

A major problem in modern electrochemistry, and one addressed at length in this book, is the fact that most electrochemical reactions of technological importance are complex ones in which electron transfer and chemical processes are closely interlinked. The central problems are:

**1.3
Reaction mechanisms**

1. Is electron transfer taking place on the bare-metal electrode or is the electrode modified by the presence of an adsorbed layer?
2. Are the important steps in the mechanism taking place in solution (i.e. are they homogeneous reactions) or are surface adsorbed intermediates involved?
3. Are there several different mechanisms involved, possibly leading to a range of products, or is the reaction dominated by one mechanism?

As an example we may consider the Kolbe reaction, the oxidation of carboxylic acid and carboxylates of the form $R-COOH$ or $R-COO^-$ to form coupled hydrocarbon products of the form R_2. Investigation of this reaction in aqueous and non-aqueous solvents has revealed that the processes taking place are very complex indeed. In general, the product R_2 is only formed at high current densities on smooth electrodes. At lower current densities, alkenes and non-dimeric products such as $R-H$ are found, and, especially in alkaline solutions, the product $R-OH$ can be formed in good yield (the Hofer–Möst reaction). Even more remarkable is the fact that if a platinum electrode is immersed in aqueous solution and the potential steadily *increased*, the first electrochemical reaction that takes place, at moderate positive potentials, is the oxidation of water to oxygen. Only above a critical potential is the oxygen evolution reaction suppressed and oxidation products from the carboxylates become predominant.

A number of earlier workers attempted to reconcile these observations by invoking a role for oxygen or peroxide, but it now seems most likely that the basic mechanism takes the form:

$$R\text{–}COO^- \longrightarrow R\text{–}COO^{\bullet} + e^- \qquad (a) \qquad E$$

$$R\text{–}COO^{\bullet} \longrightarrow R^{\bullet} + CO_2 \qquad (b) \qquad C$$

$$2R^{\bullet} \longrightarrow R_2 \qquad (c) \qquad C$$

We have used the notation E to refer to a simple electron transfer step, and C to refer to a simple chemical step. It is likely that step (a) is usually rate limiting but a kinetic analysis, using the principles discussed later in this book, is inconsistent with the discharge of the electron just onto the bare metal; it seems highly likely that a substantial coverage of $R^{\bullet}$ adsorbed on the surface is necessary both to drive reaction (c) and to inhibit the oxygen evolution process. Clearly, identification of this layer is highly desirable, but has proved extremely difficult in this case owing to the simultaneous vigorous evolution of CO_2.

The participation of surface radical species is, then, strongly suspected in the Kolbe reaction, but there are other radical reactions, such as the reductive dimerisation of CO_2, that are thought to be homogeneous, particularly in non-aqueous solvents. The basic mechanism in this latter case is thought to be:

$$CO_2 + e^- \longrightarrow CO_2^{-\bullet} \qquad E$$

$$CO_2 + CO_2^{-\bullet} \longrightarrow O\text{–}C\text{–}O\text{–}CO_2^{-\bullet} \qquad C$$

$$O\text{–}C\text{–}O\text{–}CO_2^{-\bullet} + e^- \longrightarrow CO + CO_3^{2-} \qquad E$$

This apparently simple process is, however, complicated both by an alternative route:

$$2CO_2^{-\bullet} \longrightarrow C_2O_4^{2-}$$

and by very high sensitivity to any adventitious water present:

$$CO_2^{-\bullet} + H_2O \longrightarrow HCOO^{\bullet} + OH^-$$

$$HCOO^{\bullet} + e^- \longrightarrow HCOO^-$$

in which a second ECE reaction leads to formate formation. The CO_2 reaction has become very topical in recent years, with the necessity of understanding the processes involved in CO_2 reduction being given added urgency through the greenhouse effect.

The weapons in the hands of the electrochemist when a mechanistic investigation is launched are:

1. The identification of the products, both major and minor.
2. The investigation of the *sensitivity* of the product distribution to the presence of particular reactants such as adventitious water in non-aqueous media.
3. The measurement of the *rate* of the desired reaction as a function of
    (a) potential
    (b) concentration of reagents
    (c) isotopic constitution of reagents
    (d) temperature.
4. The use of time-resolved techniques to reveal the presence of consecutive reactions.
5. The direct spectroscopic identification of intermediates.
6. The detection and identification of films formed on surfaces by spectroscopic and other means.

Some of these techniques are perhaps self-evident; it is known, for example, that methanol can be oxidised not only to CO_2 but also to formaldehyde, formic acid and even (reportedly) methyl formate in strongly acidic aqueous solutions. Clearly any mechanism must account for the formation of all of these products.

The most powerful approach, at least in principle, is the measurement of the *rate* of the desired reaction as a function of potential and reagent concentration. In essence, any reaction can be written as a set of consecutive steps; this is true even if the reaction is apparently a simple process such as the electrolyte deposition of a monovalent cation such as Ag^+, since loss of water of hydration from the cation and the (possibly assisted) transport of atoms over the surface to appropriate lattice sites are clearly consecutive processes.

A sequence of chemical reactions can be analysed theoretically in one of two ways:

(a) A steady state method, in which the change of concentration of highly reactive intermediates with time is put equal to zero;
(b) By use of the quasi-equilibrium hypothesis, in which all steps prior to the rate-determining step (rds) are postulated to be at rapid equilibrium. This is equivalent to the supposition that the rate constant of the rds is at least a factor of ten slower than that of all the antecedent steps in both directions.

In practice, whilst (a) will lead to more accurate results, these results are usually far more cumbersome, particularly for more complex reactions, and far less easy to compare with experimental data.

1.3.1   Homogeneous electrochemical mechanisms

In order to understand the methodology in some detail, we first consider homogeneous processes, where the electrochemical techniques used are well-established. Such processes are not central to this book, which is primarily concerned with *electrode* processes, but they do serve to illustrate the manner in which mechanisms can be explored. As indicated above, any step in the electrochemical mechanism must be either *chemical* (denoted by C) or *electrochemical* (denoted by E) in nature. It is not normally the case that more than one electron is transferred simultaneously, so possible sequences may be written down straightforwardly.

Leaving aside very simple inorganic single electron transfer steps, such as those discussed above, the simplest mechanism conceivable would consist of two processes, one chemical and the other electrochemical, denoted CE or EC. A simple example of a CE process would be the reduction of a weak acid, such as acetic acid:

$$CH_3COOH \underset{k_r}{\overset{k_d}{\rightleftharpoons}} CH_3COO^- + H^+ \qquad C$$

$$H^+ + e^- \longrightarrow \tfrac{1}{2}H_2 \qquad\qquad\qquad E$$

The simple EC process is less common, though the reverse of the reaction above would clearly be a case in point, and a better known example is:

$$Fe^{3+} + e^- \longrightarrow Fe^{2+} \qquad\qquad\qquad E$$
$$Fe^{2+} + H_2O_2 \longrightarrow Fe^{3+} + OH^- + OH^\bullet \qquad C$$

Investigation of this type of reaction is usually carried out by controlled convection methods which allow rather simple analytical formulae to be derived under a variety of limiting conditions. Thus, for the CE reactions above, it is possible to adjust the potential on a Pt electrode such that CH_3COOH and CH_3COO^- are both electrochemically *inactive* but H^+ is reduced to H_2 at a limiting current density determined only by the rate of the first step coupled with the transport of H^+ to the electrode surface. In fact, the CE reaction has been studied extensively by rotating disc and it has been found, as expected, that the limiting current at higher rotation rates is *less* than that found by extrapolation from lower values of ω, since the undissociated acid is swept away from the surface at high rotation rates before it has time to dissociate. If the *extrapolated* limiting current density is given by J_{ext}, then:

$$J_{lim} = J_{ext} - \frac{D^{1/6} \cdot [^-OAc]^{1/2} \cdot J_{lim} \cdot \omega^{1/2}}{1.62 \, v^{1/6}(k_d/k_r) \cdot (k_r)^{1/2}} \qquad (1.49)$$

and since the acid dissociation constant is $K_a = k_r/k_d$, this allows k_r to be determined, at least in principle.

Of more intrinsic interest are processes involving two electrons, since these constitute the great bulk of organic reductive processes, especially in protonic solvents. The reason for this is that the first electron transfer will generate an unstable radical species, whereas the second can regenerate a stable closed-shell product. One example that we have already encountered is CO_2 reduction but a case that has been well studied is that of quinone reduction as:

$$A + e^- \longrightarrow B^- \qquad E$$
$$B^- + H^+ \longrightarrow BH \qquad C$$
$$BH + e^- \longrightarrow CH^- \qquad E$$
$$CH^- + H^+ \longrightarrow CH_2 \qquad C$$

where A, BH and CH_2 are:

$$A \qquad\qquad BH \qquad\qquad CH2$$

Initially, it was thought that all two-electron reaction pathways were of this general sort, and a so-called scheme of squares was elaborated, which, for the system shown, would take the form:

$$A \underset{}{\overset{e^-}{\rightleftharpoons}} B^- \underset{}{\overset{e^-}{\rightleftharpoons}} C^{2-}$$

$$H^+ \Big\Updownarrow \qquad H^+ \Big\Updownarrow \qquad H^+ \Big\Updownarrow$$

$$AH^+ \underset{}{\overset{e^-}{\rightleftharpoons}} BH \underset{}{\overset{e^-}{\rightleftharpoons}} CH^-$$

$$H^+ \Big\Updownarrow \qquad H^+ \Big\Updownarrow \qquad H^+ \Big\Updownarrow$$

$$AH_2^{2+} \underset{}{\overset{e^-}{\rightleftharpoons}} BH_2^+ \underset{}{\overset{e^-}{\rightleftharpoons}} CH_2$$

where the exact route followed during the reduction process would depend on the pH and the electrode potential. However, in recent years it has been recognised that alternate pathways are also feasible. Savéant and co-workers have explored disproportionation reactions, and a more complete two-electron route could be written schematically as:

$$A \pm e^- \rightleftharpoons B \qquad E_1$$

$$B \xrightarrow{k_1} C \qquad C_1$$

$$C \pm e^- \rightleftharpoons F \qquad E_2$$

$$B + C \xrightarrow{k_2} A + F \qquad C_2$$

The ECE reaction is essentially just $E_1C_1E_2$, with C_2 being insignificant. If C_2 is important, then the alternative route $E_1C_1C_2$ can either have C_1 rate limiting, denoted DISP1 by Savéant, or C_2 rate limiting, denoted DISP2.

In general, DISP2 reactions can easily be distinguished from ECE or DISP1 since they involve a second-order rate-limiting step, but the distinction between ECE and DISP1 is extremely difficult and, in practice, functionally impossible for the rotating disc electrode. In fact, Compton and co-workers have established that the channel electrode, a hydrodynamic technique with a much higher range of conditions possible, is capable of distinguishing between ECE and DISP1, and have demonstrated that reduction of fluoroscein to leuco-fluorescein in aqueous solution in the pH range 9.5–10.0 is actually a DISP1 reaction:

$$F + e^- \rightleftharpoons S^-$$

$$S^- + H^+ \rightleftharpoons SH$$

$$SH + S^- \longrightarrow F + LH^-$$

where S is the semi-fluoroscein radical anion and LH^- is the singly protonated leuco-fluoroscein anion.

The deposition of metals has also been studied by a large number of electrochemical techniques. For the deposition of Cu^{2+}, for example, it is reasonable to ask whether both electrons are transported essentially simultaneously or whether an intermediate such as Cu^+ is formed in solution. Such questions, like those of the ECE problem discussed above, have usually been investigated by forced convection techniques, since the rate of flow of reactant to and away from the electrode surface gives us an important additional kinetic handle. In addition, by using a second separate electrode placed 'downstream' from the main working electrode, reasonably long-lived intermediates can be transported by the convection flow of the electrolyte to this second electrode and detected electrochemically.

The best known example of this technology is the rotating ring-disc electrode discussed in chapter 2; the 'ring' electrode in this experiment is downstream of the disc, allowing electrochemical detection of products formed at the disc surface, which then survive long enough to be swept to the ring. Thus, electrodeposition of Cu^{2+} in sulphate solution apparently occurs as a single process at an Au/Hg electrode but at potentials just below the point at which reduction begins a significant re-oxidation current can be monitored at the ring, provided the potential of the latter is sufficiently high, arising from the oxidation of Cu^+ formed at the disc. At more negative *disc* potentials, this re-oxidation ring current disappears and it is clear that the reduction of the Cu^+ formed at the disc is now so fast that none can escape to the ring. A similar experiment can be carried out for the electrochemical dissolution of indium where the intermediate In(I) was detected. In solution, the In(I) formed is oxidised further by water or protons as:

$$In(I) + 2H^+ \longrightarrow In^{3+} + H_2$$

but this reaction is slow and In(I) survives to be detected at the ring.

1.3.2   Electrochemical mechanisms with adsorbed intermediates

It is evident from the above discussion that the study of complex electrochemical mechanisms involving homogeneous processes has reached a level at which a reasonably complete analysis can be carried out. By contrast, complex processes taking place purely through surface adsorbed intermediates have proved far less easy to explore by the type of study referred to above and understanding such processes in detail has led to the development of the large array of *in situ* and *ex situ* techniques discussed in this book.

Classically, processes involving surface intermediates were investigated primarily by methods (2)–(4) above and in particular by measuring current as a function of concentration of reagents and electrode potential. A familiar example is the hydrogen evolution reaction, which may proceed by one of two possible mechanisms, both of which share a common first step:

$$H^+ + e^- \underset{M}{\rightleftharpoons} H^\cdot_{ads} \qquad 1(E)$$

$$H^+ + H^\cdot_{ads} + e^- \rightleftharpoons H_2 \qquad 2(E) \text{ (Volmer)}$$

$$2H^\cdot_{ads} \rightleftharpoons H_2 \qquad 3(C) \text{ (Tafel)}$$

This type of mechanism can be analysed by the *steady-state* method. If we consider reactions 1 and 2, and represent the appropriate rate constants by k_1, k_{-1}, etc., and the rates of the corresponding reactions by $v_1, v_{-1}...$, etc. then the rate of change of the coverage θ_H of hydrogen atoms on the surface is given by:

$$d\theta_H/dt = 0 = v_1 - v_{-1} - v_2 + v_{-2} \tag{1.50}$$

where, at steady state, the coverage of hydrogen atoms will be stationary.

Consider the rate of the first reaction in the forward direction: from equation (1.26) we have

$$v_1 = \kappa Z \cdot e^{-\lambda/4k_B T} \cdot e^{-f\beta(\Delta\phi - \Delta\phi_0)} \cdot [H^+] \cdot (1 - \theta_H) \tag{1.51}$$

where we note that $(1 - \theta_H)$ is the fraction of the surface available for the reaction to take place, and we have replaced e_0/k_B by the F/R terms in f.

If the concentration of the electrolyte is sufficient then, as we saw above, all the potential drop at the interface will take place between the metal and the OHP, and the value of $(\Delta\phi - \Delta\phi_0)$ can be replaced by $E - E^0$, where E, E^0 are potentials of the electrode with respect to a standard reference electrode. We note that E^0 is the potential at which no current flows, provided $[O]_0 = [R]_0$ in equation (1.26). We then obtain:

$$v_1 = k_1'' \cdot [H^+] \cdot (1 - \theta_H) \cdot e^{-f\beta(E-E^0)} \tag{1.52}$$

where $k_1'' = kz\, e^{-\lambda/4k_BT}$. However, E^0 is *independent* of concentration, and can be combined with the other pre-exponential factors to give a rate constant, k_1:

$$v_1 = k_1 \cdot [H^+] \cdot (1 - \theta_H) \cdot e^{-f\beta E} \tag{1.53}$$

Two things to notice are that k_1 will depend on the reference electrode chosen and secondly E^0 is the equilibrium potential *not* for the standard hydrogen electrode but rather for the electrochemical equilibrium between H^+ and $H^\cdot_{ads}$. This can be seen by explicitly writing out the equation corresponding to the reverse reaction:

$$v_{-1} = \kappa Z \cdot e^{-\lambda/4k_BT} \cdot \theta_H \cdot e^{f(1-\beta)(E-E^0)} \tag{1.54}$$

and equating (1.49) and (1.50) gives:

$$e^{-f\beta(E-E^0)} \cdot [H^+] \cdot (1 - \theta_H) = e^{f(1-\beta)(E-E^0)} \cdot \theta_H \tag{1.55}$$

which is satisfied at $[H^+] = 1$, $\theta_H = \frac{1}{2}$ when $E = E^0$. Of course, the value of E^0 will differ from that of the standard electrode potential $E^0_{H^+/H_2}$ only by a constant that will depend on the free energy of adsorption of H on the surface, and this allows us to define k_1 with respect to the NHE if we so wish.

The overpotential η is defined as $E - E_r$, where E_r is the equilibrium electrode potential at the concentration $[H^+]$ and θ_H that actually pertain in the system. In fact, since θ_H is unknown normally, we actually define E_r with reference to the values of $[H^+]$ and p_{H_2} through the Nernst equation:

$$E_r = E^0_{H^+/H_2} + (RT/F) \cdot \log_e([H^+]/[p_{H_2}]^{1/2}) \tag{1.56}$$

The rate v_1 can also be defined with respect to the overpotential, η

$$\begin{aligned}
v_1 &= k_1' \cdot [H^+](1 - \theta_H) \cdot \exp[-f\beta(E - E^0_{H^+/H_2})] \tag{1.57}\\
&= k_1' \cdot [H^+](1 - \theta_H) \cdot \exp[-f\beta(E - E_r + [RT/F]\log_e\{[H^+]/[pH_2]^{1/2}\})]\\
&= k_1' \cdot [H^+] \cdot (1 - \theta_H) \cdot ([H^+]/[p_{H_2}]^{1/2})^{-\beta} \cdot \exp[-f\beta\eta]\\
&= k_1' \cdot [H^+] \cdot (1 - \theta_H) \cdot ([H^+]/[p_{H_2}]^{1/2})^{-\beta} \cdot \exp[-f\beta\eta]\\
&= k_1' \cdot [H^+]^{1-\beta}(1 - \theta_H) \exp[-f\beta\eta] \tag{1.58}
\end{aligned}$$

assuming $p_{H_2} = 1$ atm and $k_1' \equiv k_1 \cdot \exp[f\beta(E^0 - E^0_{H^+/H_2})]$.

The importance of this will become more apparent when we consider the reaction order with respect to the concentration of H^+ below. Returning to the complete scheme and defining $\Delta E = E - E^0_{H^+/H_2}$, we have

$$v_1 = k_1[H^+]\cdot(1 - \theta_H)\exp[-f\beta\Delta E] \tag{1.59}$$

$$v_{-1} = k_{-1}\theta_H\exp[f(1 - \beta)\Delta E] \tag{1.60}$$

$$v_2 = k_2\theta_H[H^+]\exp(-f\beta'\Delta E) \tag{1.61}$$

$$v_{-2} = k_{-2}p_{H_2}(1 - \theta_H)\exp[f(1 - \beta')\Delta E] \tag{1.62}$$

provided, as above, that the supporting electrolyte concentration is sufficient to restrict the potential change at the working electrode to within the OHP. In all of (1.59)–(1.62), ΔE is the difference between the electrode potential and the suitably chosen reference which is not dependent on the concentration of H^+ or on p_{H_2}.

Using the steady-state approximation, (1.50), substituting the expressions (1.59) to (1.62) for v_1, v_{-1}, v_2 and v_{-2}, we can write

$$\theta_H = \frac{k_1[H^+]e^{-f\beta\Delta E} + p_{H_2}\cdot k_{-2}e^{f(1-\beta')\Delta E}}{k_1[H^+]e^{-f\beta\Delta E} + k_{-1}e^{f(1-\beta)\Delta E} + k_2[H^+]e^{-f\beta'\Delta E} + k_{-2}p_{H_2}e^{f(1-\beta')\Delta E}}$$

$$\tag{1.63}$$

Unfortunately, this is far too cumbersome an expression to be useful, and it is normal to seek approximations to the formula by making either of reactions 1 and 2 above rate limiting. We can also make the reasonable assumption that regardless of which reaction is rate limiting, the reverse of the second reaction can be neglected provided that we are sufficiently negative in potential for the reaction to be essentially driven just one way.

If reaction 1 is slow, then $k_2 \gg k_1, k_{-1}$, and $\theta_H \to 0$, so the overall reaction rate is now:

$$v \equiv v_1 = k_1[H^+]\exp[-f\beta\Delta E] \tag{1.64}$$

The actual current passed $I = 2FAk_1[H^+]\exp[-f\beta\Delta E]$ since two electrons are transferred for every occurrence of reaction 1. Equation (1.64) constitutes the fundamental kinetic equation for the hydrogen evolution reaction (*her*) under the conditions that the first reaction is rate limiting and that the reverse reaction can be neglected. From this equation, we can calculate the two main observables that can be measured in any electrochemical reaction. The first is the Tafel slope, defined for historical reasons as:

$$b = (\partial\eta/\partial\log_{10}I)_{[H^+]} \tag{1.65}$$

Since ΔE and η are linearly related, evidently for the situation here, with the kinetic equation given by (1.64) or equivalently from (1.58):

$$b = 2.303RT/\beta F \tag{1.66}$$

which at room temperature has the value $120\,mV$ for $\beta = \frac{1}{2}$.

The second observable is the reaction *order*, which has two commonly used definitions. The simplest definition to use, though not necessarily to measure, is:

$$p_{\Delta E} = (\partial \ln I / \partial \log_e[H^+])_{\Delta E} \tag{1.67}$$

which clearly has the value of unity for (1.64). An alternative definition is:

$$p_\eta = (\partial \ln I / \partial \log_e[H^+])_\eta \tag{1.68}$$

which clearly differs from the first definition since (1.58) differs from (1.64). In fact, for the cathodic reaction here it is easily seen that:

$$p_\eta = p_{\Delta E} - (\beta F / RT) \cdot (dE_r / d \log_e[H^+]) = 1 - \beta \tag{1.69}$$

The reader is cautioned to be very careful indeed to ascertain which definition is being used by the author of a paper.

The analysis above can also be carried through under the condition that reaction 2 is slow. From (1.63) we then have:

$$\theta_H \approx \frac{k_1[H^+]}{k_1[H^+] + k_{-1}\,e^{f\Delta E}} \tag{1.70}$$

and if $\theta_H \ll 1$, then $\theta_H \sim (k_1[H^+]/k_{-1}) \cdot \exp[-f\Delta E]$. The current is now given by the rate v_2 in (1.61), remembering two electrons pass for each time v_2 occurs:

$$I \sim 2k_2 FA\theta_H[H^+] \exp[-f\beta'\Delta E] \sim 2k_2 F(k_1/k_{-1})[H^+]^2 \exp[-f(1+\beta')\Delta E] \tag{1.71}$$

It is evident that the Tafel slope is now clearly $2.303RT/F(1 + \beta')$, which is approximately $40\,mV$ if $\beta' \sim \frac{1}{2}$. The reaction order with respect to H^+ is given by:

$$p_{\Delta E} = 2; p_\eta = p_{\Delta E} - \{(1 + \beta')F/RT\} \cdot (dE_r / d \log_e[H^+])$$
$$= 1 - \beta' \tag{1.72}$$

If reaction 2 is slow and $\theta_H \to 1$ (corresponding to $k_1 c_{H^+} \gg k_{-1} \exp[f\Delta E]$), then:

$$I \sim 2FAk_2[H^+] \exp[-f\beta'\Delta E] \tag{1.73}$$

which is essentially indistinguishable from (1.64).

We have, up to now, not considered reaction 3 at all. If reaction 1 is rate limiting, of course, we can determine nothing at all about whether it is followed by reaction 2 or reaction 3; indeed, it is a general rule that any kinetic analysis will be unable to shed light on any step following the rate-limiting step.

However, if reaction 3 is rate limiting we can deduce something useful and we will illustrate the quasi-equilibrium method by using it to derive the kinetic equation under these conditions. This method assumes that *all* reactions prior to the rate limiting step are in equilibrium. Thus, for reaction 1:

$$v_1 = v_{-1}$$

or:

$$k_1[H^+](1 - \theta_H)\exp[-f\beta\Delta E] = k_{-1}\theta_H\exp[f(1 - \beta)\Delta E] \qquad (1.74)$$

whence:

$$\frac{\theta_H}{(1 - \theta_H)} = (k_1/k_{-1})\cdot[H^+]\exp[-f\Delta E] \qquad (1.75)$$

The rate of reaction 3 is clearly $v_3 = k_3\theta_H^2$ so:

$$I \sim 2FAk_3\theta_H^2 \sim 2F(k_1/k_{-1})^2\cdot k_3[H^+]^2\cdot\exp[-2f\Delta E] \qquad (1.76)$$

whence, the Tafel slope is given by:

$$b = 2.303(RT/2F) \sim 30\,mV \text{ at } 298\,K \qquad (1.77)$$

Obviously:

$$p_{\Delta E} = 2 \text{ and interestingly } p_\eta = 0. \qquad (1.78)$$

It is evident that the above type of analysis can be carried out for a wide range of electrochemical reactions, and Tafel slopes and reaction orders can be calculated for various conditions of coverage by intermediates. In principle, therefore, given the experimental Tafel slope and order of a reaction, it should be possible to determine its mechanism by simply consulting theoretical tabulations of order and Tafel slope until a match is found. Unfortunately this is rarely a satisfactory procedure: definitive answers are the exception rather than the rule, since there are a large number of possible rate-limiting steps and limiting coverage conditions, and in addition β commonly varies from 0.3 to 0.7, giving a wide range of predicted values for all scenarios. There is also the implicit assumption above that the rate constant for all the reactions involving $H_{ads}^{\cdot}$ are assumed to be independent of coverage, and also implicit in our analysis has been the fact that either the concentrations of such solution species as H^+ are the same at the electrode surface as in the bulk or, at the least, can be estimated in some reliable way. For concentrations in excess of 0.1 M and small currents, it is certainly the case that the concentration of reagent at the electrode surface will be closely similar to that in the bulk but at more dilute solutions or higher current densities control of the concentration by convective flow of the reagents will be necessary.

The problem of the rate constant being dependent on coverage is a serious one. We have already seen that the Temkin isotherm is commonly obeyed by adsorbates, but fundamental to this isotherm is the idea that the stability of the adsorbate will *decrease* with increasing coverage. It follows that the *activation energy* will also decrease with coverage for reactions involving *loss* of adsorbate and will increase for reactions leading to an increase in coverage. We saw above that the enthalpy of adsorption increased linearly with θ (remember that the enthalpy is *negative*):

$$\Delta H_\theta = \Delta H_0 + X\theta RT$$

and the change in activation energy will normally therefore have the form:

$$U_\theta^\ddagger = U_0^\ddagger + \xi X \theta RT \qquad (1.79)$$

where $0 < \xi < 1$ and normally $0.2 < \theta < 0.8$ to ensure validity of the Temkin isotherm.

Consider a simple anodic discharge reaction:

$$X^- \rightleftharpoons X_{ads} + e^- \quad (1)$$

$$2X_{ads} \longrightarrow X_2 \qquad (2)$$

Then the rates of these reactions are:

$$v_1 = k_1(1 - \theta_X) \cdot [X^-] \cdot \exp[f\beta\Delta E] \cdot \exp[-\xi X\theta) \qquad (1.80)$$

$$v_{-1} = k_{-1} \cdot \theta_X \cdot \exp[-f(1-\beta)\Delta E] \cdot \exp[(1-\xi)X\theta] \qquad (1.81)$$

If reaction 2 is rate limiting, then from the quasi-equilibrium hypothesis $v_1 = v_{-1}$:

$$X\theta = F\Delta E/RT + \log_e(k_1/k_{-1}) + \log_e[(1-\theta_X)/\theta_X] + \log_e[X^-] \qquad (1.82)$$

If $0.2 < \theta_X < 0.8$, then the term $\log_e[(1-\theta_X)/\theta_X]$ will change very little in comparison to the linear term $X\theta$, provided, of course, that X is reasonably large. It follows that θ itself will be potential dependent, satisfying the equation:

$$X\theta \approx F\Delta E/RT + \log_e\left(\frac{k_1}{k_{-1}}\right) + \log_e[X^-] \qquad (1.83)$$

If the reverse of equation (2) above is *activated*, then $v_2 = k_2\theta_X^2 e^{2\xi X\theta}$. However, in the event that chemisorption of X_2 is *not* activated, *all* the change in enthalpy of activation will appear in the activation energy (i.e. $\xi = 1$) and $v_2 = k_2\theta_X^2 e^{2X\theta}$.

Hence, if reaction 2 is rate limiting and chemisorption is activated:

$$I \sim 2FAk_2\theta_X^2 e^{2X\xi\theta} \sim 2Fk_2' \cdot \exp[2f\xi\Delta E] \cdot [X^-]^{2\alpha} \qquad (1.84)$$

assuming that the change in I with θ is dominated by the exponential part. The Tafel slope is now $2.303RT/2\xi F \sim 60\,\text{mV}$ if $\xi \sim \frac{1}{2}$, as compared to the $30\,\text{mV}$ expected if X_{ads} obeys the Langmuir isotherm and $\theta_X \ll 1$, as can be verified by carrying out an analysis on equations (1) and (2) above, following the methodology for the hydrogen case. The major difference between the two cases would be the reaction order, which is clearly 2ξ for the Temkin case and 2 for the Langmuir case.

1.3.3   The direct detection of intermediates

The use of Tafel slopes and reaction orders to explore the mechanism of an electrochemical reaction involving adsorbed intermediates is clearly fraught

with problems, and attention has increasingly turned in the last twenty years
to the direct verification of a proposed mechanism by the identification and
study of adsorbed species. The methods by which this can be done are the
primary subject material of this book and are summarised below.

(1) *Pulse methods*:   In these techniques, the surface is prepared in a steady-
    state form with a stationary coverage of the intermediate. The current
    or potential is then rapidly altered, and the resultant potential or current
    changes monitored. From these measurements, following correction for
    base-line effects, the charge associated with the change in coverage of the
    intermediate can be obtained.
(2) *AC methods*:   Very closely related to (1) are AC methods, in which the
    AC response of the electrode under steady-state conditions is measured
    over a wide range of frequencies, and the resultant detailed data are then
    modelled. This technique is intrinsically very powerful, and the manner
    in which mechanistic information can be obtained is discussed in detail
    below.
(3) *Cyclic voltammetric methods*:   In these, the potential is swept linearly
    with time and the oxidation or reduction of the surface species can
    be followed by measuring the resultant current. Great care is needed in
    the interpretation of cyclic voltammograms and examples are given in
    chapter 2.
(4) *Physicochemical methods*:   The direct *spectroscopic* detection of inter-
    mediates has proved immensely difficult, especially in the infrared, owing
    to interference by the solvent, but increasingly powerful tools are being
    developed. These direct techniques undoubtedly offer the most convincing
    proof of a model mechanism, and they also indicate whether films on
    electrode surfaces are forming that may not be detectable electrochemically.
    A detailed description of these techniques is given in chapter 2.

This book is concerned with mechanisms; in chapter 3 some well-known electrochemical reactions are considered in detail and the mechanisms proposed
for these reactions examined in the light of modern advances. The rate at
which the subject is changing is now so fast that this section can only give
a snapshot of a subject in the throes of rapid change.

Further reading

Albery, W.J. (1975) *Electrode Kinetics*, Clarendon Press, Oxford.
Albery, W.J. and Hitchman, M. (1971) *Ring Disc Electrodes*, Clarendon Press, Oxford.
Bard, A.J. and Faulkener, L.R. (1980) *Electrochemical Methods*, John Wiley and Sons, New York.
Bockris, J.O'M. and Reddy, A.K.N. (1970) *Modern Electrochemistry*, Plenum, New York.
Compton, R.G. and Hamnett, A. (eds) (1989) *Comprehensive Chemical Kinetics*, Vol. 29, Elsevier,
    Amsterdam.
Sawyer, D.T. and Roberts, J.L. (1974) *Experimental Electrochemistry for Chemists*, John Wiley
    and Sons, New York.
The Southampton Electrochemistry Group (1990) *Instrumental Methods in Electrochemistry*,
    Ellis Horwood, Chichester.

2 Techniques giving mechanistic information

2.1
Surface specific processes—*in situ* techniques

2.1.1 Electrocapillarity

The model of the electrode/electrolyte interface proposed in the first chapter is termed the Stern model. This is now accepted as the definitive picture of the structure of the interface, primarily as a result of electrocapillarity studies.

Electrocapillarity is the measurement of the variation of the surface tension of mercury in (usually) aqueous electrolyte with applied potential. The surface tension, γ, of an interface relates to the surface free energy, G, by the expression:

$$\gamma = (\partial G / \partial A)_{T,P,\mu}, \qquad J\,m^{-2} = N\,m^{-1} \tag{2.1}$$

where A is the *area* of the electrode; plainly γ can be regarded as the free energy per unit area of an interface. Any external perturbation, such as applied potential, that alters the energetics of the interface will thus change γ. Solid metal surfaces are generally not homogeneous and only recently have reliable techniques been available for the measurement of the surface tension of solids; however, mercury is a liquid under normal conditions and so has a homogeneous and *reproducible* surface that can be formed clean. In addition, the surface tension of mercury is easy to measure as a function of the applied potential and mercury is a polarisable electrode, i.e. it has a wide range of potential over which no faradaic current will flow (in the absence of electroactive species). Altering the potential across the interface then simply charges it without any electron transfer complicating matters. Mercury is thus an ideal choice as the substrate in electrochemical experiments involving the measurement of γ.

A typical electrocapillarity system is shown in Figure 2.1(a). The mercury reservoir provides a source of clean mercury to feed a capillary tube; the height of mercury in this tube can be varied such that the mass of the Hg column exactly balances the surface tension between the mercury and the capillary walls, see Figure 2.1(b). A voltage V is applied across the mercury in the capillary and a second electrode which is non-polarisable (i.e. the interface will not sustain a change in the potential dropped across it), such as the normal hydrogen electrode, NHE. The potential distribution across the two interfaces is shown in Figure 2.1(c). As can be seen:

$$V = (\phi_{M,q} - \phi_S) + (\phi_S - \phi_{Pt}) \tag{2.2a}$$

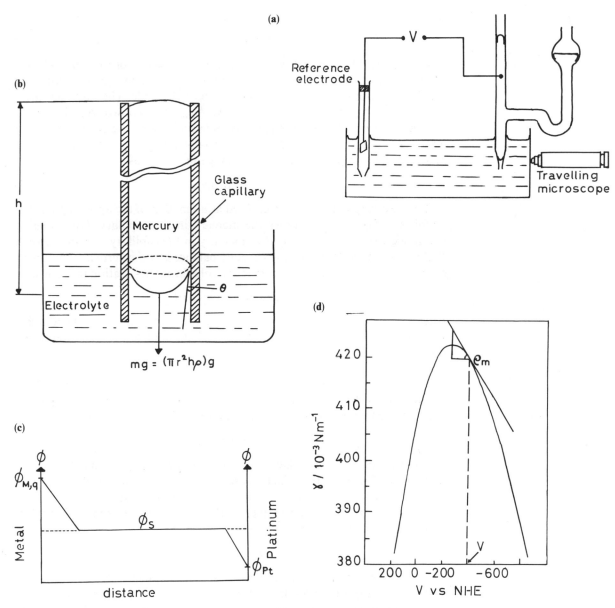

Figure 2.1 (a) A schematic representation of the apparatus employed in an electrocapillarity experiment. (b) A schematic representation of the mercury/electrolyte interface in an electro-capillarity experiment. The height of the mercury column, of mass m and density ρ, is h, the radius of the capillary is r, and the contact angle between the mercury and the capillary wall is θ. (c) A simplified schematic representation of the potential distribution across the metal/electrolyte interface and across the platinum/electrolyte interface of an NHE reference electrode. (d) A plot of the surface tension of a mercury drop electrode in contact with 1 M HCl as a function of potential. The surface charge density, ρ_M, on the mercury at any potential can be obtained as the slope of the curve at that potential. After *Modern Electrochemistry*, J.O'M. Bockris and A.K.N. Reddy, Plenum, New York, 1970, Vol. 2, p. 702.

or:

$$V = \Delta\phi + \Delta\phi_{ref} \qquad (2.2b)$$

where $\phi_{M,q}$ is the potential on the electrode as a result of the charged ions at the interface, ϕ_S is the potential in the bulk solvent and ϕ_{Pt} is the potential on the platinum surface of the NHE reference electrode. The potential drop of interest is $\Delta\phi$. Unfortunately, because $\Delta\phi_{ref}$ is unknown, $\Delta\phi$ cannot be related directly to the measured applied potential V. However, as the reference electrode is non-polarisable, $\Delta\phi_{ref}$ is essentially independent of V; any change in V will then result in an identical change in $\Delta\phi$. Thus:

$$dV = d(\Delta\phi) \qquad (2.3)$$

The surface tension can be measured as a function of the applied potential as follows. Figure 2.1(b) shows the mercury in the capillary. On changing the applied potential, the height of mercury in the capillary is adjusted such that the weight of the mercury is exactly balanced by the vertical component of the surface tension, and the mercury column is thus stationary. Thus:

$$mass = -\gamma 2\pi r \cos\theta$$

where $2\pi r$ is the contact length, r is the radius of the capillary tube and θ is the contact angle between the mercury and the wall of the tube. Replacing the mass:

$$h\pi r^2 \rho = -\gamma 2\pi r \cos\theta$$

where ρ is the density of mercury and h the height of the column. Allowing $\theta \approx 180°$ for mercury/glass and rearranging:

$$\gamma = h\rho r/2 \qquad (2.4)$$

A typical plot of the surface tension of mercury vs. the applied potential is shown in Figure 2.1(d), which shows the γ/V plot for a 1M HCl solution. Clearly, the form of the plot is an inverted parabola, suggesting that $\gamma \propto -V^2$.

A first step in attempting to explain the shape of the plot in Figure 2.1(d) is provided by the *Lippmann equation*:

$$(\partial\gamma/\partial V)_\mu = -\sigma_M \qquad (2.5)$$

This relates the change in surface tension with the applied potential at constant electrolyte composition to the surface charge density on the metal, σ_M.

In order to extract the variation of γ with V from equation (2.5), link this with the potential drop across the mercury interface, $\Delta\phi$, and so make sense of the experimentally obtained γ vs. V plots, the potential dependence of σ_M is required; this is only possible within the framework of a model.

The earliest model was provided by Helmholtz, and is shown schematically in Figure 2.2(a) for a negatively charged electrode. Immediately next to the electrode is a layer of water molecules, attracted via their dipoles. As a result of coulombic attraction, cations approach the electrode as close as their

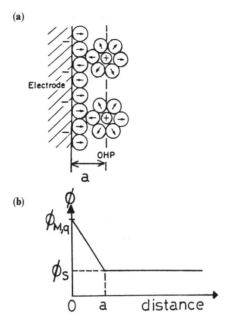

Figure 2.2 (a) The structure of the electrode/electrolyte interface, assuming a single layer of solvated ions adjacent to the electrode. The distance of closest approach of the ions to the electrode is a, and the ion sheet forms the outer Helmholtz plane (OHP). (b) The variation of the potential as a function of the distance from the metal surface for the interface shown in (a).

solvation sheaths, and the layer of 'adsorbed' water molecules, allow; this distance of nearest approach is a. The Helmholtz model predicts that the total potential difference between the electrode and the bulk solution, $\Delta\phi = \phi_{M,q} - \phi_S$, is dropped across this Helmholtz layer, as shown in Figure 2.2(b). In other words, the nearest cations to the electrode form a sheet of charge parallel to the electrode surface, of equal and opposite charge to σ_M, effectively forming one half of a parallel-plate capacitor.

Taking the analogy with a capacitor one step further, the potential dropped across its plates is proportional to the charge density on the plates, σ:

$$\Delta\phi \propto \sigma \tag{2.6}$$

the constant of proportionality is the capacitance C,

$$C \cdot \Delta\phi = \sigma \tag{2.7}$$

and is *independent of $\Delta\phi$ or σ*. The capacitance is related to the separation of the plates, d, by:

$$C = \varepsilon_0 \varepsilon / d \tag{2.8}$$

where ε is the relative permittivity (sometimes termed the dielectric constant) of the medium between the plates and ε_0 is the absolute permittivity of a vacuum ($8.854 \times 10^{-12}\,\mathrm{F\,m^{-1}}$).

Thus, in terms of the Helmholtz model equation (2.7) becomes:

$$\Delta\phi = \sigma_M/C_H \tag{2.9}$$

or:

$$\Delta\phi = \sigma_M \cdot a/\varepsilon_0\varepsilon \tag{2.10}$$

where σ_M is the charge on the metal, $-\sigma_M$ the charge on the Helmholtz plane of ions and ε is that for water, $c.$ 80.

Thus, a relationship has been established between $\Delta\phi$ and σ_M. Now, equation (2.5) stated:

$$d\gamma = -\sigma_M \cdot dV \tag{2.5}$$

and, from equation (2.3), $dV = d(\Delta\phi)$, thus:

$$d\gamma = -\sigma_M \cdot d(\Delta\phi) \tag{2.11}$$

From equations (2.9) and (2.11), therefore:

$$d\gamma = -C_H(\Delta\phi) \cdot d(\Delta\phi) \tag{2.12}$$

and:

$$\int d\gamma = -C_H \cdot \int (\Delta\phi) \cdot d(\Delta\phi) \tag{2.13}$$

$$\gamma = -[C_H \cdot (\Delta\phi)^2/2] + \text{constant} \tag{2.14}$$

Thus, we have a relationship between γ and the potential drop across the mercury/electrolyte interface, as desired; in addition, the relationship is parabolic. However, a link must now be established between the unmeasurable $\Delta\phi$, and the measured applied potential.

The relationship between $\Delta\phi$ and V can be found by a consideration of Figure 2.3 which shows a typical plot of γ vs. $\Delta\phi$ on the basis of equation (2.14). Thus, at the maximum on the curve, $d\gamma/d(\Delta\phi) = 0 = \sigma_M$; the potential at γ_{max} corresponds to the *point of zero charge*, the pzc. Since the pzc of an electrode

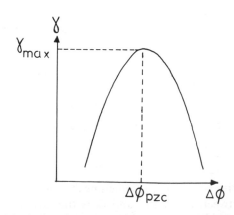

Figure 2.3   A plot of γ vs. potential according to equation (2.15).

is an extremely important parameter, the potential of the electrode with respect to the pzc has a major influence on the nature of species attracted to the electrode surface. At potentials positive of the pzc, anions are attracted to the surface; at potentials negative of the pzc, cations will be attracted. This argument does not take into account the high polarisability of anions and the role of dispersion forces (see below) but is sufficient at this point. Neutral molecules would be expected to adsorb most strongly at the pzc, since such adsorption requires the displacement of water molecules and ions from the surface region; the removal of the dipolar water molecules and charged ions will be much easier near the pzc.

Now, at the point of zero charge, equation (2.9) implies that $\Delta\phi = 0$; i.e. that the pzc corresponds to a potential drop across the interface of zero and, from equation (2.2), that $\phi_M = \phi_S$. This is not found in practice owing to the layer of water molecules at the electrode surface that are present even at the pzc. These water dipoles give rise to an additional contribution to $\Delta\phi$, see Figure 2.4(a). This additional potential drop, $\Delta\phi_D$, will change sign according to the orientation of the water dipoles at the electrode, and equation (2.2) can thus be re-written as:

$$V = \Delta\phi_D + (\phi_{M,q} - \phi_S) + (\phi_S - \phi_{Pt})$$

or,

$$V = \Delta\phi + \Delta\phi_{ref}$$

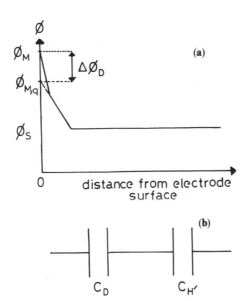

Figure 2.4   (a) The dependence of the potential as a function of the distance from the electrode surface, taking into account the presence of adsorbed water dipoles. (b) The interface in (a) represented in terms of two capacitors. C_D is the dipole capacitance ($= \varepsilon\varepsilon_0/[a - r_{H_2O}]$) and $C_{H'}$ is the original Helmholtz capacitance ($= \varepsilon\varepsilon_0/r_{H_2O}$).

where $\Delta\phi = [\Delta\phi_D + (\phi_{M,q} - \phi_S)]$. The contribution of the water dipoles cannot be separately measured explicitly. In the simplest model, they are assumed to be static at potentials away from the pzc, but to 'flip over' at the pzc. Hence, at potentials away from the pzc, we can again write:

$$dV = d(\Delta\phi)$$

At the pzc,

$$\Delta\phi_{pzc} = \Delta\phi_{D,q=0} - \phi_S$$

where $\Delta\phi_{D,q=0}$ is the only potential drop across the mercury interface. A more correct expression of equation (2.13) would be to replace $\Delta\phi$ by $(\Delta\phi - \Delta\phi_{pzc})$, i.e. $\Delta(\Delta\phi)$. From equation (2.2) we can write:

$$\Delta V = \Delta(\Delta\phi) \tag{2.15}$$

where

$$\Delta V = (V - V_{pzc}) \tag{2.16}$$

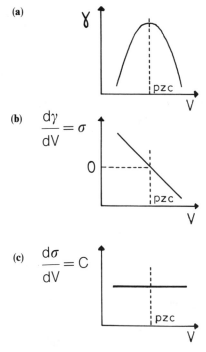

Figure 2.5   (a) Schematic plot of surface tension vs. potential according to equation (2.15) and assuming that cations and anions behave in identical manners. (b) Plot of surface charge density vs. potential obtained *via* the differentiation of the curve in (a). (c) Plot of differential capacitance vs. potential obtained *via* the differentiation of the curve in (b).

Replacing (2.16) in equation (2.13), and remembering that at $\gamma = \gamma_{max}$, $\Delta\phi = \Delta\phi_{pzc}$, $V = V_{pzc}$:

$$\int_{\gamma}^{\gamma_{max}} d\gamma = -C_H \int_{V}^{V_{pzc}} [V - V_{pzc}] \cdot dV \qquad (2.17)$$

giving:

$$\gamma = \gamma_{max} - C_H(V - V_{pzc})^2/2 \qquad (2.18)$$

The modified Helmholtz capacitance, C_H, is now the total of two capacitors in series, (see Figure 2.4(b)):

$$1/C_H = r_{H_2O}/\varepsilon\varepsilon_0 + (a - r_{H_2O})/\varepsilon\varepsilon_0 \qquad (2.19)$$

where r_{H_2O} is the radius of the water molecule; the first term represents the contribution from the water dipoles, (C_D), and the second the capacitance of the Helmholtz layer of ions, $(C_{H'})$.

Thus, we now have a reasonable model of the interface in terms of the classical Helmholtz model that can explain the parabolic dependence of γ on the applied potential. The various plots predicted by equation (2.18) are shown in Figures 2.5(a) to (c). The variation in the surface tension of the mercury electrode with the applied potential should obey equation (2.18). Obtaining the slope of this curve at each potential V (i.e. differentiating equation (2.18)), gives the charge on the electrode, σ_M, at that potential. A

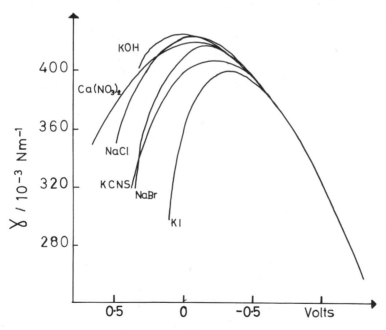

Figure 2.6 Plots of surface tension vs. potential for mercury in contact with various electrolytes at 18°C. The potential axis is plotted with respect to the pzc for NaF. From the work of D.C. Grahame (1947).

plot of σ_M vs. V should be a straight line with σ_M passing through 0 at $V = V_{pzc}$. The slope of this line $(d^2\gamma/dV^2)$, gives the Helmholtz capacitance of the interface, which should be independent of V.

Figure 2.6 shows typical electrocapillarity curves for a range of electrolytes at a mercury electrode. Several observations on Figure 2.6 can be made. First, the identity of the cation seems to have very little effect on the negative potential side of the electrocapillarity curve. However, the identity of the anion can give rise to marked deviations from the behaviour predicted by the Helmholtz model. Such deviations appear as an increasing curvature of the $\gamma/\Delta\phi$ plot in the positive potential region and a shifting of the pzc to more negative potentials. Considering first only the former observation, this increasing curvature implies an increase in the slope of the σ_M vs. $\Delta\phi$ plot and thus an increase in the capacitance of the double layer having anions in the Helmholtz plane. This, in turn, implies that anions can be present nearer than a distance dictated by the closest approach of the solvation sheath. Such an increase in C_H is indicative of the energy balance that exists between the loss of stabilisation of an ion on shedding some, or all, of its solvating water, and the gain in electrostatic stabilisation afforded by the ion approaching as near as possible to the oppositely charged electrode. An important contribution to this balance is the strength of the solvation of the ion; in general, the smaller cations are more strongly solvated than the larger anions, hence the insensitivity of the electrocapillarity curve in the negative potential region to the identity of the cation. However, anions show a wide variation in solvation energy, from the small, non-polarisable F^- ion which is strongly solvated, to I^- which is large and polarisable and weakly solvated. The Helmholtz model can quite easily take account of this difference between the cationic and anionic contributions to the interfacial capacitance by allowing the γ vs. V curve to be comprised of two half-parabolae, joined at the pzc. This gives rise to the corresponding plots of σ_M and C_H shown in Figures 2.7(a) to (c).

The second observation, that of the shift in the pzc to more negative potentials, shows the presence of anions not just nearer a negatively charged electrode than expected but *in contact* with the electrode, even when it has a negative charge. The potential is having to be made much more negative than that required to replace *solution* anions by cations. Hence, the concept of 'specific adsorption' must be introduced to modify the Helmholtz model in order to account for these deviations; again, it is a question of the energy balance between the loss of stabilisation caused by the removal of the water of solvation, the gain in energy afforded by the formation of the bond between ion and electrode and the energy change caused by removing the surface water. This time the interaction between the ion and the electrode may be electrostatic or, in the case of anions at negatively charged electrodes, may be via Van der Waals forces. Ions, solvent dipoles and neutral molecules may all be adsorbed at an electrode surface, irrespective of the potential of the electrode, via Van der Waals forces, with the adsorption weakened or

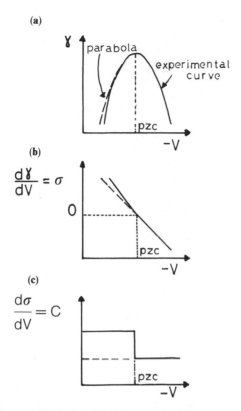

Figure 2.7 (a) Experimental (full line) and ideally parabolic (dashed line) electrocapillary curves and the corresponding (b) charge vs. potential, and (c) capacity vs. potential curves.

strengthened by potential changes at the electrode. For anions at a positively charged electrode the strongest interaction will be due to the coulombic attraction; however, Van der Waals forces may be strong enough to allow negatively charged ions to be adsorbed at a negatively charged metal. Again, the solvation energy is an extremely important factor and hence specific adsorption is predominantly observed for anions.

As a result of the above considerations, the Helmholtz model of the interface now shows two planes of interest (see Figure 2.8). The inner Helmholtz plane (IHP) has the solvent molecules and specifically adsorbed ions (usually anions); the outer Helmholtz plane (OHP), the solvated ions, both cations and anions. It can be seen from Figure 2.8 that the dielectric in the capacitor space now comprises two sorts of water: that specifically adsorbed at the electrode surface and that lying between the two Helmholtz planes. Continuing the analogy with capacitance, these two forms of water act as the dielectric in two capacitors connected in series.

As was seen above, the Helmholtz capacitance, C_H, is given by:

$$1/C_H = r_{H_2O}/\varepsilon\varepsilon_0 + (a - r_{H_2O})/\varepsilon\varepsilon_0 \qquad (2.19)$$

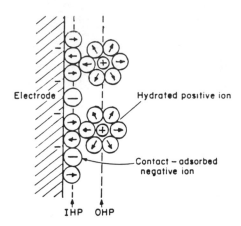

Figure 2.8   The 'double layer' Helmholtz model of the electrode/electrolyte interface.

To obtain an idea of the magnitude of C_H that may be expected from equation (2.19), consider a negatively charged metal: the radius of a typical cation is *c*. 2Å, and the diameter of water about the same, $\varepsilon_{H_2O} = 80$, $\varepsilon_0 = 8.854 \times 10^{-12}\,F\,m^{-1}$. This gives a double layer capacitance of *c*. $1.2\,F\,m^{-2}$; in practice, the observed values are 0.05 to $0.5\,F\,m^{-2}$. Thus, the theoretical Helmholtz capacitance is much higher than that found experimentally, at first sight indicating a serious flaw in the revised model. However, the problem can be traced to the value of ε_{H_2O} used. The value of 80 employed for the dielectric constant of water assumes that the molecules are free to rotate and so re-orient on changing the applied electric field. The water molecules at the electrode surface will not be able to change orientation at all and those between the IHP and OHP have only limited freedom. Calculations suggest that the adsorbed water has $\varepsilon \approx 6$ and the water between the Helmholtz planes has $\varepsilon \approx 30$; feeding these values into equation (2.19) gives $C_H \approx 0.27\,F\,m^{-2}$, in fairly good agreement with experiment.

At this point it should be noted that specifically adsorbed anions cannot easily be accounted for by the Helmholtz model in terms of their contribution to C_H. Nevertheless, the Helmholtz model of two layers of charge in the interfacial region, the 'double layer' model, seems to be reasonably good as far as explaining the general shape of the γ vs. V plots and the magnitude of the double layer capacitance. However, more serious difficulties arise on considering the experimentally observed plots of the interfacial capacitance vs. applied potential. Figure 2.9 shows capacitance vs. potential plots for the Hg/aqueous NaF interface at various electrolyte concentrations and is typical of the experimentally observed curves in the absence of specific adsorption.

Several important observations can be made on Figure 2.9. As was discussed above, the plot of C_H vs. V should be a step function, indicating that away from the pzc the capacitance is independent of potential. In addition, equation (2.19) does not include any dependence on concentration. From

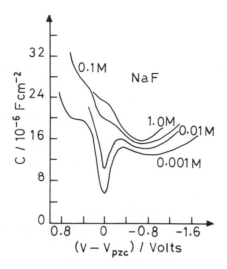

Figure 2.9 Plots of differential capacitance vs. potential at 25°C for mercury in contact with various concentrations of aqueous sodium fluoride. The potential axis is plotted with respect to the pzc. From the work of D.C. Grahame (1947).

Figure 2.9, it can be seen that the interfacial capacitance *does* show a dependence on concentration, particularly at low concentrations. In addition, whilst there is some evidence of the expected step function away from the pzc, the capacitance is *not independent of V*. Finally, and most destructive, the Helmholtz model most certainly cannot explain the pronounced minimum in the plot at the pzc at low concentration. The first consequence of Figure 2.9 is that it is no longer correct to consider that differentiating the γ vs. V plot twice with respect to V gives the absolute double layer capacitance C_H where C_H is independent of concentration and potential, and only depends on the radius of the solvated and/or unsolvated ion. This implies that the $d\gamma/dV$ (i.e. σ_M) vs. V plot is not two straight lines joined at the pzc. Thus, in practice, the experimentally obtained capacitance is $(d\sigma_M/dV)$, and is termed the differential capacitance. (The value quoted above of 0.05–$0.5\,F\,m^{-2}$ for the double-layer was in terms of differential capacitance.) A particular value of (dq_M/dV) is obtained, and is valid, *only at a particular electrolyte concentration and potential*. This admits the experimentally observed dependence of the double layer capacity on V and concentration. All subsequent calculations thus use differential capacitances specific to a particular concentration and potential.

Another important and fundamental consequence of Figure 2.9 is that the Helmholtz theory cannot cope with a concentration dependence of the differential capacitance nor can it be modified in such a way as to incorporate a minimum in $(d\sigma_M/dV)$ at low concentrations of electrolyte. Instead, a second theory must be invoked, having a dependence on electrolyte concentration but in such a way that it is only important at low concentration. A first step along this particular road is provided by the Gouy–Chapman theory. As for the Helmholtz model, this model also allows the positive charge on the metal to be balanced by an equal and opposite charge in solution but reconsiders where this charge actually resides in the electrolyte. In the Helmholtz model, the balancing solution charges are fixed in the OHP and IHP, neglecting altogether the effects of thermal motion. The Gouy–Chapman model moves to the opposite extreme and considers the charge density near the electrode to be a balance between thermal motion attempting to randomise the concentration of ions and the coulombic attraction exerted by the charged electrode attraction oppositely charged ions. Gouy–Chapman *does not allow for an inner Helmholtz plane at all*; the nearest approach of the ions to the electrode is dictated by the thickness of their solvation sheaths. This leads to the concept of a diffuse double layer with ions of both charges in a spatially extended region near the electrode surface. Ions of charge opposite to that on the electrode are present in excess, due to the coulombic attraction, whilst ions of the same charge are present in a lower concentration. This picture represents a 'snapshot' of the actual situation, with all of the ions in constant motion. Thus, there is a falling off of the nett charge density from the OHP out into the bulk; in the bulk the concentrations of negatively and positively charged ions are equal, hence the nett charge density is zero, see Figure 2.10.

If N_i^0 is the number of ions of type i and charge z_i per unit volume present in the bulk at equilibrium, then the Boltzmann equation leads to the following expression for the number of ions of type i at a distance x from the electrode:

$$N_i = N_i^0 \exp[-z_i e_0 \phi(x)/kT] \tag{2.20}$$

where e_0 is the charge on an electron and $\phi(x)$ the potential at distance x. *As can be seen from Figure 2.10, this equation is valid only for $x \geq a$, since the Gouy–Chapman model only allows a minimum distance of approach of*

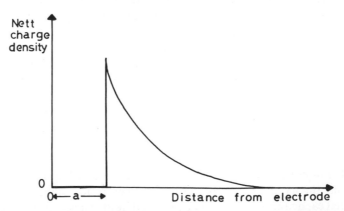

Figure 2.10   Schematic representation of the variation of the excess, or nett, surface charge density as a function of the distance away from the electrode surface, according to the Gouy–Chapman theory. The distance of nearest approach of the ions, with their associated solvation sheaths, is *a*.

an ion of type i of a, where a is the hydrated diameter of i. Thus, letting $\xi = x - a$, and remembering that $N_i = N_i^0$ when $\phi(x) = \phi_S$, then $\phi(x) = \phi(\xi) - \phi_S$ and:

$$N_i(\xi) = N_i^0 \exp\{-[z_i e_0(\phi(\xi) - \phi_S)]/kT\} \qquad (2.21)$$

From the above equation, it can be seen that the number of ions of charge opposite to that on the electrode decreases exponentially with potential from an excess at the OHP $(x = a)$, to that of the bulk.

The charge density at a point ξ is simply:

$$q(\xi) = \Sigma_i N_i(\xi) z_i e_0 \qquad (2.22)$$

from equation (2.21), therefore:

$$q(\xi) = \Sigma_i z_i e_0 N_i^0 \exp\{-[z_i e_0(\phi(\xi) - \phi_S)]/kT\} \qquad (2.23)$$

The charge density at a point ξ can also be described in terms of the Poisson equation (again with $\phi(x) = \phi(\xi) - \phi_S$):

$$q(\xi) = -\varepsilon\varepsilon_0[d^2\phi(\xi)/d\xi^2] \qquad (2.24)$$

Thus, combining equations (2.23) and (2.24):

$$[d^2\phi(\xi)/d\xi^2] = -[\Sigma_i z_i e_0 N_i^0 \exp\{-[z_i e_0(\phi(\xi) - \phi_S)]/kT\}]/\varepsilon\varepsilon_0 \qquad (2.25)$$

The exponential term can be expanded as:

$$\Sigma_i z_i e_0 N_i^0 - \Sigma_i z_i e_0 N_i^0 [z_i e_0(\phi(\xi) - \phi_S)]/kT$$

However, electroneutrality tells us that in a given volume in the bulk of the solution, the sum of all of the charges must be zero, hence the first term in the above expression is zero and therefore we can write:

$$-[z_i e_0(\phi(\xi) - \phi_S)]/kT$$

N_i^0/N_A, where N_A is Avogadro's number, is just the concentration of species i, c_i; and the ionic strength of the solution, $I = 0.5\Sigma_i c_i z_i^2$. Thus, we can rewrite equation (2.25) as:

$$[d^2\phi(\xi)/d\xi^2] = (2000 e_0^2 N_A/\varepsilon\varepsilon_0 kT)I(\phi(\xi) - \phi_S) \qquad (2.26)$$

(the factor of 1000 takes into account the fact that the concentration is in $mol\,dm^{-3}$, instead of the more correct $mol\,m^{-3}$) or:

$$[d^2\phi(\xi)/d\xi^2] = \kappa^2[\phi(\xi) - \phi_S] \qquad (2.27)$$

with

$$\kappa^2 = (2000 e_0^2 N_A/\varepsilon\varepsilon_0 kT)I \qquad (2.28)$$

Integrating equation (2.27) twice, with the boundary conditions that ϕ can vary between $\phi(\xi)$ at $\xi = 0$, i.e. ϕ_{OHP} to ϕ_S, gives:

$$\phi(\xi) = \phi_S + (\phi_{OHP} - \phi_S)e^{-\kappa\xi}, \qquad (2.29)$$

which shows the expected exponential fall-off in potential with distance from the outer Helmholtz plane; i.e. as $x \to \infty$, the exponential term tends to zero and $\phi(\xi) \to \phi_S$. The constant κ is of particular importance as its inverse, $1/\kappa$, is a measure of the thickness of the *diffuse* double layer and is termed the shielding, or Debye, length. From equation (2.29),

$$(\phi(\xi) - \phi_S)/(\phi_{OHP} - \phi_S) = e^{-\kappa\xi}$$

from which it can be seen that κ^{-1} is the distance from the OHP to a point in the diffuse layer where the potential has dropped to $1/e$, (i.e. 67%), of its total change from the OHP to the bulk of the solution. Effectively, the Debye length is the average distance from the electrode that the second plate of the 'diffuse layer' capacitor can be placed.

It can be shown that the differential capacitance of the *diffuse layer*, according to Gouy–Chapman theory, C_{GC}, is given by:

$$C_{GC} = (\varepsilon\varepsilon_0/\kappa^{-1})\cosh(z_i e_0[\phi(\xi) - \phi_S]/kT) \qquad (2.30)$$

A cosh plot resembles a very steep parabola for small values of $(\phi(\xi) - \phi_s)$, as shown in Figure 2.11, the concentration dependence of C_{GC} arising through κ (equation (2.28)). At the pzc, the only drop across the mercury interface is due to the water dipoles; and we can write $\phi(\xi) = \phi_S$. Hence, from equation (2.30), $C_{GC} = \varepsilon\varepsilon_0/\kappa^{-1}$ and we can read off the value of the Debye length from the differential capacitance plot.

The data of Figure 2.11 bear some relationship to Figure 2.9 at low

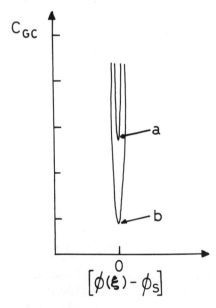

Figure 2.11 According to the Gouy–Chapman theory, the capacity of the electrode/electrolyte interface should be a cosh function of the potential difference across it (see text). Concentration of electrolyte in (b) > than that in (a).

concentrations but the theory is obviously flawed at higher concentrations. In addition, the differential capacities predicted by the Gouy–Chapman theory agree reasonably well with those observed experimentally at low concentrations, however, they are anything up to an order of magnitude too low in concentrated electrolyte (e.g. 1.0 M). Thus, the Gouy–Chapman model can predict the sharp minimum in the differential capacitance vs. potential plot and the Helmholtz theory can explain the 'wings' to potentials either side of the pzc. It was left to Stern to point out that the most realistic model was a combination of both models. The Stern model effectively considers the total potential drop between the electrode and the bulk of the solution, $\Delta\phi$, as the sum of two differences:

$$\Delta\phi = (\phi_M - \phi_{OHP}) + (\phi_{OHP} - \phi_S)$$

where $\phi_M = (\phi_{M,q} + \Delta\phi_D)$, or:

$$\Delta\phi = \Delta\phi_H + \Delta\phi_{GC} \tag{2.31}$$

where $\Delta\phi_H$ is the potential drop across the Helmholtz layers and $\Delta\phi_{GC}$ is the potential dropped across the diffuse double layer.

Thus, the overall double layer capacitance is given by:

$$1/C_{DL} = 1/C_H + 1/C_{GC} \tag{2.32}$$

and:

$$1/C_{DL} \approx [r_{H_2O}/\varepsilon_D\varepsilon_0 + (a - r_{H_2O})/\varepsilon_{IHP-OHP}\varepsilon_0] + (1/\kappa)/\varepsilon\varepsilon_0 \tag{2.33}$$

where ε_D is the dielectric constant for adsorbed water, ε that for bulk water, and $\varepsilon_{IHP-OHP}$ the constant for water molecules between the two Helmholtz planes.

Equation (2.33) now defines the double layer in the final model of the structure of the electrolyte near the electrode: specifically adsorbed ions and solvent in the IHP, solvated ions forming a plane parallel to the electrode in the OHP and a diffuse layer of ions having an excess of ions charged opposite to that on the electrode. The excess charge density in the latter region decays exponentially with distance away from the OHP. In addition, the Stern model allows some prediction of the relative importance of the diffuse vs. Helmholtz layers as a function of concentration. Table 2.1 shows

Table 2.1 Debye lengths for various electrolytes calculated from equation (2.28) in water

Concentration (mol dm^{-3})	κ^{-1}(Å) electrolyte type		
	1:1	1:2/2:1	2:2
10^{-4}	304	176	152
10^{-3}	96	55.5	48.1
10^{-2}	30.4	17.6	15.2
10^{-1}	9.6	5.5	4.8

some values of κ^{-1} calculated from equation (2.28) for aqueous solutions of a strong electrolyte.

It can be seen from the table that, in dilute solutions, the diffuse layer may extend some hundreds of angstroms out from the electrode. In contrast, in more concentrated solutions, i.e. 0.1 M, the diffuse layer thickness decreases to $< 10\,\text{Å}$; not much more than the thickness of the Helmholtz layer. As C_H has no concentration dependence it remains constant on changing the concentration; however, from equations (2.22) and (2.23), C_{GC} decreases as the concentration of the electrolyte increases. Thus, at low concentration:

$$1/C_{GC} \gg 1/C_H$$

and thus:

$$1/C_{DL} \approx 1/C_{GC}, \qquad C_{DL} \approx C_{GC}$$

Similarly, in sufficiently concentrated solutions, the double layer capacity is effectively that of the Helmholtz layer; in this regime, the analogy between the double layer and two capacitors in series works very well. In effect, at higher concentrations, most of the solution charge is squeezed into the two Helmholtz planes or confined very close to them. Very little charge is thus scattered diffusely into the Gouy–Chapman diffuse layer. At low enough concentrations the solution charge is scattered under the simultaneous influence of thermal and coulombic forces; the only evidence of a Helmholtz plane being defined by the distance of nearest approach of solvated ions to the electrode surface.

Largely through the painstaking work of Grahame in the 1940s, electrocapillarity effectively established the first experimental basis for the now accepted double layer theory. The basic picture of the electrode/electrolyte interface was thus in place.

It would seem, at first glance, that the place of electrocapillarity lies in the history of electrochemistry rather than its future, since its application appears limited only to the mercury/electrolyte interface. However, the work of Sato and colleagues (1986, 1987, 1991) has very definitely placed the technique, or at least a development of it, very firmly at the frontiers again.

In terms of understanding the mercury/electrolyte interface, it is clear from the above discussion that the measurement of the surface free energy (in terms of the surface tension), is central. If the electrocapillarity technique could be applied to solid electrodes, then it is capable of supplying information extremely difficult to obtain by any other technique. Sato has indeed developed a technique to measure the surface tension of a metal electrode which he terms piezoelectric surface stress measurement and is based upon the previous work of Gokhshtein (1970).

The working electrode, in the form of a rectangle of thin foil, c. 20 μm thick, is glued to a piezo-electric ceramic element of the same lateral dimensions. Both the piezo element and the edge of the working electrode are then isolated from the electrolyte with inert epoxy cement. The essence of the piezo-electric effect is that if the ceramic element is deformed in any way, a potential difference develops across it, directly related to the deformation. Thus, if the working electrode potential is ramped, as in cyclic voltammetry (see next section), this induces changes in the surface free energy of the electrode which, in turn, cause the electrode surface area to expand or contract. This deformation is transmitted, via the glue, to the piezo-electric element which is then forced to deform. The magnitude of the potential drop induced across the element is then measured and is directly proportional to the original change in the surface free energy of the working electrode. As may be imagined, the changes being measured are extremely small and may easily be lost. Consequently, it is the *derivative* of the surface free energy change per unit area with respect to the working electrode potential (dγ/dE), that is actually measured. Thus, a 5 mV sinusoidal wave at 320 Hz is superimposed on the potential ramp. This, in turn, induces an AC voltage across the piezo element. By employing a lock-in amplifier only that voltage across the piezo

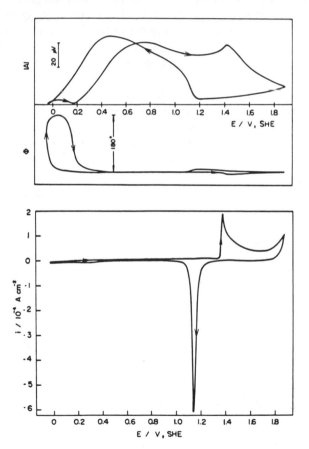

Figure 2.12 (Upper) piezoelectric signal curve and (lower) cyclic voltammogram of a gold electrode in pH 3.0 0.5 M Na_2SO_4/H_2SO_4 electrolyte. The magnitude of the voltage modulation was 5 mV at 200 Hz, with a potential scan rate of 33.3 mV s^{-1}. From Seo, Jiang and Sato (1987).

element that changes *at the same frequency* as the small potential modulation is detected, will be measured. As will be discussed further below, such a lock-in detection system is extremely sensitive and is often employed where the expected response is small. The amplitude $|a|$ of the induced voltage across the piezo element is directly proportional to $(d\gamma/dE)$. In addition, there is a phase difference ϕ between the induced voltage and the imposed signal on the working electrode. At potentials below the pzc, $\phi = 0$ (i.e. the signals are in phase), however, at potentials greater than the pzc, $\phi = 180°$.

The dramatic change in ϕ is extremely useful, as the $(d\gamma/dE)$ vs. E plots obtained by this method are often extremely complicated; i.e. although the pzc is indicated by a minimum in $(d\gamma/dE)$, there may be several such peaks in the plot. However, only that associated with the pzc will also show a $180°$ change in ϕ.

The results obtained by the method of Sato come in the form of plots of $|a|$ and ϕ against potential. Since $|a| \propto (d\gamma/dE)$, then integrating the plot of $|a|$ vs. E around the pzc (shown by the $180°$ change in ϕ), gives the electrocapillarity curve of $\Delta\gamma$ vs. E, providing the proportionality constant between $|a|$ and $(d\gamma/dE)$ can be determined. This latter is normally straightforward since $d\gamma/dE = |\sigma|$, which can be determined independently.

Seo *et al.* (1987) employed the piezoelectric surface stress technique to study the behaviour of gold on electrochemical cycling in aqueous sulphate solution. They obtained plots of $|a|$ and ϕ vs. E at a range of pH from 0.3 to 14 and a typical example of these is shown in Figure 2.12 which shows the plots obtained at pH 3. It should be emphasised that the data in Figure 2.12 were obtained dynamically by simultaneously scanning the potential in the

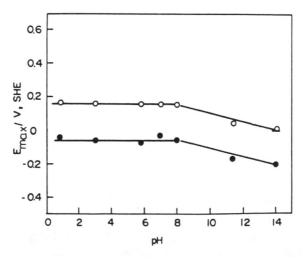

Figure 2.13   pH dependence of the electrocapillary maximum, E_{max}. The solutions were 0.5 M Na_2SO_4/H_2SO_4, except for the pH 14 electrolyte which was 1 M NaOH. The open circles represent data obtained from the anodic scan and the filled circles from the cathodic scan. From Seo, Jiang and Sato (1987).

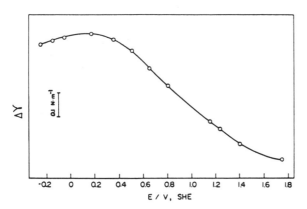

Figure 2.14 Electrocapillary curve of a gold electrode in pH 3.0 sulphate electrolyte. From Seo, Jiang and Sato (1987).

positive direction from -0.05 V vs. SHE to $+1.85$ V vs. SHE and back. The expected change in the phase of ϕ can be seen in the positive-going scan at $+0.2$ V and -0.05 V in the negative-going direction. A plot of this phase change vs. pH is shown in Figure 2.13. In agreement with expectation, the phase change marking the pzc is *not* pH dependent in acid solution but shows marked pH dependence in alkali since OH^- ions are known to adsorb strongly on gold. The electrocapillarity curve at pH 3 is shown in Figure 2.14; it is derived from the integration of the $|a|$ vs. E curve.

2.1.2 Linear sweep voltammetry and cyclic voltammetry

Linear sweep voltammetry (LSV) and cyclic voltammetry (CV) were first reported in 1938 and described theoretically in 1948 by Randles and Sevcik. In effect rather than monitoring the response of an electrochemical system to a small amplitude periodic potential change, as in AC impedance, in LSV and CV a large periodic potential change is imposed on the system. In LSV, the potential is ramped between two chosen limits at a steady rate v and the current monitored; a linear sweep voltammogram is then just a plot of current vs. the (time-dependent) potential. Cyclic voltammetry is exactly the same as LSV except that the potential is swept back and forth between the two chosen limits one or more times and the current monitored continuously. Both techniques have grown enormously in popularity over the past few decades, so much so that obtaining a CV or LSV of a new electrochemical system is often the first experiment performed by the electrochemist, giving invaluable information as to the presence of electroactive species in solution or at the electrode surface.

The equipment necessary to carry out LSV and CV measurements is shown in Figure 2.15. The electrochemical cell usually consists of a vessel that can be sealed to prevent air entering the solution, with inlet and outlet ports to

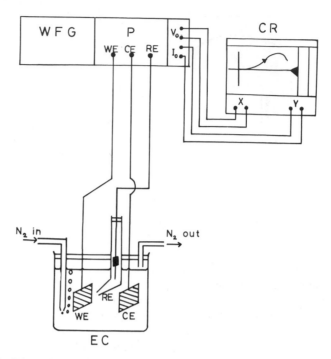

Figure 2.15 Schematic representation of the equipment necessary to perform linear sweep voltammetry (LSV) or cyclic voltammetry (CV). WFG waveform generator, P potentiostat, CR chart recorder, EC electrochemical cell, WE working electrode, CE counter electrode, RE reference electrode.

allow the saturating of the solution with an inert gas, N_2 or Ar. The removal of O_2 is usually necessary to prevent currents due to the reduction of O_2 interfering with the response from the system under study. The standard cell configuration consists of three electrodes immersed in the electrolyte: the working electrode (WE), counter electrode (CE), and reference electrode (RE). The potential at the WE is monitored and controlled very precisely with respect to the RE via the potentiostat; this may be controlled in turn via interfacing with a computer. The desired waveform is imposed on the potential at the WE by a waveform generator; the current flowing between the WE and CE is usually measured as the potential drop V across a resistor R (from which $I = V/R$), the latter connected in series with the two electrodes. The resulting I/V trace is then either plotted out directly via an XY chart recorder or, where possible, retained in a computer to allow any desired data manipulation prior to a hard copy being taken.

Figures 2.16(a) and (b) show the linear sweep voltammogram and cyclic voltammogram that would be expected from an electroactive adsorbed species assuming that:

(i) At time $t = 0$, the electrode surface has the maximum coverage of the oxidised form, O.

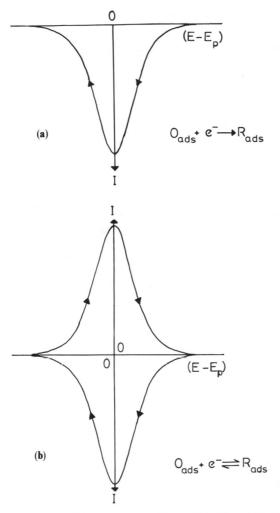

Figure 2.16 The current/potential curves expected from (a) LSV and (b) CV studies on the reversible redox reaction: $O_{ads} + e^- \rightleftharpoons R_{ads}$. E_p is the potential at which the maxima (and minima) in the current are observed.

(ii) That the electrochemical reaction can be represented by:

$$O_{ads} + ne^- \longleftrightarrow R_{ads}$$

and is kinetically fast enough to maintain Nernstian (i.e. reversible) behaviour at each point in the CV or LSV scan such that the concentrations of O and R *near the electrode* obey the Nernst equation.

(iii) That the oxidised and reduced species are both strongly adsorbed and have the same enthalpy of adsorption.

(iv) There is no O or R in solution.

The shape of the voltammograms in Figures 2.16(a) and (b) can be understood by the following treatment. Consider the cathodic scan in the CV; i.e. the reaction is as written in (ii). If E_i is the initial potential, at which essentially only O is present on the surface, the potential at any time t is given by:

$$E = E_i + vt \tag{2.34}$$

where v is the scan rate, dE/dt. As defined, v is negative for the positive scan. Neglecting the activities of the surface species, E can be expressed approximately in terms of the Nernst equation:

$$E = E_0 + [RT/nF] \cdot \log_e [\theta_{ox}/\theta_{red}] \tag{2.35}$$

where θ_{ox} and θ_{red} are the coverages of the oxidised and reduced forms at any time t. $\theta_{red} = (1 - \theta_{ox})$, thus:

$$E = E_0 + [RT/nF] \cdot \log_e [\theta_{ox}/(1 - \theta_{ox})] \tag{2.36}$$

From equation (2.36), it can be seen that $E = E_0$ when $\theta_{ox} = 1/2$; this defines E_0, it is the potential at which the surface coverage by the oxidised form is 1/2, *it is not the thermodynamic reduction potential of the O/R couple*. Differentiating (2.36):

$$dE/d\theta_{ox} = [RT/nF] \cdot [1/\theta_{ox} + 1/(1 - \theta_{ox})]$$
$$= [RT/nF] \cdot [1/\theta_{ox}(1 - \theta_{ox})]$$
$$[d\theta_{ox}/dE] = [nF/RT] \cdot [\theta_{ox}(1 - \theta_{ox})] \tag{2.37}$$

From Faraday's laws of electrolysis, the charge passed at any time t, Q, is the current in amperes, I, multiplied by the time in seconds, t; in terms of the surface coverage this is:

$$Q = N_T n e_0 (1 - \theta_{ox})$$

where N_T is the total number of O and R species present on the surface, e_0 is the electronic charge and we have assumed that E_i is sufficiently positive of E^0 for θ_{ox} to be essentially unity. In addition, since:

$$Q = \int_0^t I \, dt$$

$$dQ/dt = I = -N_T n e_0 (d\theta_{ox}/dt), \text{ or:}$$

$$dQ = I \, dt = -N_T n e_0 \cdot d\theta_{ox} \tag{2.38}$$

The negative sign indicates that the current is *cathodic*; i.e. that a *reduction* is taking place. From equations (2.37) and (2.38), since $(d\theta/dE) = (d\theta/dt)/(dE/dt)$, it follows that:

$$I = -[n^2 F N_T e_0 \cdot v/RT] \theta_{ox}(1 - \theta_{ox}) \tag{2.39}$$

Differentiating to find the position of the minimum (or maximum) and the value of the peak current:

$$dI/d\theta_{ox} = -[n^2 F N_T e_0 \cdot v/RT](1 - 2\theta_{ox})$$

which is zero when $\theta_{ox} = 1/2$; i.e. when $E = E_0$. Inserting $\theta = 1/2$ into equation (2.39) gives *the peak current*, I_p, as:

$$I_p = -[n^2 F N_T e_0 \cdot v/4RT] \qquad (2.40)$$

Thus, the peak current is directly proportional to the scan rate and a plot of I_p vs. v gives the total amount, N_T, of species present on the surface. In order to increase I_p, and thus the sensitivity of the technique, it is only necessary to increase the scan rate. This is the first indication of one of the major advantages of CV and LSV; they can be used to detect extremely low amounts of electroactive species. However, this is a double-edged sword; not only are the techniques very sensitive to the presence of an analyte, but their sensitivity to any impurities in the electrolyte demands the use of extremely pure electrolytes. From equation (2.36):

$$\theta_{ox}/(1 - \theta_{ox}) = \exp[nf(E - E_0)]$$

re-arranging:

$$\theta_{ox} = \exp[nf(E - E_0)]/(1 + \exp[nf(E - E_0)]) \qquad (2.41)$$

Replacing equation (2.41) in (2.39) and writing $a = \exp[nf(E - E_0)]$

$$I = -[a/(1 + a)^2] \cdot vn^2 f N_T e_0 \qquad (2.42)$$

which is the equation of the cathodic sweep in the CV in Figure 2.16(b).

For an ideal Nernstian reaction, the peak potentials of the cathodic and anodic sweeps will be the same, and equal to E_0. The width of the cathodic (or anodic) wave at half peak height, $\Delta E_{1/2}$, can be found by replacing I by $I_p/2 = -[n^2 f N_T e_0 v/8]$ in equation (2.39) and so obtaining the two values for θ. Using equation (2.36) then gives a $\Delta E_{1/2}$ of *c.* $0.09/n$ V, at 298 K. In practice, $\Delta E_{1/2}$ is never as small as this ideal value as a result of adsorbate–adsorbate interactions; these intermolecular interactions cause a smearing out of the observed redox potential, and a full treatment of this and other complications can be found in the standard texts referred to at the end of chapter 1.

The most important experimentally measured quantity is the total charge passed during the reduction, Q_T, given by:

$$Q_T = \int_0^{t_{max}} I \, dt \qquad (2.44)$$

From equation (2.34), $dE = v \, dt$; $E = E_i$ at $t = 0$ and $E = E_f$ at $t = t_{max}$, thus:

$$Q_T = [1/v] \cdot \int_{E_i}^{E_f} I \, dE \qquad (2.45)$$

The integral is just the area, A, under the CV, thus:

$$vQ = A \qquad (2.46)$$

This is another indication of the large potential sensitivity of the CV and LSV methods. Equation (2.46) shows that, in order to increase the measurable area under the voltammogram, the scan rate need only be increased. This will be applicable so long as the kinetics of the surface electron transfer reaction are sufficiently fast.

A simple example of the redox behaviour of surface-bound species can be seen in Figure 2.17, which shows the behaviour of a bare platinum electrode in N_2-saturated aqueous sulphuric acid when a 'saw tooth' potential is applied. There are two clearly resolved redox processes between 0.0 V and 0.4 V, and these are known to correspond to the formation and removal of weakly and strongly bound hydride, respectively (see section on the platinum CV in chapter 3). The peak currents of the cathodic and anodic reactions for these processes occur at the same potential indicating that the processes are not kinetically limited and are behaving in essentially an ideal Nernstian fashion. The weakly bound hydride is thought to be simply H atoms adsorbed on top of the surface Pt atoms, such that they are still exposed to the

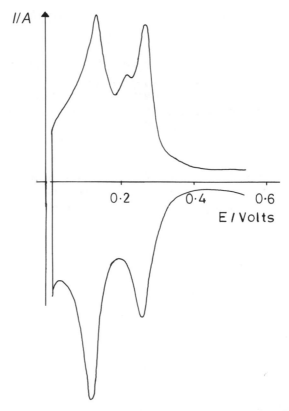

Figure 2.17 Cyclic voltammogram of a Pt electrode immersed in N_2-saturated aqueous sulphuric acid showing the hydrogen adsorption and desorption peaks.

electrolyte, whilst the strongly bound form is adsorbed below the surface layer of Pt atoms with its electron in the conduction band of the metal.

A question arises as to what happens if the Nernstian approximation breaks down. Under these circumstances, we must use the proper equations for the kinetics of electron transfer discussed in chapter 1. The simplest case is that of a completely *irreversible* system, where only oxidation (or reduction) is possible and a single electron is transferred, i.e. consider the process:

$$R_{ads} \longrightarrow O + e^-,$$

in which the potential is ramped up from a value at which only adsorbed R is present on the surface. The current passed at any potential will be given simply by the Butler–Volmer expression, modified to take into account the fact that it concerns only surface species:

$$I = I_0 \theta_{red} \exp[(1 - \beta)f(E - E^0)] \tag{2.48}$$

with, as above, $E = E_i + vt$ and I_0 is the exchange current density.

$$Q = N_T e_0 (1 - \theta_{red})$$
$$(dQ/dt) = I = -N_T e_0 (d\theta_{red}/dt)$$
$$(d\theta_{red}/dt) = -(I_0 \theta_{red}/N_T e_0)\exp[(1 - \beta)f(E - E^0)]$$

which can be re-written as:

$$(d\theta_{red}/dt) = -(I_0 \theta_{red}/N_T e_0)\exp[(1 - \beta)f(E_i + vt - E^0)] \tag{2.49}$$

Integrating equation (2.49), and inserting equation (2.48) gives:

$$I = I_0 \exp[(1 - \beta)f(E - E^0)] \cdot \exp[-\gamma(\xi - 1)] \tag{2.50}$$

where:

$$\xi = \exp[(1 - \beta)f(E - E_i)]$$

and:

$$\gamma = (I_0 RT/N_T e_0(1 - \beta)vF) \cdot \exp[(1 - \beta)f(E_i - E^0)]$$

By differentiating equation (2.50) with respect to E it can be shown that the potential, E_{max}, at which the current reaches a maximum is given by:

$$E_{max} = E^0 + (RT/(1 - \beta)F)\log_e[(N_T e_0(1 - \beta)fv)/I_0] \tag{2.51}$$

From equation (2.51) it can be seen that the slower the intrinsic kinetics of the reaction, (i.e. the smaller I_0), the more E_{max} is displaced from E^0. It is also clear that E_{max} displaces as $\log_e v$ with increasing scan rate. In principle, a plot of E_{max} vs. $\log_e v$ allows us to determine β and I_0.

A classic example of an irreversible reaction is the electrochemical oxidation of adsorbed CO. This reaction has been studied by a number of workers. Typically, CO is allowed to adsorb at a platinum electrode, after which the electrode potential is swept up to a value where the adsorbed CO is oxidised to CO_2:

$$CO_{ads} + H_2O \longrightarrow CO_2 + 2H^+ + 2e^-$$

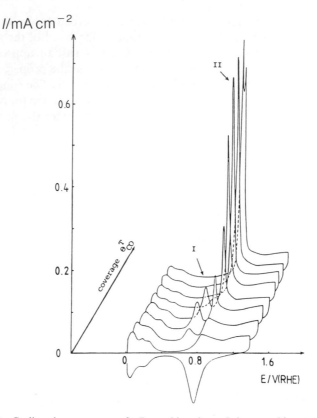

Figure 2.18 Cyclic voltammograms of a Pt working electrode immersed in aqueous perchloric acid (in the absence of CO) showing the oxidation peaks of adsorbed CO for different degrees of coverage θ_{CO}^T. The scan rate was $50\,mV\,s^{-1}$. The adsorption was effected by exposing the platinum working electrode to CO-saturated electrolyte for a sufficient length of time to give the coverage required. From Beden *et al.* (1985).

Some representative voltammogramms from the work of Beden *et al.* (1985) are shown in Figure 2.18.

The authors found that as CO is more strongly adsorbed than hydrogen, the introduction of CO into the electrochemical cell was immediately accompanied by its adsorption at the platinum electrode. This was shown by the decrease in the charge under the hydride adsorption features, and also the appearance of the oxidative stripping peaks I and II.

The first peak, I, grows in as the coverage increases from 0, and increases with increasing coverage before disappearing at high CO concentrations. The second, much sharper peak, II, appears at 0.85 V, and increases steadily in intensity until saturation coverage is attained. (At very high coverage, this feature is split; an interesting observation that the authors were unable to explain at the time.)

The authors interpreted the voltammograms in terms of at least two energetically different adsorbed states of CO on the surface. They then assumed that peak I was attributed to the oxidation of bridged CO, $Pt_2C=O$ (CO_B), and peak II to linearly adsorbed $C\equiv O$, (CO_L). Two electrons per Pt site are needed to oxidise CO_L, and one to oxidise CO_B; hence, the authors could calculate the coverages of the two forms, θ_B and θ_L, as follows:

$$\theta_L = Q_{CO}^L/2Q_H^0$$

and:

$$\theta_B = Q_{CO}^B/Q_H^0$$

where Q_{CO}^i is the charge under the stripping peak attributed to the type i CO, and Q_H^0 is the charge under the oxidative wave corresponding to the stripping of the monolayer of hydride off clean platinum, see Figure 2.19(b). The total coverage of adsorbed CO, θ_{CO}^T, was defined as:

$$\theta_{CO}^T = \theta_{CO}^L + \theta_{CO}^B = 1 - \theta_H,$$

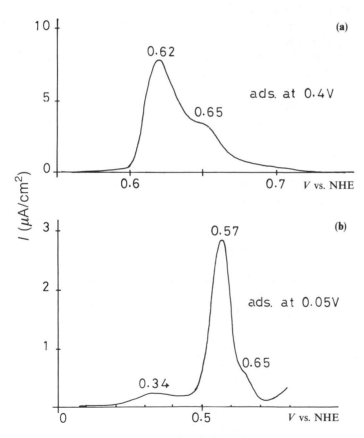

Figure 2.19   Linear sweep voltammograms of a platinum electrode immersed in N_2-saturated 0.5 M H_2SO_4, showing the anodic stripping of adsorbed CO. The CO was adsorbed from the CO-saturated electrolyte for 10 minutes at the designated potential. The scan rate was $1\,mV\,s^{-1}$. The adsorption potential was (a) 0.05 V and (b) 0.4 V vs. NHE. Note the different electrode potential scales for the two plots. From Kunimatsu *et al.* (1986).

where θ_H is the coverage of the Pt surface by hydride in the presence of the adsorbed CO, equal to Q_H/Q_H^0, with Q_H the area under the corresponding hydride stripping wave. The authors then used *in situ* IR and UV-visible spectroscopy to confirm the identities of the CO species responsible for the observed electrochemistry.

The potential at which adsorption occurs also has a major effect upon the nature of the surface species. Thus, Figures 2.19(a) and (b) show LSVs for the oxidation of the adsorbed CO on a platinum electrode in 0.5 M H_2SO_4, from the work of Kunimatsu *et al.* (1986). In Figure 2.19(a), the CO was adsorbed at 0.4 V vs. NHE and in 2.19(b) the adsorption was carried out at 0.05 V. In both cases, the dissolved CO was removed via N_2-purging after the adsorption. As can be seen from the figures, adsorption at a potential at which the surface already has hydride present (see Figure 2.17), results in CO species that are somewhat easier to oxidise than those formed on the bare Pt. By integrating the curves in Figures 2.19(a) and (b) to obtain the charge passed, it was found that the charge required to strip off the CO_{ads} was roughly the same.

These two studies quite nicely show the strength and weakness of voltammetry; the strength lies in the ease and unambiguity with which the technique declares the existence of a reaction, and its high sensitivity in doing so. The weakness of voltammetry is that it shows that 'something is happening'; however, precisely what is often beyond the capability of the technique. By making an educated guess as to the identity of any surface species, the dimunition of the hydride structure can then be employed to produce a model of the adsorption, in terms of the fractional coverage by each of the adsorbed species; clearly, this approach is open to great error. The ideal means of employing LSV or CV in such a problem is to use them to determine the potentials at which the onset of reaction occurs, and then to use another more suitable *in situ* technique to provide information on the identity of the species present, the sites they occupy on the surface, etc.

Cyclic voltammetry also finds application in the investigation of somewhat thicker layers on electrodes, such as electronically conducting polymers. There has been an explosion of interest in these polymers, centred around the promise of a new generation of materials for the electronics and optical industries of the 21st century. The initiator of this interest was a paper by Shirakawa *et al.* (1977), reporting the chemical preparation of films of polyacetylene (PA). More importantly, they also reported that the shiny, coherent films so formed could be doped via charge-transfer reactions with redox agents to give conductivities in the metallic regime. This led eventually to the discovery of a range of polymers based on benzenoid and heterocyclic monomer units, polyparaphenylene, polyaniline, polypyrrole, polythiophene, etc., that could be *electrochemically* polymerised to form coherent semi-conducting films on the electrode. These films can then be electrochemically cycled between oxidised and neutral ('doped' and 'undoped') forms; in the neutral form they are generally insulating and switching causes the polymer

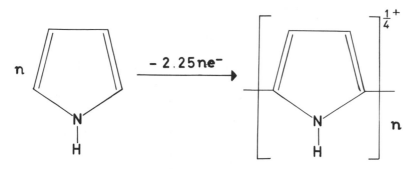

Figure 2.20   The anodic polymerisation of pyrrole.

to become semiconducting, a process normally accompanied by a colour change.

A typical example of such a polymer is polypyrrole. The exact mechanism by which the electropolymerisation of pyrrole occurs remains a source of controversy; however, by assuming 100% growth efficiency, then it can be calculated that 2.25 electrons are removed per monomer unit. Only two electrons are required to polymerise the monomer; however, the film is formed in a highly oxidised state, corresponding to one electron per four units, see Figure 2.20.

The thickness of the film can then be calculated from the charge, Q, passed during growth, via the relationship:

$$L = QW/2.25F\sigma A$$

where A is the area of the electrode, W is the molecular weight of the monomer and σ its density (usually taken to be about $1.5\,g/cm^3$). L is the thickness of the film assuming no swelling due to the presence of electrolyte; to estimate the electrolyte content, this figure can be compared to that obtained from other methods, e.g. ellipsometry (see below), which give the actual value.

The thickness of the 'pure' film can also be obtained by measuring the area under the oxidative or reductive waves of the CV of a conducting polymer, providing the extent of doping of the film (i.e. the number of charges in the film per monomer unit) at the anodic limit is known.

A typical CV of polypyrrole is shown in Figure 2.21 for a 40 nm film on Pt in 1 M $NaClO_4$.

Clearly, there is considerable asymmetry between the cathodic and anodic waves in Figure 2.21. More precisely, the exact structure of the cyclic voltammograms of such conducting polymers has attracted much speculation as to the processes responsible. Two important theories that have been advanced are those of Feldberg (1984), and Heinze et al. (1987). The Feldberg theory postulates that the CV is the result of a combination of a redox wave and a capacitive component, the redox wave being responsible for the peak in the anodic sweep; the generation of the charge carriers during the oxidation

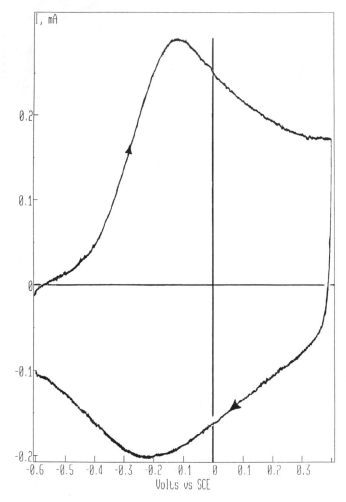

Figure 2.21   A cyclic voltammogram of a 40 nm polypyrrole film on platinum in N_2-saturated
aqueous 1.0 M $NaClO_4$.

of the polymer then results in major morphological changes, such that a
huge capacitance develops. In contrast, the theory of Heinze *et al.* explains
the shape of the CV in terms of a number of redox waves resulting from the
film comprising a range of conjugation lengths. The peak in the oxidative
wave was attributed to morphological changes; these changes stabilise the
system such that the redox energies of the charge carriers are lowered.

   Again, voltammetry alone is insufficient to shed any light on the electrochemistry of these systems. As will be seen in the next chapter, the application
of a range of *in situ* techniques was required before any major advances in
understanding were obtained.

   Thus, the information yielded by voltammetry in the study of adsorption on electrodes can be summarised thus:

(1) The presence of adsorption, with high sensitivity.
(2) The potential at which adsorption occurs (if the process is potential-dependent) and that at which stripping commences.
(3) The fractional coverage of the surface by adsorbed species.
(4) The number of distinct adsorbates.

As has already been stated, voltammetry is usually the predecessor to other techniques capable of furnishing more detailed information. However, the importance of the technique should not be underrated; it is generally the very first experiment performed by an electrochemist on a new system. Without the valuable information furnished by voltammetry, an electrochemist would waste a large amount of time having to use other, less straightforward, means of determining the basic parameters of the system.

2.1.3 Scanning tunnelling microscopy

The late 1980s saw the introduction into electrochemistry of a major new technique, scanning tunnelling microscopy (STM), which allows real-space (atomic) imaging of the structural and electronic properties of both bare and adsorbate-covered surfaces. The technique had originally been exploited at the gas/solid interface, but it was later realised that it could be employed in liquids. As a result, it has rapidly found application in electrochemistry.

STM is based on the phenomenon of quantum mechanical tunnelling. A sharp tip ($<1\,\mu m$ in diameter) of a chemically inert metal (W or Pt–Ir) is brought close enough to the surface of a conducting sample for the electronic wavefunctions of tip and sample to overlap; tunnelling can then occur and is maintained by the presence of a potential difference between the two. The subsequent variation of the tunnelling current, or height of the tip above the surface, as the tip is scanned across the surface of the sample is then used to obtain a topographical image of the surface *with atomic resolution, both vertically (in the z direction) and laterally*. In order for the technique to have real application, tunnelling should take place at a reasonable potential difference (the 'bias voltage', V_b). Typical bias voltages are $2\,mV$ to $2\,V$, with tunnelling currents, i_t, of 10^{-10} to 10^{-7} amps.

The tip is first brought near to the sample via a coarse Z positioner, such as a controlled approach piezo-electric motor. This provides a $10\,\text{Å}$ step size and so is capable of very precise and controlled movement. As soon as a tunnelling current is detected, the tip is stopped and the fine control system is operated. Thus the tip is mounted on the end of a hollow, cylindrical piezo crystal, a 'tube scanner' (a piezo-electric crystal can be made to deform under the influence of a voltage applied across it). The tube scanner has four strip electrodes on the outside, providing movement in the x and y directions (lateral movement), and one electrode covering the whole of the inside. A

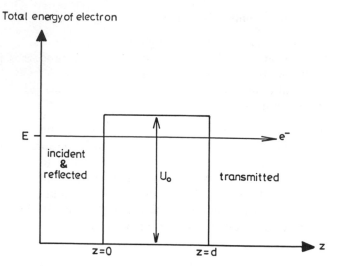

Figure 2.22   Schematic representation of an electron, of total energy E, tunnelling through a rectangular barrier of height U_0. From Christensen (1992).

combination of all five electrodes provides the height adjustment. This system takes over the cautious advance of the tip until the preset tunnelling current is achieved; usually corresponding to a tip–sample separation of c. 5–20 Å. The tube scanner is then used to raster the tip across the surface, with two modes of operation generally being employed to obtain the image: constant current and constant height.

Before proceeding to consider the mechanics of the data collection in more detail, it may be helpful to consider the factors governing the magnitude of the tunnelling current i_t. By considering an electron tunnelling through a rectangular potential barrier of height U_0 (see Figure 2.22) it is relatively straightforward to show that the tunnelling current is given by:

$$i_t \approx D_S(E_{F,S})D_T(E_{F,T})V_b \exp(-2\kappa d) \qquad (2.52)$$

where $D_S(E_{F,S})$ and $D_T(E_{F,T})$ are the density of states of the sample and tip at their respective Fermi level energies (i.e. the energy of the topmost filled level), d is the tip–sample separation, V_b is the bias voltage ($V_b = V_S - V_T$) and the exponential term is the probability of tunnelling through the potential barrier, with κ given by:

$$\kappa = [2m_e(U_0 + E_e)/\hbar^2]^{1/2} \qquad (2.53)$$

where E_e is the energy of the electron. The tunnelling process for metallic tip and sample separated by a vacuum can be represented schematically as in Figures 2.23(a) and (b). Figure 2.23(a) depicts the case where tip and sample are joined only by a wire, i.e. $V_b = 0$. Under UHV conditions, the bias potential, and hence the energy of the electrons, is usually very small with respect to U_0, and U_0 can be approximated to the mean of the two barrier heights:

$$U_0 \approx [\phi_T + \phi_S]/2 = \bar{\phi}$$

where ϕ_T and ϕ_S are the work functions of tip and sample. Under these conditions, where the barrier is trapezoidal, κ can be written as:

$$\kappa \approx [2m_e \bar{\phi}\hbar^2]^{1/2} \qquad (2.54)$$

Hence, for a given sample and at a given position, i_t varies directly as V_b and exponentially with the tip–sample separation. As V_b is varied, so the energy of the sample varies with respect to that of the tip.

Figures 2.23(a) and (b) should be a little more complicated than those

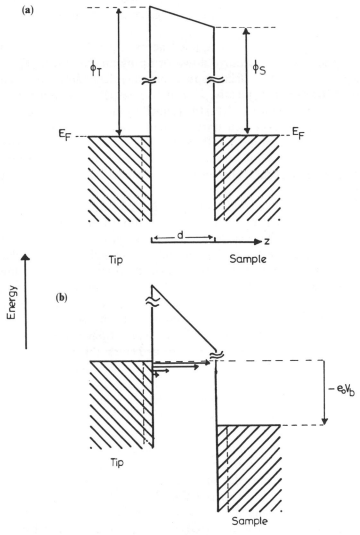

Figure 2.23 (a) The energy levels of tip and sample when connected by a wire. E_F is the Fermi level, d the tip/sample separation, and ϕ_S and ϕ_T the work functions of the sample and tip, respectively. (b) As for (a) except depicting the effect on the system of an imposed bias voltage, V_b. The bias voltage is defined as $V_{sample} - V_{tip}$. From Christensen (1992).

representing simple band theory since the latter deals only with the energy levels of the bulk material and takes no account of the inhomogeneity of the surface. Each point on the surface will have associated with it a distribution of energy levels and a plot of the number of such levels having an energy (E) against energy E constitutes the local density of states (LDOS). The LDOS of the tip are not shown in Figure 2.23 for clarity. These states may be occupied or unoccupied, entirely analogous to molecules having filled and unfilled molecular orbitals, though fortunately, for a metal, the relevant LDOS terms in equation (2.52) (i.e. $D_S(E_{F,S})$ and $D_T(E_{F,T})$) show very little spatial dependence; the surface is largely homogeneous on an atomic scale and the electronic wavefunctions are generally localised at the atomic positions.

If a potential, V_{bias}, is now applied across tip and sample, the relative energy of the sample will move down or up by an amount $|e_0 V_{bias}|$ (where e_0 is the charge on the electron in coulombs) depending on whether the polarity of the sample is positive or negative, respectively. When the sample is biased positive (Figure 2.23(b)) the tunnelling current arises from electrons tunnelling from the occupied states of the tip into unoccupied states on the sample. At negative sample bias the situation is reversed and electrons tunnel from occupied states of the sample to unoccupied states of the tip. Obviously, tunnelling cannot occur between occupied states on both the sample and the tip.

Returning to the two modes of operation commonly employed in STM: in the constant current mode, the tip is positioned at some point (x, y) and moved across the surface such that y is constant and x varies. During this lateral movement, the tunnelling current i_t is sensed. Feedback electronics then change the height of the tip (z) in order to maintain a constant i_t. At the end of the scan, the tip is displaced in the y direction, and the process repeated. The image is then a 3-dimensional map of (x, y, z) and the scanning process is termed 'rastering', being analogous to the process occurring in a television tube. In the constant height mode, the tip is maintained at a fixed height z and scanned across the surface as before; this time i_t is monitored and the image is a 3-dimensional plot of (x, y, i_t). In each case rapid variations in z or i_t result as the tip passes over surface features such as atoms. These two processes are represented schematically in Figures 2.24(a) and (b).

One of the prerequisites for electrochemical STM is rapid imaging to avoid thermal drifts (see below) and the constant height mode allows for faster imaging since the electronics of the device can respond faster than the z translator. However, this method must be applied with care and is only desirable when atomically flat surfaces are being imaged since tip crash into an 'atomic mountain' is a real possibility. In both modes, the tube scanner must be calibrated using a surface of known topography, such as highly ordered pyrolytic graphite (HOPG), if accurate topographical information is to be obtained.

For metal samples at tip–sample separations where $d \gg 1/\kappa$ (i.e. $d \geqslant 10\,\text{Å}$)

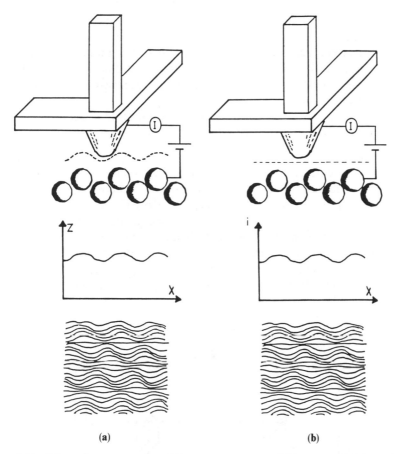

Figure 2.24   Schematic representation of (a) constant current and (b) constant height modes of operation of an STM. From Christensen (1992).

the approximation can be made that the LDOS functions and κ can be thought of as essentially constant: i_t is then exponentially proportional to d and the STM image is a simple topograph of the sample surface. This is aided by the fact that the exponential nature of the tunnelling process tends to select one or two atoms lying proud of the mean surface of the tip. As a result the sample is imaged by these atoms. This is a key point. Typical tip diameters are of the order of $1\,\mu m$ [$1000\,\text{Å}$]; if this selection did not occur, the technique would not function with anywhere near the resolution it does. For a typical value of ($[\phi_T + \phi_S]/2$) of $4\,\text{eV}$, $\kappa = 1\,\text{Å}^{-1}$ and the tunnelling current decreases by an order of magnitude for every $1\,\text{Å}$ increase in the tip–sample separation. On the other hand, if the current is kept constant to within 2%, the gap d remains constant to within $0.01\,\text{Å}$. This is the basis for interpreting the image as a simple topograph.

   This exponential dependence leads to very high vertical resolution, of the order of 0.1 Å, rendering the imaging of monatomic steps on samples such as platinum (diameter *c.* 2.746 Å) relatively straightforward. By the early 1990s, the lateral resolution of STM was such that atomic resolution was readily achieved. However, this requires an exceptional stability and noise-free operation (see discussion below). More routine measurements are usually performed at lateral resolutions of *c.* 5 Å.

The problems to overcome in order to attain atomic resolution with an STM can be divided into three areas:

1. *Vibrational isolation.* In order to maintain the tip a few atomic diameters above the sample, the STM must be totally isolated from the vibrations of the building it is in. Originally this was achieved by levitating the instrument on superconducting magnets. More recently, it was found that suspending the instrument by springs or elasticated ropes was sufficient, and rather less expensive, particularly if used in conjunction with a series of isolating plates. An arrangement of this type is shown in Figure 2.25, in which the STM head and sample cell sit on a series of metal plates separated by rubber 'O' rings. The whole assembly is then suspended by elasticated ropes or springs from a frame. In order to prevent acoustic

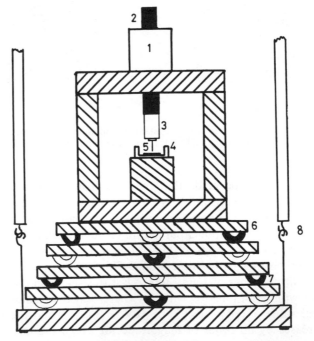

Figure 2.25 Schematic representation of the STM head and electrochemical assembly. (1) Inchworm motor, (2) Inchworm, (3) Faraday cage around tube scanner, (4) Teflon electrochemical cell, (5) working electrode (i.e. sample), (6) stainless steel plates, (7) halved rubber 'O' rings, (8) elasticated ropes attached to baseplate. The counter and reference electrodes and the various electrical connections are not shown for clarity. From Christensen (1992).

vibrations the 'head' and sample cell are protected by a glass bell jar which also allows control of the atmosphere above the electrochemical cell.

2. *Electronic noise.*   For a noise amplitude of 0.1 Å only atoms separated by $\geqslant 4$ Å can be resolved. Thus, the electronics must be carefully shielded, with the pre-amplifier stage placed as close to the tip as possible and the STM itself placed in a Faraday cage.

3. *Thermal drift.*   It is essential that the STM can acquire images rapidly (i.e. $\ll 10$ seconds) since thermal drifts due to electrolyte cooling, expanding electrodes, etc. will degrade lateral resolution over longer collection times.

The first STM experiments were performed under UHV conditions, and so the bias potential was simply applied as a difference across the tip and sample. However, introducing an electrolyte above the sample brought with it some particular problems. It is no longer sufficient simply to apply a bias voltage equal to the potential difference between tip and sample as this means that the potentials of the tip and sample are undefined with respect to any fixed reference, a wholly undesirable situation. Consequently, modern electrochemical STM systems operate under bipotentiostatic control with the tip and sample controlled and monitored independently with respect to the reference electrode. The bias potential is then still given by $(V_S - V_T)$, but V_T and V_S are now potentials with respect to the reference electrode.

There is a second, rather important, practical problem inherent in operating an STM *in situ*: maintaining a fixed bias voltage if the surface of the sample is to be imaged as a function of its potential, since STM images of samples other than metals, or of metals coated with adsorbates, are often strongly dependent on V_b (see below). This means that V_T must be altered exactly in step with V_S if V_b is to be maintained. This is then only possible if both V_S and V_T remain within the potential window of the solvent, i.e. at potentials within the limits set by the destruction of the solvent. In addition, there is a further constraint on V_T, as it must not be in a region where faradaic reactions such as O_2 reduction can inject noise into i_t. In order to reduce this latter problem, STM tips are usually coated in an inert material, such as black vacuum wax, to reduce the exposed area. If the tip is sufficiently sharp, only the very end of the tip will break through the surface tension of the wax and so be exposed to solution.

In general STM tips are manufactured through electrochemical etching. One such method involves forming a 'soap bubble' of 2 M NaOH in a Pt/Ir loop and inserting a length of tungsten wire through the bubble. An AC potential of *c.* 1–15 V peak-to-peak is then applied between the tip wire and the loop at 50 Hz. The wire is then etched to breaking point, at which time the weight of the wire helps to pull it out into a very fine point.

As was noted above, STM images are often strongly dependent both on V_{bias} and on the type of feature being imaged at the surface. As a result, the image is often not a simple topographical 'map' of the surface features. Consider, for example, a surface having an adsorbed atom with an excess of

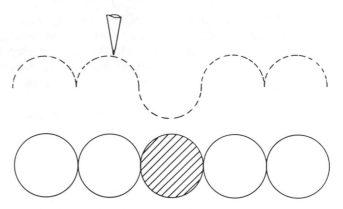

Figure 2.26   The expected trajectory that would be followed by an STM tip in constant current mode over a metal sample. The surface has one alien atom more electronegative than its neighbours. From Christensen (1992).

electrons. Under such circumstances, when the sample is negatively biased, electrons will flow readily from this atom to the tip. However, when the tip is positively biased, the sample atom will not readily accept any more electrons. This is shown in Figure 2.26. When the sample is negatively biased, tunnelling occurs readily and i_t increases (constant z mode) or the tip is retracted (constant i_t mode). If the sample is positively charged, tunnelling does not occur so readily so the tip is moved towards the sample to try and restore i_t (in constant current mode) or i_t decreases (constant z mode).

An elegant example of the ability of STM to link observed electrochemistry with topographic changes at the atomic level is provided by the work of Itaya *et al.* (1990), which concerned the application of STM to study the effects of potential cycling on the topography of a Pt(111) electrode. This follows the general interest over recent years in the study of well-defined highly ordered crystals and of platinum in particular because of its fundamental technological importance as the foremost electrocatalyst. Polycrystalline surfaces are too complicated for such studies whereas the well-defined, homogeneous and essentially flat nature of a single-crystal surface provides a readily understood starting point. An important observation which links single-crystal Pt(111) electrochemistry with that of the polycrystalline material arises from cyclic voltammetry. Cyclic voltammetry involves the repeated linear ramping of the potential of the sample electrode between two preset limits and simultaneously monitoring the current that passes. If Pt(111) is cycled in this manner between a potential above that at which hydrogen evolution takes place and one in the double layer region, as shown in Figure 2.27(a), then the surface is stable; the various features in the voltammogram merely indicate the formation and removal of surface hydrides. However, if the electrode is cycled up to a potential where surface oxides are formed, and back, then the cyclic voltammogram is observed to rapidly revert to that expected for polycrystalline Pt, see Figure 2.27(b).

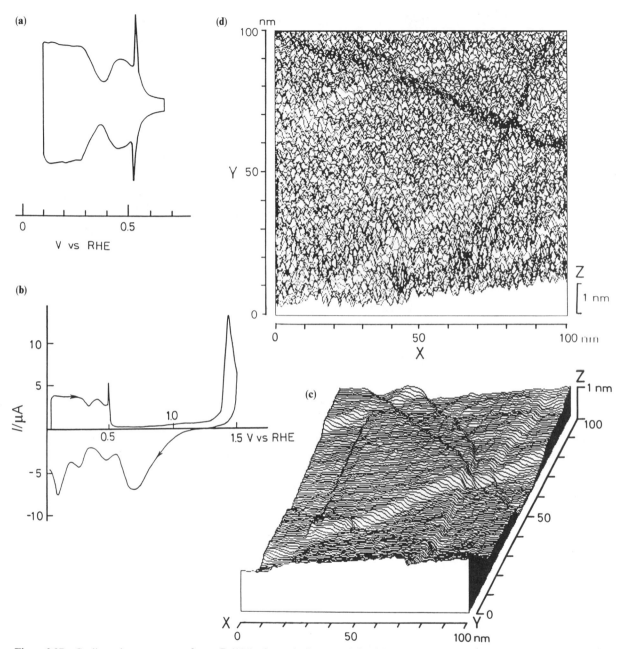

Figure 2.27 Cyclic voltammograms for a Pt(111) electrode immersed in 0.05 M H$_2$SO$_4$: (a) 0.1 V to 0.65 V vs. RHE, (b) 0.05 V to 1.5 V. (c) STM image of a 100 nm × 100 nm Pt(111) facet surface obtained in 0.5 M H$_2$SO$_4$. The electrode potentials of the Pt sample and Pt tip electrodes were 0.95 V and 0.9 V, respectively. The tunnelling current was 2 nA. Scan speed was 200 nm s^{-1}. (d) STM image of a 100 nm × 100 nm area of the Pt(111) surface obtained after 5 potential cycles between 0.05 V and 1.5 V vs. RHE. The conditions were as in (c). From Itaya *et al.* (1990).

Itaya and colleagues decided to employ STM to see if any correlation existed between the surface topography and the electrochemistry in Figures 2.27(a) and (b). The images were obtained at constant current with the tip and sample controlled independently.

Figure 2.27(c) shows a token STM image of a 1000 Å × 1000 Å area of the Pt(111) surface. The image was collected at 0.95 V vs. the reversible hydrogen electrode, RHE, i.e. in the double layer region (see Figures 2.27(a) and (b)). As can be seen, the surface is dominated by steps of c. 2.3 Å, consistent with monatomic height (the diameter of Pt is 2.38 Å). The steps cross at c. 60°, such that the terraces so formed are triangular in shape, as expected for a surface with threefold symmetry.

Figure 2.27(d) shows an image of a 1000 Å × 1000 Å area of the surface collected at 0.95 V after five potential cycles between 1.5 V and 0.05 V. After one potential cycle, noisy signals appeared on the terraces although even after five cycles the original monatomic steps can still clearly be distinguished and the position of these steps remain unchanged. This observation is very important as it indicates that the only change induced by the potential cycling is the formation of the disordered structures on the terraces. From Figure 2.27(d), it can be seen that the disordered structures on the terraces are *randomly oriented* islands, c. 30–50 Å in diameter, and one or two atoms high.

The authors observed that neither the location of the steps nor the disordered structures on the terraces changed with time in the absence of further potential cycling. However, they did observe that an image collected at 1.5 V (i.e. well into the region where the surface is covered with oxide) during the first cycle shows no differences from that in Figure 2.27(c), strongly suggesting that it is the *reduction* of this oxide surface, rather than its formation, that produces the structures observed in Figure 2.27(d).

The authors noted that the islands formed on the terraces are only one or two atoms high, strongly suggesting that formation of the oxide is accompanied by place-exchange, see p. 248. Hence, they concluded that the roughening of the Pt(111) surface is via this place-exchange mechanism. Once the oxide layer is stripped off, the Pt atoms left behind as a result of the place-exchange (in the form of adatoms) do not go back into their original positions. This results in the observed topographical changes. It does not seem unreasonable to correlate the formation of these randomly oriented small islands with the appearance of the polycrystalline platinum electrochemistry.

The 'eyewitness' nature of the work of Itaya and colleagues speaks volumes for the potential of STM in electrochemistry.

2.1.4   Scanning tunnelling spectroscopy

As was noted in the previous section, the images obtained in STM are often strongly dependent on the magnitude and sign of the bias voltage, as can be seen from a consideration of equation (2.52). Employing STM as a purely topographic tool assumes that $D_S(E_{F,S})$ and $D_T(E_{F,T})$ are uniform across the surface, i.e. they show no dependence on position. As was discussed above, this is a reasonable approximation for a clean metal surface. However, for a metal on which is an adsorbate, or for non metals, $D_S(E_{F,S})$ may show a marked spatial variation, as exemplified in Figure 2.26.

At the simplest level, the presence of such atoms as a function of position could be probed very simply by careful choice of potential. Thus, for metals the LDOS are generally manifest as contours of constant state density across the surface, which closely follow the topographic atom positions. For semiconductors, however, marked LDOS differences may develop. Thus the energy of a state and its spatial distribution may depend on the chemical identity and position of the surface atoms. By choosing V_{bias}, it is possible to probe occupied and unoccupied states, and this allows for selective atom imaging, as first reported by Feenstra and co-workers (1987). The authors studied GaAs(110), in which the surface consists of equal numbers of Ga and As atoms. Charge transfer from Ga to As results in an occupied state centred at the As atoms and an empty state centred at the Ga atoms. It also results in a small vertical displacement of the Ga atoms with respect to the As atoms. By applying a $V_{bias} = -1.9$ V (vs. sample), the STM selectively imaged the Ga atoms; with a bias voltage of $+1.9$ V, the As atoms were imaged. By combining the bias-dependent images with theoretical calculations, the tilt angle between the surface Ga and As atoms was quantitatively determined.

Using a slightly more sophisticated approach, the presence of the alien atom in Figure 2.26 could be verified by holding the tip at a constant height over the surface, ramping the bias potential and monitoring i_t. For locations other than that over the alien atom, we may expect a smooth increase in i_t with V_b (for a homogeneous and uniform density of states). However, over the alien atom, this smooth increase will have a ripple superimposed on it where the tunnelling current is reduced due to the filled LDOS, see Figure 2.28. This is the basis of scanning tunnelling spectroscopy; V_S or V_T is ramped and the tunnelling current monitored.

In order to ameliorate the sharply sloping background obtained in an STS spectrum, the data are often presented as di_t/dV_b vs. V_b, i.e. the data are either numerically differentiated after collection or V_b has a small modulation applied on top of the ramp, and the differential di_t/dV_b is measured directly as a function of V_b. The ripples due to the presence of LDOS are now manifest as clear peaks in the differential plot. di_t/dV_b vs. V_b curves are often referred to as conductance plots and directly reflect the spatial distribution of the surface electronic states; they may be used to identify the energy of a state and its associated width. If V is the bias potential at which the onset of a ripple in the i/V plot occurs, or the onset of the corresponding peak in the di/dV plot, then the energy of the localised surface state is $e_0 \times V$. Some caution must be exercised in interpreting the differential plots, however, since

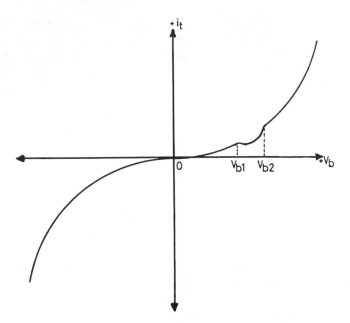

Figure 2.28   Schematic representation of the STS spectrum expected if the tip is poised over the alien atom depicted in Figure 2.26. The energy of the filled LDOS represented by the alien atom lies between V_{b1} and V_{b2}.

changes in the tunnelling probability can dominate i_t if, for example, $e^{-2\kappa d}$ shows a dependence on V_{bias}.

An effective approach to obtaining STS spectra that has been employed *ex situ* is that of Hamers *et al.* (1986) which reverts to the simpler approach of measuring the $i–V$ plot, followed by numerical differentiation if required. CITS, current imaging tunnelling spectroscopy, is essentially a mixture of STM and STS. In this technique, the feedback loop used for constant-current STM is gated in such a way that it is only switched in for some fraction of the time (e.g. 3/10). When the feedback loop is active, a constant bias voltage, $V_{c,bias}$, is applied to give tunnelling, and thus topographic imaging. When the feedback loop is switched out, the tip is held stationary above the sample at a *constant height*. The bias voltage, $V_{r,bias}$, is then linearly ramped between two values and the tunnelling current, $i_{r,t}$, measured as a function of $V_{r,bias}$. The bias voltage is then returned to $V_{c,bias}$, the feedback loop activated and topographic imaging continued. The $i–V$ curve is acquired very rapidly, e.g. in 4.5×10^{-4} s, compared to the speed of the topographic scanning, e.g. 4000 Å/s.

The high speed of the $i–V$ data collection eliminates vertical and lateral drift of the tip position, and the surface topography and the spatially resolved $i–V$ characteristics are determined *simultaneously* at each point in the x scan.

As a result of the constraints placed on V_b by the potential window of the solvent, and faradaic noise at the tip, obtaining STS spectra *in situ* has been via rather more simple methods, and generally over more restricted ranges of V_b than those that can be obtained under UHV conditions. One such approach is tip-current voltammetry (TCV) which was first reported by Uosaki and Koinuma (1992) and was employed by them to study surface states on n-doped GaAs. n-GaAs is of great interest to device manufacturers as a result of the extremely high mobility of the electrons responsible for the conduction mechanism and consequent extremely rapid response time. One of the most important characteristics of such a semiconductor is the number of states present at its surface, since such surface states can have a major, even dominant, effect on the electrochemistry and photoelectrochemistry of the semiconductor. Thus, surface states can provide recombination centres for photoexcited electron/hole pairs and so reduce the efficiency of photo-electrochemical energy conversion. They can also act as an electron mediator between the semiconductor and species in solution and so facilitate electron transfer. It has been suggested that the behaviour of these states is a function of their energy and so their identification, mapping as a function of position, and the determination of their energy distribution is an extremely important goal.

The technique of TCV is relatively straightforward; the potential of the semiconductor sample is ramped relatively slowly (c. 20 mV/s) and the tunnelling current is monitored, with the tip, via the feedback loop, able to adjust its height above the surface in response to any deviation in the tunnelling

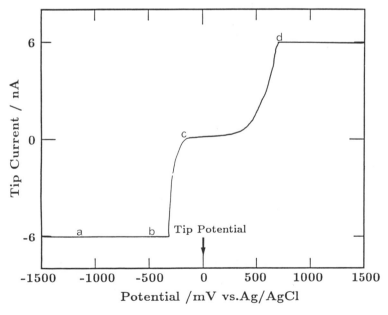

Figure 2.29 Tip current as a function of the potential of an etched n-GaAs electrode in 2 M HClO$_4$. Tip potential: 0 V vs. Ag/AgCl. Preset tunnelling current: 6 nA. From Uosaki and Koinuma (1992).

current from its preset value. Thus, Figure 2.29 shows the TCV of an etched, highly doped n-GaAs electrode in aqueous perchloric acid. The preset tunnelling current was 6 nA and the potential of the tip was 0 V.

As can be seen from Figure 2.29, there are two potential regions in which the feedback loop can maintain the preset tunnelling current: < -250 mV and > 700 mV. In intermediate regions, the tunnelling current cannot be sustained and the tip moves continually towards the surface until it crashes. The authors reported that the flat band potential, E_{FB}, of n-GaAs is -950 mV under the conditions of the experiment and so were able to draw the energy diagrams shown in Figure 2.30 to represent the various regions of the TCV in Figure 2.29.

Thus, when $E_S < E_{FB}$, i.e. region (a) in Figure 2.29, there are enough carriers (electrons) at the surface to sustain the tunnelling current: the process is essentially the same as when imaging a metal sample. STM images obtained in this potential region are stable and essentially unchanging.

When $E_S > E_{FB}$, region (b), a 'depletion' layer forms in the semiconductor due to the bending of the bands under the influence of the electric field. Increasing the potential increases this band bending and so increases the effective barrier to tunnelling it represents. However, the high doping level

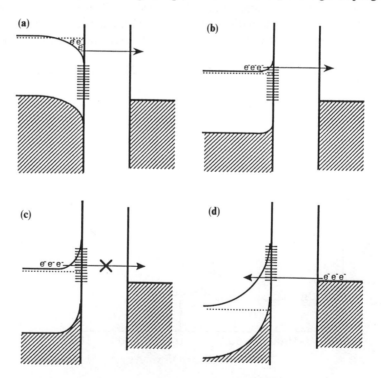

Figure 2.30 The energetics of the n-GaAs/electrolyte/tip interface at various potentials. (a) $E_S < E_{FB}$, (b) $E_S > E_{FB}$, (c) $E_S \gg E_{FB}$, (d) tunnelling occurs from tip to sample. From Vosaki and Koinuma (1992).

means that the band bending occurs over a relatively short distance into the semiconductor and the barrier is thus narrow enough to permit electrons to tunnel out, at least until $-250\,mV$.

When $E_S \gg E_{FB}$, region (c), the barrier becomes too thick to permit the electrons to tunnel out and the tip crashes.

Eventually, in region (d), tunnelling occurs from the tip to the sample. Although the depletion layer is still thick, the effective thickness of the barrier in this region is actually reduced and the presence of the surface states plays a dominant role in maintaining the tunnelling current in this region.

These authors provided further evidence in support of their conclusions by performing TCV experiments on n-GaAs samples that had been treated with (a) $RuCl_3$ and (b) ammonium sulphide. Ru is known to increase the surface state density and sulphur is known to remove surface states.

Figure 2.31 shows the TCV for the Ru-treated surface. Again, there are two potential regions where the tunnelling current could be maintained, $< -100\,mV$ and $> +100\,mV$. Clearly, however, the middle region, wherein the tunnelling current cannot be sustained, is much narrower due to the increase in the surface state density and the concomitant increase in the tunnelling probability. Figure 2.32 shows the dominant role of surface states with respect to tunnelling at potentials in the higher region; removal of the surface states, via treatment with ammonium sulphide, results in the thin depletion layer being an effective barrier to tunnelling.

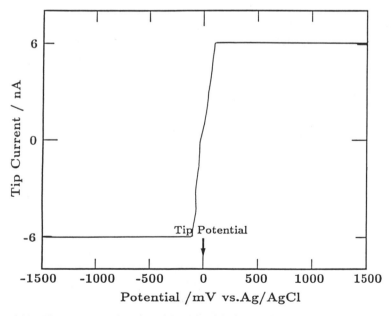

Figure 2.31 Tip current as a function of the potential of an $RuCl_3$-treated n-GaAs electrode in 2 M $HClO_4$. Tip potential: 0 V vs. Ag/AgCl. Preset tunnelling current: 6 nA. From Uosaki and Koinuma (1992).

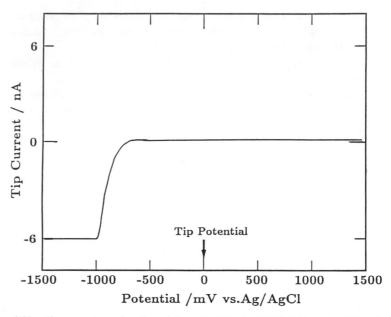

Figure 2.32   Tip current as a function of the potential of a $(NH_4)_2S$-treated n-GaAs electrode in $2M$ $HClO_4$. Tip potential: $0V$ vs. Ag/AgCl. Preset tunnelling current: $6nA$. From Uosaki and Koinuma (1992).

STS, both *in situ* and *ex situ*, is making a major impact in the areas of surface science and electrochemistry, particularly in the study of the semiconductor/vacuum and semiconductor/electrolyte interfaces.

2.1.5   Atomic force microscopy

The success of STM has resulted in the development of a whole variety of related scanning probe microscopes, the most important of which is atomic force microscopy, AFM, also known as scanning force microscopy. AFM was first reported in 1986 by Binnig, Quate and Gerber.

As with STM, AFM is an atomic-scale probe. However, unlike STM, the technique does not rely on quantum mechanical tunnelling. The AFM probe detects the interatomic forces between the tip and surface atoms on the sample and thus provides images that are closer to simple topographs. In addition, it can be used to image non-conducting substrates. Because of the complications involved in applying AFM *in situ*, the technique was only really employed in electrochemistry after about 1989 (however, where it has been applied it has produced extremely interesting, and even dramatic, results). Consequently, only a brief description will be given here; further information on the technique can be found in any of a number of excellent review articles, some of which are given in the further reading list for this section at the end of the chapter.

The operation of an AFM is shown schematically in Figure 2.33. A sharp probe tip is mounted on a spring (in the form of a cantilever) and the tip tracked across the surface of the sample. In most of the instruments in use tracking is achieved by moving the sample and maintaining the fragile tip mounting stationary. Interatomic forces, both attractive and repulsive, between atoms at the tip and on the surface cause the spring to deflect. This deflection

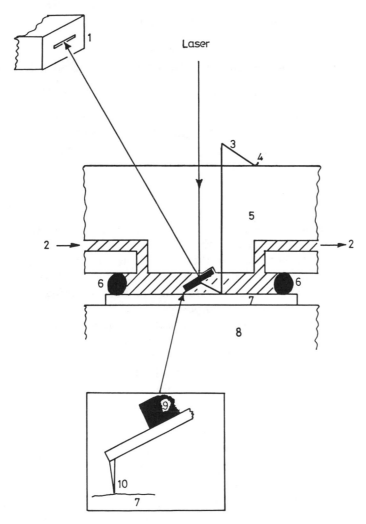

Figure 2.33 Schematic representation of an AFM electrochemical cell and its mode of operation. (1) photodiode, (2) electrolyte solution inlet/outlet, (3) spring clip, (4) cantilever holder, (5) glass cell body, (6) 'O' ring, (7) sample, (8) x, y, z translator, (9) mirror and (10) tip. After Manne *et al.* (1991).

is related to the magnitude of the force experienced by the cantilever spring, having a spring constant k. The AFM image, in its simplest form, is thus a plot of tip deflection against lateral position.

Several types of interaction can be probed with AFM: (i) Van der Waals forces and ionic repulsion, (ii) magnetic and electrostatic forces, (iii) adhesion and frictional forces and (iv) the elastic and plastic properties of the surface. In terms of the interactions relevant to electrochemistry, only those interactions typified in (i) will be considered.

The interatomic force between two atoms a distance R apart can be described in terms of the potential energy of the system $P(R)$, one widely applicable form of which is the Lennard–Jones potential:

$$P(R) = [C_1/R^{12}] - [C_2/R^6], \tag{2.55}$$

where C_1 and C_2 are constants. The force $\mathbf{F}(\mathbf{R})$ at any point R is then

$$\mathbf{F}(\mathbf{R}) = -[\mathrm{d}P(R)/\mathrm{d}R] \cdot \mathbf{F_U}(\mathbf{R}) = \{[12C_1/R^{13}] - [6C_2/R^7]\} \cdot \mathbf{F_U}(\mathbf{R});$$

where $\mathbf{F_U}(\mathbf{R})$ is the unit vector in the R direction and takes into account that force is a vector quantity while potential is a scalar.

The second term in equation (2.55) describes the long range Van der Waals or dispersive (attractive) forces. The first term describes the much shorter range repulsive forces experienced when the electron clouds on two atoms come into contact; the repulsion increases rapidly with decreasing distance, with the atoms behaving almost as hard spheres.

AFM has been used to image surfaces by probing both the attractive and repulsive forces experienced by the tip as a result of its proximity to the sample surface. In both modes, the probe tip is mounted on a cantilever spring. Three main designs have been employed: metal foil with a splinter of diamond, a shaped tungsten wire that acts both as spring and tip, and microfabricated tip/cantilever composites.

In order to render the AFM as insensitive as possible to low frequency noise, e.g. building vibrations, the spring constant of the cantilever, k, should be as low as possible, typically 0.1–$20\,\mathrm{N/m}$, and the resonant frequency of the spring, v_R, should be as high as possible. This latter requirement can be seen from the following simple calculation: if a is the vibration amplitude induced in response to an imposed vibration of frequency v and amplitude a_0, then $a \approx a_0[v/v_R]^2$. Typically, a building vibration has $a_0 \approx 1\,\mu\mathrm{m}$ and $v = 20\,\mathrm{Hz}$. For a resonance frequency of $10\,\mathrm{kHz}$, this gives a noise amplitude of $0.04\,\mathrm{\AA}$. This will cause very little interference and thus atomic resolution imaging is possible with very little vibration isolation, in contrast to STM.

It has been found that the relation $v_R \approx \sqrt{[k/m]}$, where m is the mass of the spring, holds for all of the cantilever geometries commonly employed. Hence, the lighter the cantilever and the lower k, the higher the resonance frequency of the lever and the lower the noise. Very light mass, and a great simplicity of design, was achieved with the application of microfabrication techniques to the production of the cantilever/tip system. AFM cantilevers of silicon oxide and nitride have been manufactured and employed successfully. Obviously, the sharper the tip, the higher the resolution and SiO_2 micro-cantilevers have proved to be sharp at the edge (due to microroughness caused by protruding microtips) all the way down to a few angstroms. Thus they can be used on *very flat* samples simply by pointing the cantilever at an angle towards the surface. On more uneven surfaces, all-in-one SiO_2 and Si_3N_4 cantilevers with integrated tips have been manufactured.

The possibility of sensing either the attractive or repulsive atomic forces leads to two common modes of operation in AFM. In the attractive mode the tip is poised above the surface and the cantilever vibrated (using a piezo crystal) at a frequency slightly different from v_R. The amplitude of vibration is typically *c.* 20 Å, hence the tip–surface separation must be relatively large, decreasing the *lateral* resolution. The attractive force felt by the tip, $\mathbf{F(R)}$, is detected as the force gradient, $\mathbf{F(R)}^{\cdot} = d\mathbf{F(R)}/dR = \Delta v/[v_R/2k]$, where Δv, the change in resonant frequency, is measured via the change in the vibration amplitude. The image is then a plot of force vs. position.

In the repulsive mode of operation, the AFM tip is in physical contact with the surface, and is best suited to AFMs having tunnelling or direct optical beam feedback. Imaging has been used in two ways. The first is exactly analogous to that for the attractive mode: the tip is vibrated at a frequency of several hundred hertz, with a peak to peak amplitude of 5–20 Å, the exact conditions depending on the resonance frequency and force constant of the cantilever spring. The amplitude of the vibration of the tip is reduced by hard contact with the surface and profiling is achieved by keeping the amplitude at a fixed value via the movement of the sample up or down using an (x, y, z) translator (such as a piezo crystal). Alternatively, the force is monitored at constant tip height (cf. the constant current and constant height modes in STM). The second repulsive method is the simplest AFM approach: the tip is maintained against the surface with a spring and cantilever arrange-ment and the motion of the tip in the z direction as the sample is moved via its attachment to an (x, y) translator is monitored.

The methods of detecting the deflection of the cantilever arm should now be considered. In effect, all that is required is a signal that varies rapidly with the movement of the AFM tip. This can then be related to the movement in the z direction and plotted vs. position to give the image, or a feedback loop can be used to change the sample height to maintain a fixed deflection. The method employed in the original paper on AFM was to detect the spring deflection via tunnelling to a STM tip mounted behind the cantilever arm and this approach is the one employed by most groups starting in AFM. Several optical techniques, including interferometry, have been applied to monitor the cantilever deflection. The most attractive (and simple) of which is the use of a position-sensitive detector to measure the deflection of a laser beam after its reflection from a mirror mounted on the back of the cantilever arm, see Figure 2.33. Such optical techniques have several advantages over tunnelling: (i) less dependence on the roughness of the cantilever surface, (ii) the tunnelling approach is prone to contamination of the STM tip, and (iii) there is less force imposed on the AFM arm by the laser photons than by the STM tip, resulting in more reliable operation.

If AFM is to be a non-destructive tool, then the tracking force should not be large enough to damage the sample. Calculations suggest that this force should not exceed 10^{-11} N for biological surfaces (i.e. adsorbed biological species) and 10^{-9} N for 'hard' surfaces. Other calculations show that a force of c. 10^{-8} N causes large elastic compressive deformation on graphite surfaces, with forces over 5×10^{-8} N causing the tip to puncture the surface. Thus, a polymerised monolayer of n-(aminoethyl)-10, 12-tricosadiynamide has been imaged with a tracking force of c. 10^{-8} N, barely enough to prevent damage to the monolayer. The monolayer was finally damaged after tens of images were obtained. Even though it is clear that caution must be exercised in obtaining AFM images, the number of such images of surfaces, and species adsorbed at surfaces, is increasing rapidly.

As was mentioned above, reports on the application of AFM in $situ$ in electrochemistry are few. One such report concerns the study of the under-potential deposition (upd) of copper on gold and reinforces the need for caution when employing repulsive AFM to the study of adsorbed species.

Underpotential deposition is a much-studied phenomenon in electro-chemistry and is the electrochemical reduction of a metal cation to form a monolayer or submonolayer of the corresponding metal at the surface of an electrode. The critical point is that deposition occurs at a potential higher than that dictated by the reversible potential of the metal/metal cation couple, suggesting that such a upd layer is energetically quite different from the bulk metal. However, subsequent deposition on a upd monolayer occurs at the expected potential, and the resulting surface is typical of the bulk metal.

The upd process at noble metal surfaces is of considerable interest for several reasons:

1. as a model of deposition in general;
2. upd surfaces show increased resistance to poisoning by oxidation products; and thus are of interest with respect to fuel cell and biosensor development;
3. the distinctive chemical properties of the upd surface.

A representative example of the upd process is copper on gold and an extremely illuminating study of this system using repulsive AFM was reported by Manne et al. (1991). The authors employed a commercially available AFM, the essentials of which are shown in Figure 2.33. The reference electrode was a copper wire in contact with the electrolyte at the outlet of the cell. The counter electrode was the stainless steel spring clip holding the AFM cantilever in place. The working electrode was a 100 nm thick evaporated Au film (which is known to expose mainly the Au(111) surface) mounted on an (x, y) translator.

Figure 2.34 shows the Au(111) surface in 0.1 M $HClO_4$/1×10^{-3} M $CuClO_4$ at $+0.7$ V, prior to the deposition of the Cu. The (x, y) displacement of the

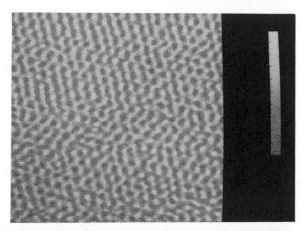

Figure 2.34   AFM image of the Au(111) surface in 0.1 M $HClO_4$/1 × 10^{-3} M $CuClO_4$ at 0.7 V prior to copper deposition. The reference electrode was a copper wire in contact with the same electrolyte. The Au–Au spacing is 2.9 Å. From Manne *et al.* (1991).

translator was calibrated using the known atom–atom spacing of the close-packed Au(111) lattice of 2.9 Å. The image was the same when 0.1 M H_2SO_4/ 1 × 10^{-3} M $CuSO_4$ was employed as the electrolyte; a surprising observation in the light of the fact that SO_4^{2-} is thought to be strongly adsorbed at this potential. The authors postulated that either sulphate is not as strongly adsorbed as was previously thought or the AFM tip 'pushes away' the adsorbed SO_4^{2-} during its passage across the surface.

Figure 2.35 shows an image collected at −0.1 V in the perchlorate solution after the bulk deposition of several monolayers of Cu. The Cu–Cu distance

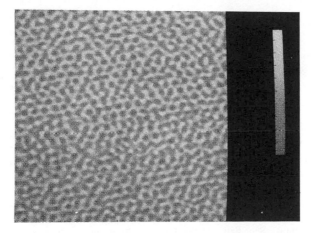

Figure 2.35   AFM image of the Au(111) surface after bulk deposition of several monolayers of Cu. The potential of the gold is −0.1 V, and the Cu–Cu spacing is 2.6 Å. From Manne *et al.* (1991).

in Figure 2.35 was found to be 2.6 Å, which was taken as evidence of the Cu atoms simply sitting 'on top' of the underlying gold atoms. Sweeping the potential up to +0.11 V stripped off all but the initially deposited upd layer. At this point, the authors observed that the nature of the electrolyte had a major effect on the structure of the adsorbed Cu. Thus the Cu–Cu spacing in the perchlorate electrolyte was the same as for the Au(111) surface at 0.7 V, i.e. 2.9 Å. However, the Cu lattice was no longer exactly overlying, i.e. commensurate with, the underlying gold lattice. It appeared that the Cu lattice direction was rotated by $30° \pm 10°$ relative to the gold. This is schematically represented in Figure 2.36(a). The authors attributed this rearrangement to the strain of incorporating the larger copper atoms (Cu radius 1.75 Å, Au 1.42 Å) into a close-packed structure exactly over the Au(111) surface. In 0.1 M H_2SO_4 at +0.144 V the upd Cu layer formed a much more open lattice with a Cu–Cu spacing of 4.9 Å; an observation supported by STM. This more open structure is shown schematically in Figure 2.36(b). The authors postulated that the more open structure found in the sulphuric acid electrolyte was due to the stabilisation of the Cu monolayer by co-adsorption with SO_4^{2-}, although they had not observed this co-adsorbed sulphate directly. This was supported by radiochemical data which showed that, prior to Cu upd on gold, there is little or no sulphate on the gold surface. Immediately following the deposition, substantial sulphate adsorption occurs.

On sweeping the potential back to +0.7 V, the upd Cu was removed and the original Au(111) surface regained, showing unequivocally that the upd of Cu on Au(111) in either electrolyte is completely reversible, contrary to some previous speculation.

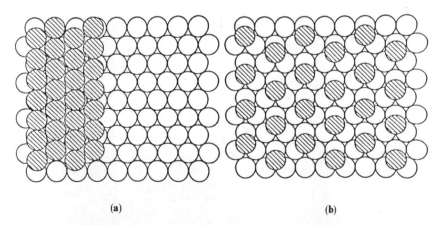

(a)   (b)

Figure 2.36(a)   Schematic representation of the incommensurate close-packed overlayer of Cu on Au formed in the perchlorate electrolyte The open circles are the gold atoms. Only part of the monolayer is shown in order to exhibit the overlayer–underlayer orientation. (b) Schematic representation of the more open lattice formed in the sulphuric acid electrolyte. From Manne *et al.* (1991).

Thus, AFM is a very powerful complement to STM and has the great advantage of not involving quantum-mechanical tunnelling. This means that non-conducting samples can be imaged, both in terms of the surface and the adsorbate, and that images can be obtained when faradaic current is flowing, as in the study discussed above. The latter could not be done with STM as a faradaic current has a disastrous effect on the tunnelling current, hence AFM offers the very real possibility of real-time imaging, in the same way that *in situ* Fourier transform infrared (FTIR) offers the capability to monitor an electrochemical process before, during and after a potential step has induced change (see section 2.1.6). With the initial difficulties rapidly being overcome, AFM is certain to become one of the major *in situ* tools of the future in electrochemistry.

2.1.6. Infrared spectroscopy

It has long been realised that infrared (IR) spectroscopy would be an ideal tool if applied *in situ* since it can provide information on molecular composition and symmetry, bond lengths and force constants. In addition, it can be used to determine the orientation of adsorbed species by means of the surface selection rule described below. However, IR spectroscopy does not possess the spatial resolution of STM or STS, though it does supply the simplest means of obtaining the spatially averaged *molecular* information.

Two major problems long defeated attempts to employ IR *in situ*:

1. *Solvent absorption*: all common solvents, and most especially water, absorb IR radiation very strongly.
2. *Sensitivity*: how to detect the tiny absorptions from a monolayer of adsorbed species when conventional detectors were noisy and conventional IR sources weak.

Some idea of the problem can be gleaned from Figure 2.37, which shows a transmittance spectrum of a *c.* 5 μm thick layer of water. On the same scale the IR absorptions expected from a monolayer of an adsorbed species may be of the same order as the thickness of the pen trace.

One possibility of circumventing the problem of the solvent absorption is to use Raman spectroscopy, where the probing light is in the visible, and this approach is detailed in section 2.1.7. However, the difficulties experienced with the application of Raman to the electrode/electrolyte interface (*vide infra*), refocused attention on the seductive simplicity of IR spectroscopy, particularly as the technique had proved invaluable in the study of species at the gas/solid and vacuum/solid interfaces.

The techniques eventually developed for the application of IR spectroscopy all employ one of three approaches to overcome the problem of the strong solvent absorption and then can be subdivided according to the means by which the sensitivity of the particular technique is increased. We first divide

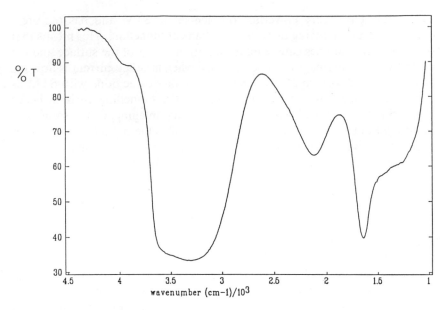

Figure 2.37   Infrared transmittance spectrum of a *c.* 5 μm-thick layer of water.

the *in situ* IR techniques into *transmission, internal reflectance* and *external reflectance.*

2.1.6.1   Transmission.   This approach is the simplest, in which a thin layer of electrolyte is formed between the infrared transparent windows of the cell via a spacer which is usually the working electrode itself, e.g. a gold minigrid or a slice of reticulated vitreous carbon. Because of its simplicity, this was one of the earliest methods, however, it is very restricted when employed in electrochemistry as most changes occur in the Helmholtz layer which is only a few molecules thick. Hence, even when several minigrids are used, the method is insensitive to adsorbed species. In addition, the restricted pathlength required places stringent demands on the cell design, and the technique is effectively limited to non-aqueous solvents. Consequently, *in situ* transmission IR is not the technique chosen to study surface processes at an electrode; its application to the study of solution species is described below.

2.1.6.2   Internal reflectance (attenuated total reflectance ATR).   The internal reflectance or, more usually, attenuated total reflectance (ATR), technique depends on the total reflectance of an IR beam at the internal face of an IR-transparent crystal of high IR refractive index, as shown in Figure 2.38. Medium 1 is a prism of such a material (for example, Si, Ge or KRS-5 [thallous bromide–iodide]), medium 2 is a thin coating of a metal (Au, Pt, Fe) which forms the working electrode and medium 3 is the electrolyte. The

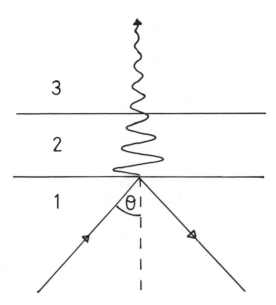

Figure 2.38 Schematic representation of the evanescent wave resulting from the total internal reflection of an electromagnetic ray at an interface.

technique operates via *critical internal reflection*. In brief, the refractive index of a material, $\hat{n}_j$, is given by:

$$\hat{n}_j = n_j - ik_j \tag{2.56}$$

where n_j is the real part of the refractive index for medium j, $i = \sqrt{[-1]}$ and $k_j = \alpha_j \lambda / 4\pi$, with α_j = the adsorption coefficient of medium j. If medium 3 in Figure 2.38 is non-absorbing, and $n_3 < n_1$, then for angles of incidence $\theta > \theta_c$ with:

$$\theta_c = \sin^{-1}[n_3/n_1] \tag{2.57}$$

then the incident ray is totally reflected within the crystal. At the point of incidence, the incident and reflected waves superimpose to form a standing, non-propagating (evanescent) wave that decays exponentially out from the crystal, as shown in the figure. The distance in medium 3 over which the mean square electric field strength of the evanescent wave decays to $1/e$ of its value at the boundary with medium 2, is called the penetration depth (τ) and is given by:

$$\tau = \lambda / - [4\pi \cdot \mathrm{Im}\, \xi_3] \tag{2.58}$$

where $\xi_3 = (\hat{n}_3^2 - n_1^2 \sin^2 \theta)^{1/2}$, and Im requires that the coefficient of the imaginary term in ξ_3 is inserted.

For the situation more usually found in practice, $n_3 < n_1$; if $k_3 = 0$, then $\xi_3 = (n_3^2 - n_1^2 \sin^2 \theta)^{1/2}$. For a Ge prism ($n_1 = 4$) in contact with air ($n_3 = 1$):

$$\xi_3 = [1^2 - 16\sin^2(45°)]^{1/2} = \pm 2.65i \qquad (2.59)$$

taking the negative root this gives $\tau = \lambda/-4\pi[-2.65] \approx \lambda/33$.

For absorbing media of low k, τ is of the order of $\lambda/10$. At $1000\,\text{cm}^{-1}$ this corresponds to a penetration depth of $1\,\mu\text{m}$. Clearly, this reduced pathlength overcomes the problem of strong solvent absorption. However, the very small pathlength does require high concentrations of reactant species, though improved sensitivity can be obtained when adsorbed material is investigated. This sensitivity problem can also be offset to some degree by employing thin crystals in which the IR ray suffers many internal reflections and the effective pathlength, l_{eff}, is given by

$$l_{\text{eff}} = N\tau \qquad (2.60)$$

where N is the number of internal reflections where the evanescent wave probes the electrolyte. N can simply be calculated from a knowledge of the crystal dimensions and geometry.

Lock-in detection in dispersive instruments or multiple scans on a Fourier transform infrared spectrometer have all been employed in order to enhance the sensitivity of the internal reflectance approach to *in situ* IR. In all these approaches the electrochemical cell is of a relatively simple design, again using the standard three-electrode arrangement (see Figure 2.39). In the 'lock-in' detection technique, the potential at the working electrode is modulated as a square wave and the detector 'locked in' to this frequency. In this way only the signal modulated at this frequency is measured, the detector being scanned across the wavelength scale, via a monochromator, until the full spectrum is collected.

If the crystal is a semiconductor such as Si, it can be used as the working electrode itself and this was the means employed in the early experiments. However, the limited number of suitable electrode materials available was a severe restriction and attempts to use metal-coated crystals suffered severely from the low sensitivity caused by the attenuation of the IR beam by the coating. In addition, 'lock-in' detection is mainly limited to those electrochemical systems capable of responding sufficiently rapidly to the imposed potential modulation.

The capabilities of the ATR technique were greatly enhanced with the advent of FTIR. These instruments have greatly improved optical throughput and speed, and consequently are capable of large signal-to-noise ratio without the need for advanced detection techniques. A full description of the principle and operation of these instruments is beyond the scope of this book; the interested reader is referred to any one of the excellent texts on this subject (see for example Griffiths and deHaseth, 1986). In this method, a spectrum (consisting of many co-added and averaged scans) is collected at the base potential and used as background. The potential is then stepped, a second

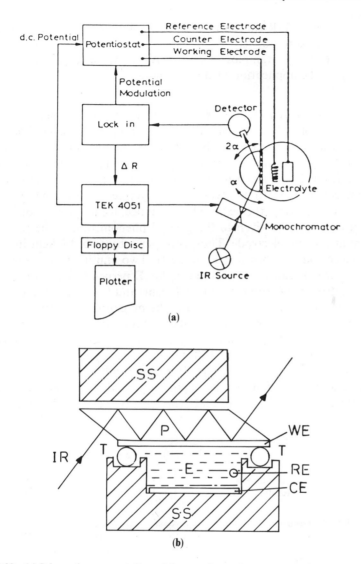

Figure 2.39 (a) Schematic representation of the experimental arrangement for attenuated total reflection of infrared radiation in an electrochemical cell. (b) Schematic representation of the ATR cell design commonly employed in *in situ* IR ATR experiments. SS = stainless steel cell body, usually coated with teflon; P = Ge or Si prism; WE = working electrode, evaporated or sputtered onto prism; CE = platinum counter electrode; RE = reference electrode; T = teflon or viton 'O' ring seals; E = electrolyte.

spectrum is collected and ratioed to the first. The signal-to-noise ratio increases as $\sqrt{N}$, where N is the number of individual scans; each scan can take as little as 15 ms, compared to several minutes for a dispersive instrument. Such a single step experiment imposes no restriction on the type of electrochemical systems that can be investigated though if the electrochemistry is truly reversible, FTIR cannot match the sensitivity of the lock-in detection methods referred to above.

There are many examples of *in situ* ATR in the literature which ably demonstrate the application of the technique.

2.1.6.3 External reflectance. The most commonly applied *in situ* IR techniques involve the external reflectance approach. These methods seek to minimise the strong solvent absorption by simply pressing a reflective working electrode against the IR transparent window of the electrochemical cell. The result is a thin layer of electrolyte trapped between electrode and window; usually 1 to 50 μm. A typical 'thin layer' cell is shown in Figure 2.40.

Before considering the external reflectance techniques in detail, some thought must first be given to the general phenomenon of the reflection of IR light at a metal electrode. Thus, consider a beam of IR light incident on a metal electrode, as shown in Figure 2.41. Two limiting polarisations of the electric vector **E** can be considered: S, for **E** perpendicular to the *plane* of reflection (from the German word Senkrecht) and P, for parallel. At the point of incidence a standing wave is set up, the magnitude of which depends on the phase shift that the incident ray suffers on reflection. This, in turn, depends on the angle of incidence, θ, and the polarisation of the light. The phase shift for S-polarised light is close to 180° for all angles of incidence, provided the reflection takes place at a highly absorbing substrate. Also, for such a substrate, almost all of the incident light is reflected. At the point of incidence, the electric vectors of incident and reflected rays very nearly cancel out for

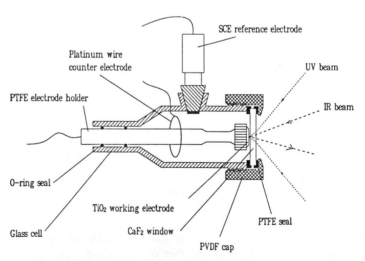

Figure 2.40 Schematic representation of the external reflectance cell design commonly employed in *in situ* IR experiments. If the working electrode is a semiconductor, then the semiconductor/electrolyte interface can be studied under illumination with, for example, UV light by directing the beam perpendicular to the IR beam, as shown.

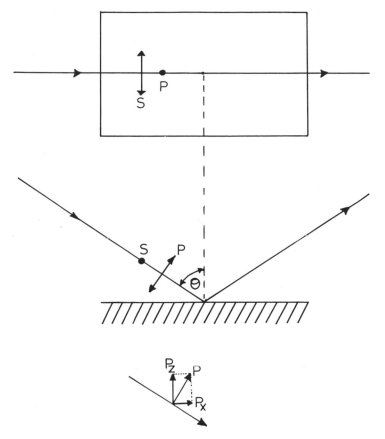

Figure 2.41 The direction of the electric vector for light polarised in the S- and P-directions before being incident at a planar surface at an angle θ, and the resolution of the electric vector of the P-polarised light into its components parallel and perpendicular to the reflective surface.

all θ and the standing wave thus has a near-zero intensity (see Figure 2.42). If the electric field has no amplitude at the surface, it cannot interact with any adsorbed species. Thus, we would expect very little absorption from a layer on the surface, with the incident light polarised perpendicular to the plane of reflection.

P-polarised light can be reduced to two components: P_x polarised parallel to the surface and P_z polarised perpendicular to the surface. The P_x component also suffers a 180° phase change on reflection for all θ and is thus 'blind' to any surface species. However, the phase shift for the P_z component changes rapidly with θ. This results in the ratio of the standing wave/incident ray mean electric field strength, $\langle E_S^2 \rangle / \langle E_I^2 \rangle$, varying with θ as shown in Figure 2.42.

The above discussion has two important implications:

1. S-polarised light is effectively blind to adsorbed and near-surface species at all angles of incidence.

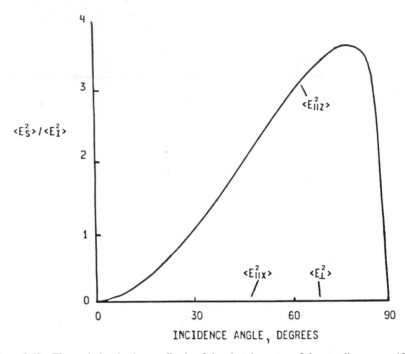

Figure 2.42   The variation in the amplitude of the electric vector of the standing wave, $\langle E_s^2 \rangle$, ratioed to the intensity of the incident ray, $\langle E_1^2 \rangle$, as a function of the angle of incidence at a reflecting surface, $n = 3.0$ and $k = 30$, in contact with a medium with $n = 1.0$ and $k = 0$. From J.K. Foley, C. Korzeniewski, J.J. Daschbach and S. Pons in *Electroanalytical Chemistry. A Series of Advances*, A.J. Bard (ed.), Vol. 14, Marcel Dekker, New York, 1986.

2. P-polarised light can interact with adsorbed species; however, this is possible only under a constraint. Thus, it is known that the interaction of a dipole oscillator on, or near, a metal surface with incident IR radiation depends on the orientation of the dipole relative to the surface. The radiation cannot give rise to oscillating electric fields parallel to the surface and so can interact only with those normal modes of a molecule for which the dipole change on vibration has a finite component perpendicular to the surface. Therefore the intensity of an observed absorption band will vary with the molecular orientation, being zero when the dipole change is parallel to the surface and a maximum when perpendicular. This is the basis of the *surface selection rule* (SSR) as reported by Greenler in 1966.

The surface selection rule operates in addition to the normal IR selection rules in determining which vibrational modes are observed. As a result of the SSR the relative intensities of the fundamental IR adsorption bands of an adsorbed species can be used to give information on the orientation of the species with respect to the surface. Both S- and P-polarised light interact equally with the randomly oriented solution species.

In employing a thin-layer configuration the external reflectance approach reduces the problem of the strong solvent absorption in two ways. Firstly, this configuration yields a solution layer only a few microns thick. Secondly, exact calculations employing the Fresnel reflection equations show that the radiation absorbed by an aqueous layer $c.$ $1\,\mu m$ thick in contact with a reflective electrode is attenuated to a lesser extent than would be predicted by the Beer–Lambert law.

There are some consequences of employing the thin layer configuration that must be borne in mind:

1. The thin layer is a source of uncompensated resistance. This can lead to a voltage drop, V_{IR}, between the reflective working electrode and the reference electrode. V_{IR} is given by:

$$V_{IR} = I \cdot R \qquad (2.61)$$

where I is the current flowing across the working electrode/electrolyte interface, and R is the resistance of the electrolyte thin layer. Thus, the true potential of the working electrode, V_T, differs from that measured, V_M, according to:

$$V_T = V_M - V_{IR} \qquad (2.62)$$

The importance of this can be assessed by running a cyclic voltammogram with the electrode pulled back from the window, and comparing it to one collected with the electrode pushed against the window. The influence of IR drop is discussed in a little more detail in section 2.2.7.

2. As a result of the thin layer resistance, the time constant of the cell is relatively long, e.g. $5 \times 10^{-3}\,s$ for a 1 M electrolyte. Again, this is no real problem, particularly in the (effectively) steady-state IR methods.
3. The thin layer arrangement results in very long diffusion times for species from the bulk of the solution and thus the thin layer can become depleted of reactant species. This can be used to some advantage as will be seen in the discussion of CO_2 reduction in the section on near-electrode investigations.

The three most commonly applied external reflectance techniques can be considered in terms of the means employed to overcome the sensitivity problem. Both electrically modulated infrared spectroscopy (EMIRS) and *in situ* FTIR use potential modulation while polarization modulation infrared reflection absorption spectroscopy (PM-IRRAS) takes advantage of the surface selection rule to enhance surface sensitivity.

2.1.6.4 Electrically modulated infrared spectroscopy, (EMIRS). In all three external reflectance approaches the signal processing technique serves two purposes: (a) to remove the contributions to the reflected ray that do not change, e.g. the detector response, the source emission envelope, the solvent, window and beamsplitter absorptions, etc., and (b) to enhance the very low signal-to-noise ratios encountered as a result of the weak absorptions of the electrogenerated species.

In the EMIRS (and *in situ* FTIR) technique, the potential of the working electrode is changed from a base value, V_b, at which the reflectivity of the electrode is $R(\bar{v})_b$, to a value V_w, where the reflectivity is $R(\bar{v})_w$. Spectra are usually plotted in the form $(\Delta R/R)$ vs. $\bar{v}$, where:

$$(\Delta R/R) = [R(\bar{v})_w - R(\bar{v})_b]/R(\bar{v})_b \qquad (2.63)$$

This results in a *difference* spectrum where features may have $+$ or $-$ ($\Delta R/R$) according to whether they are gained or lost at V_w with respect to V_b (*vide infra*).

A schematic diagram of the EMIRS instrumentation is shown in Figure 2.43. EMIRS employs high-throughput dispersive optics and a lock-in detection system, as discussed in the section on ATR. Thus, infrared light is reflected off the polished working electrode while the electrode potential is modulated as a square wave of frequency c. 10 Hz between the base potential V_b, and the working potential V_w. The wavelength range of interest is then slowly scanned. Only that component of the reflected light having the same modulation frequency and phase as the applied potential square wave is detected

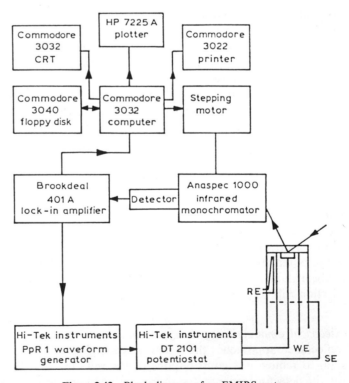

Figure 2.43 Block diagram of an EMIRS system.

and amplified. Hence, the sensitivity of the technique is increased up to that necessary for the detection of potential-induced absorption changes of sub-monolayer quantities of adsorbate ($c. \Delta R/R = 10^{-4}$). The sensitivity may be further enhanced by averaging successive scans and a polariser is placed in the IR beam to remove the superfluous S-polarised component of the light.

The intensity difference, $\Delta R/R$, observed in a potential-modulation IR experiment may result from several sources:

1. Electroreflectance effects arising from changing electron densities at the electrode surface as a result of the applied potential.
2. Potential-induced changes in absorption due to adsorption or electron transfer reactions.
3. Potential-dependent shifts in IR band frequencies of adsorbed species.
4. Potential-dependent changes in the orientation of adsorbed species.

The first example of the application of the EMIRS technique was reported by Beden *et al.* (1981) and concerned the electro-oxidation of methanol.

As will be discussed further below, a major focus of electrochemical studies has been the investigation of the electro-oxidation of small organic fuel molecules, with respect to the development of fuel cell technology. The metal anodes in such reactions are rapidly poisoned by the formation of strongly adsorbed species. In the case of the methanol fuel cell considerable controversy existed over whether triply bonded $\equiv$COH or linearly bonded $-$C$\equiv$O were the poisoning species. The paper by Beden and co-workers (1981) effectively solved this long-lasting controversy, reaching conclusions that have stood the test of time. Thus, Figure 2.44(a) shows an EMIRS spectrum collected from a Pt electrode immersed in $1\,M\ H_2SO_4/0.5\,M\ CH_3OH$. The potential was modulated between 0.05 V and 0.45 V vs. NHE.

On the basis of data obtained from the gas–solid interface, the weak band near $1850\,cm^{-1}$ was attributed to the loss of bridged C=O (Pt_2C=O) and the intense dipolar feature centred near $2070\,cm^{-1}$ was attributed to linearly-adsorbed $-$C$\equiv$O. The dipolar nature of the latter band was tentatively attributed to a potential-dependent frequency shift as opposed to a coverage-dependent shift, the justification for the former hypothesis coming in the first PM-IRRAS paper, as discussed below. The potential-dependence of the C$\equiv$O stretching frequency of linearly adsorbed CO has been interpreted as the result of differing degrees of electron back-donation from the Pt d-orbitals into the $2\pi^*$ orbitals on the C$\equiv$O and as an electric field or Stark effect. In the former case, as the potential is stepped positive, less mixing of the d and $2\pi^*$ orbitals occurs. The resultant decrease in back-donation into the anti-bonding orbitals on the CO strengthens the C$\equiv$O bond and hence increases $\bar{v}_{CO}$.

The spectra in Figure 2.44(b) show the dependence of the EMIRS response on the amplitude of the potential modulation. These were reported to indicate a decrease in coverage by adsorbed species on entering the region of sustained methanol oxidation, as would be expected.

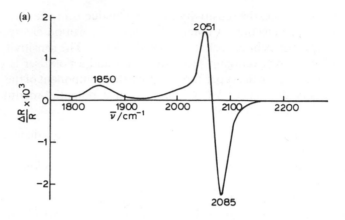

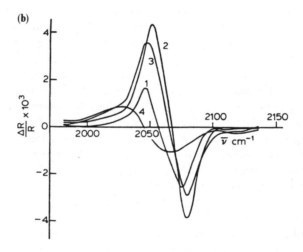

Figure 2.44 (a) Reflectance spectrum from a platinum electrode immersed in 1 M H_2SO_4 + 0.5 M methanol. Modulation was from 0.05 V to 0.45 V vs. NHE at 8.5 Hz and a scan rate of 0.0127 $\mu m\,s^{-1}$. (b) The dependence of the spectrum in (a) on the modulation amplitude. (1) 0.05 V − 0.30 V, (2) 0.05 V − 0.55 V, (3) 0.05 V − 0.80 V and (4) 0.05 V − 0.95 V. Scan rate 3.18 × 10^{-3} $\mu m\,s^{-1}$. After Beden *et al.* (1981).

This early paper concluded that $-C{\equiv}O$ is the dominant strongly adsorbed species (poison) and it is present at high coverage. Some Pt_2CO is also present, but there is no evidence of Pt_3COH under the experimental conditions employed (non steady-state, potential perturbation at 8.5 Hz and with the dissociative chemisorption of methanol slow). The identity of the poisoning species formed on bulk Pt during the electro-oxidation of methanol is no longer in doubt, largely due to this pioneering work.

2.1.6.5 Polarisation modulation infrared reflection-absorption spectroscopy (PM-IRRAS or IRRAS). Potential modulation IR studies rely on switching the potential at a reflective electrode between 'rest' and 'active' states, generating *difference* spectra. However, the EMIRS technique has several drawbacks; the relatively fast potential modulation requires that only fast and reversible electrochemical process are investigated; the absorption due to irreversibly chemisorbed species would be gradually eliminated by the rapid perturbation. Secondly, there is some concern that rapid modulation between two potentials may, to some extent, in itself induce reactions to occur.

Hence, another approach is required that will enable the study of a wider range of electrochemical processes. Two other techniques are potentially capable of fulfilling this role: PM-IRRAS and *in situ* FTIR.

As with EMIRS, PM-IRRAS is a lock-in detection technique but, unlike EMIRS, it is not the potential that is modulated but the polarisation of the incident light. The PM-IRRAS technique takes advantage of the different response of adsorbed molecules to incident IR light polarised S or P. A schematic representation of the apparatus used in the PM-IRRAS technique is shown in Figure 2.45. As was discussed above, adsorbed species will absorb only P-polarised light, subject to the orientational constraint. In order to modulate the polarisation of the IR beam rapidly between the S- and P-directions, a photoelastic modulator is employed. This consists of a non-centrosymmetric crystal oscillating at a fixed frequency and driven by an oscillating electric field. This gives rise to an AC voltage at the detector proportional to $(I_P - I_S)$, the difference in reflected intensity of the two polarisations. Some allowance must then be made for the attenuation factor, i.e. the polarisation-dependent absorption and reflection characteristics of the optical system. This has been found to be equivalent to $(I_P + I_S)^{-1}$ as a function of frequency, $\bar{v}$. The normalised difference $(I_P - I_S)/(I_P + I_S)$ is then

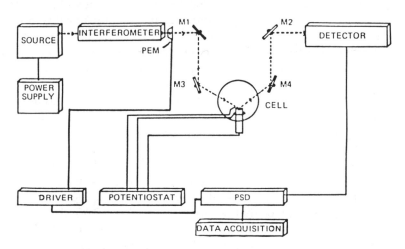

Figure 2.45 Block diagram of an FT-IRRAS system.

plotted as a function of $\bar{v}$ to give the spectrum. Since I_P is absorbed by the surface species and I_S is not, the signal contains the absolute spectrum of the adsorbed species. It follows that PM-IRRAS apparently has the advantage of supplying *absolute* spectra, i.e. spectra collected at a single potential.

A great deal of success was attendant on the early application of PM-IRRAS to the gas/solid interface. Golden *et al.* (1981) reported the development of instrumentation, using conventional dispersive optics, able to record detailed infrared reflection–absorption spectra from molecules adsorbed on single-crystal Pt without any interference from the gas-phase species.

In 1982, Russell *et al.* applied the PM-IRRAS, or IRRAS as they termed it, technique *in situ*. In the gas/solid studies the sensitivity of the PM-IRRAS technique was enhanced by operating at high angles of incidence in order to maximise the interaction between the standing wave at the surface and the adsorbate, as predicted by Greenler (see above). In order to achieve this in the *in situ* (IRRAS) studies, a prismatic window is employed. Figure 2.46 shows the angle of incidence at a reflective electrode in aqueous electrolyte calculated for (a) a plate window with a 65° angle of incidence on the window and (b) a prismatic window having a 65° bevel. The refractive index of water was taken as 1.33, i.e. with $k = 0$, for simplicity. Clearly, the prismatic window is necessary to take full advantage of the Greenler effect.

Russell and co-workers used the IRRAS technique to verify the assignment of the bipolar band observed in the EMIRS experiment. In the EMIRS experiment, the fact that the technique yields *difference* spectra rendered the assignment of the bipolar nature of the band to a potential-dependent frequency shift somewhat indirect. Thus, the capability of the IRRAS technique to deliver absolute spectra could be used to some effect. Figure 2.47(a) shows IRRAS spectra collected from a Pt electrode immersed in CO-saturated 1 M $HClO_4$ at various potentials. This clearly shows the potential-dependent nature of the frequency of the $-C\equiv O$ band. Figure 2.47(b) shows the spectrum collected at 0.05 V in Figure 2.47(a) subtracted from that taken at 0.45 V and Figure 2.47(c) shows the corresponding EMIRS spectrum taken with the potential modulated between the same two values (see Figure 2.44(a)). The agreement between the results obtained from the two, obviously complementary, techniques was very good, differing only in the band intensities and width. These effects can be attributed to the then-limited resolution of the IRRAS technique ($30\,cm^{-1}$ at $2100\,cm^{-1}$, which at that time was rather poor). Clearly, the technique had elegantly proven the principal conclusions of the EMIRS experiment.

A close inspection of Figure 2.47(a) reveals a marked curvature in the baseline which disappears on taking the difference spectrum shown in Figure 2.47(b). This curvature marks the presence of underlying water absorption and points to the fact that whilst in principle IRRAS should be an absolute technique, in practice the situation is more complicated.

As was discussed above, the surface selection rule arises because of the reduction in magnitude of the mean square electric field strength, $\langle E_S^2 \rangle$, of

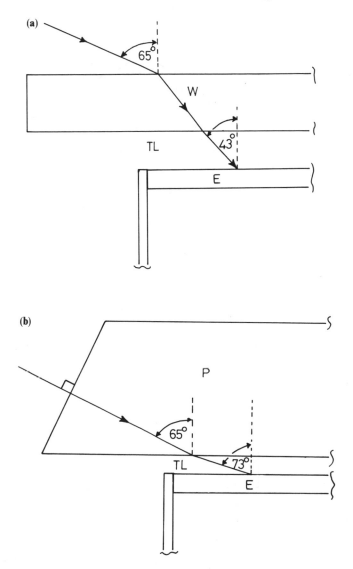

Figure 2.46 Schematic diagram showing the angle of incidence at the electrode for (a) an angle of incidence of 65° at a plate window (W), and (b) the incoming ray incident normal to the face of a prismatic window (P), having a bevel of 65°, assuming $n_{solvent} = 1.33$, $k_{solvent} = 0$, $n_{window} = 1.4$, $k_{window} = 0$, $n_{air} = 1$, and $k_{air} = 0$. TL = thin electrolyte layer, E = electrode.

the standing wave from S-polarised light reflected at a metal surface, due to a phase change of 180° between incident and reflected rays. However, this reduction in the magnitude of $\langle E_S^2 \rangle$ occurs not only at the surface, but also for an appreciable distance into the solvent layer. Hence, $(I_P - I_S)$ contains information not just on adsorbed species but also on a layer of solution

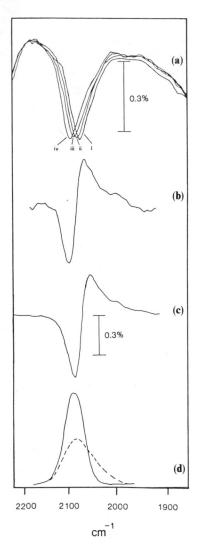

Figure 2.47 (a) IRRAS spectra of CO adsorbed on Pt in 1.0 M HClO$_4$ saturated with CO. The electrode potential was held constant at (i) 50 mV vs. NHE, (ii) 250 mV, (iii) 450 mV and (iv) 650 mV. (b) Difference spectra resulting from subtraction of the IRRAS spectra (iii) and (i). (c) EMIRS spectrum of Pt electrode in CO-saturated 1 M HClO$_4$ modulated between + 50 mV and 450 mV. (d) A schematic representation of spectra at two potentials which could produce an EMIRS spectrum similar to that shown in (c). The IRRAS spectra in (a) rule out this possibility. After Russell *et al.* (1982). Copyright 1982 American Chemical Society.

at least 0.1 μm thick. Clearly, at the gas/solid interface the effect of this is negligible. However, for highly absorbing aqueous layers this is much more important, as can be seen in Figure 2.48. This figure shows an FT-IRRAS spectrum of the near-electrode region for CO adsorbed on Pt in 0.5 M H$_2$SO$_4$ at 0.4 V vs. NHE. The intense water absorption across the full mid-IR region

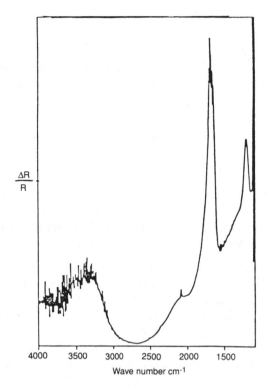

$$\frac{\Delta R}{R}$$

Wave number cm^{-1}

Figure 2.48 FT-IRRAS spectrum of the electrode/electrolyte interface for CO adsorbed on a smooth platinum electrode in 0.5 M H$_2$SO$_4$ at 0.4 V vs. NHE. From W.G. Golden, K. Kunimatsu and H. Seki, *J. Phys. Chem.*, **88** (1984) 1275. Copyright 1984 American Chemical Society.

is plainly seen and dwarfs the tiny (but still relatively strong by *in situ* IR standards) CO absorption near 2040 cm^{-1}.

Later IRRAS papers resorted to taking the *difference* between IRRAS spectra recorded at different potentials in order to remove this unchanging water absorption (see Figure 2.47(b)). This rather annuls the 'absolute' advantage of IRRAS over the less costly *in situ* FTIR method. Such absolute spectra are only obtainable of species that absorb in the spectral region relatively free from strong solvent absorptions, e.g. 2000 cm^{-1}–2800 cm^{-1}. As a result, IRRAS studies have largely been confined to the investigation of adsorbed CO species or adsorbed C≡N. As with ATR, the advent of FTIR proved to have a major effect on increasing the sensitivity of PM-IRRAS.

2.1.6.6 In-situ Fourier transform infrared spectroscopy. The final technique in this section concerns the FTIR approach which is based quite simply on the far greater throughput and speed of an FTIR spectrometer compared to a dispersive instrument. *In situ* FTIR has several acronyms depending on the exact method used. In general, as in the EMIRS technique, the FTIR-based methods rely upon potential-modulation to remove unchanging contributions to the reflectance spectrum. However, normal lock-in detection cannot be employed with an FTIR spectrometer since, by the nature of the FTIR approach, the intensity of the light striking the detector is already modulated by the Michelson interferometer, the device at the heart of the FTIR. The frequency of this modulation, $F(\bar{v})$, depends on the frequency of the IR light, $\bar{v}$ (cm^{-1}), and the velocity of the mirror, v (cm s^{-1}), according to:

$$F(\bar{v}) = v\bar{v}\,\text{s}^{-1} \tag{2.64}$$

In order to employ a lock-in detection technique, as in EMIRS, the modulation frequency of the potential at the electrode would have to be at least an order of magnitude greater than $F(\bar{v})$. Thus, the potential modulation would have to be c. 100 kHz; too great to allow sufficient relaxation time for most electrochemical processes to respond. Instead, a slow modulation or single-step approach is employed, as follows:

1. The potential is repeatedly stepped between working and reference values, as in an EMIRS experiment. However, a number of scans are collected, co-added and averaged during each step to give 'working' spectra, S_w, or reference spectra, S_b, and the frequency of the modulation is much slower than in an EMIRS experiment. All of the working spectra are co-added and averaged at the end of the experiment, to give S_w, and normalised to the co-added and averaged reference spectra, S_b, according to:

$$\Delta R/R = (S_w - S_b)/S_b,$$

thus, as with EMIRS, the spectra are plots of $\Delta R/R$ vs. $\bar{v}$ and peaks pointing up, to $+\Delta R/R$, correspond to the loss of features at the potential E_w, etc. The number of scans at each step, N, is chosen to give the required signal-to-noise ratio. This approach, termed 'potential difference infrared spectroscopy' (PDIR) by Corrigan and Weaver (1986), effectively annuls any experimental drift and/or interference from atmospheric CO_2 and water vapour. However, it is restricted to the study of reversible electrochemical systems, although the response time is not so limiting as in EMIRS.

2. In the second, more commonly-employed, approach a spectrum, S_b, of N co-added and averaged scans (i.e. individual spectra) are collected at the base potential, E_b. The potential is then stepped to successively higher (or lower) values E_n, and further spectra, S_n, taken. As above the spectra are plots of $\Delta R/R$ vs. $\bar{v}$, where:

$$\Delta R/R = (S_n - S_b)/S_b.$$

This approach is very straightforward, is not restricted to reversible systems, and so has been employed to study a wide range of electrochemical systems. Scan times are very low, even as low as 15 ms, hence N can be large without experimental drift becoming a problem. Sensitive spectra with stable baselines are thus quickly and simply obtained.

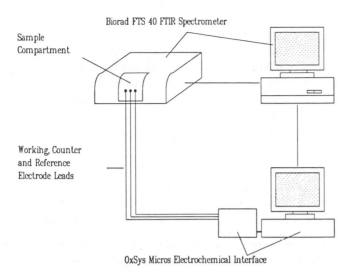

Biorad FTS 40 FTIR Spectrometer

Sample
Compartment

Working, Counter
and Reference
Electrode Leads

OxSys Micros Electrochemical Interface

Figure 2.49   Block diagram of the equipment required to perform *in situ* FTIR.

A schematic representation of the instrumentation used in the *in situ* FTIR technique is shown in Figure 2.49. As can be seen from the figure, the instrumentation is much simpler than that required to perform EMIRS or PM-IRRAS measurements.

An elegant example of the application of *in situ* FTIR to the study of absorbed species is provided by the work of Corrigan and Weaver (1986). As well as being extremely interesting in terms of the system under study, the work of these authors represents an important step towards the further development of *in situ* FTIR. If the approach is to attain its full potential, then it must be capable of providing *quantitative* as well as qualitative information.

Corrigan and Weaver employed the PDIR approach to study the potential-dependent adsorption of azide, N_3^-, at a silver electrode. The potential was switched between the reference value, -0.97 V vs. SCE (where adsorption is known to be limited) and the working potential every 30–60 scans, i.e. up to a minute per step, to a total of *c.* 1000 scans. The high number of scans was required in order to obtain the required S/N ratio; hence the PDIR technique was employed to minimise instrumental drift. Since the electro-chemical process under study was totally reversible on the timescale of the experiment, the PDIR technique was a viable option.

Figure 2.50 shows IR spectra of a silver electrode immersed in aqueous perchlorate containing N_3^-, obtained via the PDIR approach with successively higher working electrode potentials. As can be seen from the figure, a loss feature is observed near 2048 cm^{-1} at potentials $\geqslant -0.72$ V. However, a gain feature is not observed until potentials $\geqslant -0.02$ V.

The loss feature is at the frequency at which the asymmetric stretch of solution N_3^- is expected to absorb and this was the assignment made by the authors. The fact that the azide lost from the thin layer is not replenished via diffusion in from the bulk of the solution is not surprising given the restrictive nature of the thin layer configuration. In fact, the authors performed an experiment in which the diffusion of N_3^- into the thin layer was monitored as a function of time and was found to be first-order, following a $\log_e t$ dependence, and with a half-life of 55 minutes.

The gain feature was attributed to azide adsorbed 'end on' at the electrode. This assignment was based on two factors:

(i) It shows a potential-dependent frequency, increasing from $2074\,\text{cm}^{-1}$ to $2083\,\text{cm}^{-1}$ as the potential is increased from $-0.02\,\text{V}$ to $0.18\,\text{V}$.

(ii) The common observation that azide coordinated 'end on' to a number of metal cations absorbs in the range $2080\,\text{cm}^{-1}$–$2090\,\text{cm}^{-1}$.

The solution loss feature in Figure 2.50 suggests that adsorption is taking place at all the potentials referred to in the figure. Hence, the absence of a

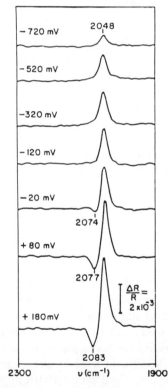

Figure 2.50 Potential difference infrared (PDIR) spectra for adsorbed azide at the silver–aqueous interface in the asymmetric N–N–N stretch region. The reference (base) potential was $-970\,\text{mV}$ vs. SCE; sample potentials as indicated. The solution contained 0.01 M NaN_3/0.49 M $NaClO_4$. The spectra are the average of 1024 interferometer scans at each potential. From Corrigan and Weaver (1986). Copyright 1986 American Chemical Society.

gain feature at potentials < -0.02 V suggests that, at these lower potentials, adsorption is taking place in a manner that renders the azide invisible to the probing IR beam. On the basis of the surface selection rule, the authors postulated that the azide adsorbs flat at lower coverages. This allows the favourable interaction between the π orbitals of the linear N_3^- anions and the orbitals of the metal surface. As the potential, and the coverage, increases, packing considerations then force the adsorbate into the vertical orientation.

The authors obtained an extinction coefficient for the solution azide absorption at 2048 cm^{-1} by measuring the absorbance of a known concentration of the azide in a 0.018 mm pathlength transmittance cell. Using this extinction coefficient with the integrated band intensities of the 2048 cm^{-1} band in Figure 2.50, they were able to calculate the surface azide concentration, Γ, as a function of potential and this is shown in Figure 2.51. For comparison, the figure also shows the values of Γ calculated from differential capacitance measurements (see Hupp *et al.* (1983) and Larkin *et al.* (1982) for a description of the differential capacitance technique). As can be seen, the agreement between the two different techniques is good.

The work of Corrigan and Weaver clearly shows the power of *in situ* FTIR in terms both of the excellent quality of the spectra that can be obtained with relatively uncomplicated instrumentation and of the fact that it points the way ahead for *in situ* infrared spectroscopy.

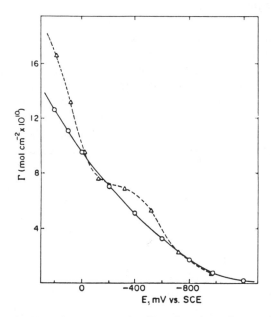

Figure 2.51 Plots of azide surface concentration, Γ, against electrode potential, E, for the silver electrode in 0.01 M NaN$_3$/0.49 M NaClO$_4$. The solid curve is as determined by differential-capacitance potential measurements, and the dashed curve from the integrated infrared intensities of the positive-going bands in Figure 2.50. Copyright 1986 American Chemical Society.

2.1.7   The Raman spectroscopies

Before describing the application of Raman to the study of the electrode/ electrolyte interface, a brief recap of the theory of Raman spectroscopy may be helpful.

When the oscillating electric field of an incident light ray interacts with a molecule, a small oscillating dipole moment is induced in the molecule as a consequence of its polarisability, α. Polarisability itself is a measure of the change in the dipole moment of a molecule induced by an electric field, and in the simplest case, where the electric field E and induced dipole moment μ are in the same direction:

$$\mu = \alpha E$$

For lower molecular symmetries, μ and E must be treated as vectors and may not be in the same direction. Under these circumstances, the polarisability has the form of a tensor.

The rapidly varying electric field of the incident electromagnetic radiation can, therefore, cause a rapid fluctuation in the dipole moment of the molecule. The magnitude of this oscillation depends on the polarisability of the molecule but is generally small.

It is known that such an oscillating dipole radiates light at the same frequency at which it oscillates. The emitted light need not be in the same direction as the incident ray, and the effect is that a small proportion of the incident light is scattered by the molecule. When the frequency of the incident light is unchanged by the scattering process, i.e. the scattering is elastic, the phenomenon is known as Rayleigh scattering. However, the molecules themselves may also undergo vibrational excitation, at frequencies dictated by their normal modes of vibration. If the molecular symmetry is such that the incident electromagnetic radiation can induce an electronic dipole change in the molecule and a vibrational transition can couple to the electronic dipole change, then a Raman transition may take place. In general, the coupling is weak and the magnitude of this inelastic scattering is small, being produced by only $c.$ 1 incident photon in 10^{10}. Thus this, the normal Raman (NR) effect yields relatively weak emission bands.

A schematic representation of the mechanisms involved in Raman transitions is shown in Figure 2.52. The process may be considered in terms of an initial excitation to a 'virtual' state between the ground and first excited electronic states, followed by a re-emission of the photon as decay takes place from the virtual state back to the electronic ground state. If both absorption and decay take place to the $v'' = 0$ vibrational level of the ground state, then the scattering is clearly elastic and the resultant band is the intense Rayleigh line, which dominates Raman spectra. Absorption from $v' = 0$ followed by emission to $v'' = 1$ yields the Stokes line with $\Delta v = 1$, corresponding to the excitation of a particular vibrational mode to its first excited state and resulting in its fundamental absorption. Transition from $v' = 1$ to $v'' = 0$

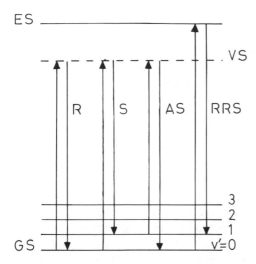

Figure 2.52 Schematic representation of the transitions giving rise to the Raman effect. GS = ground electronic state, ES = excited electronic state, VS = virtual electronic state, R = Rayleigh scattering, S = transitions giving rise to Stokes lines, AS = transitions giving rise to Anti-Stokes lines, RRS = transitions giving rise to resonance Raman.

leads to the anti-Stokes line with $\Delta v = -1$, corresponding to the de-excitation of this vibrationally 'hot' state. For a polyatomic molecule, the vibrational Raman selection rule is $\Delta v = \pm 1$; transitions to levels with $\Delta v = \pm 2, \pm 3$, etc. are only observed if the overtone can couple strongly to another fundamental or if the normal mode itself possesses appreciable anharmonicity.

For a vibration having a frequency $\bar{v}$, the Stokes and anti-Stokes spectra correspond to bands at $\bar{v}_0 - \bar{v}$ and $\bar{v}_0 + \bar{v}$. The spectra are usually presented as the intensity of scattered light vs. the shift, $|\bar{v}_0 - \bar{v}|$, in cm^{-1}.

The ratio of the intensity of anti-Stokes and Stokes lines is primarily determined by the Boltzmann population of the excited vibrational states. For mid-IR frequencies this fractional population is very low ($\approx 10^{-4}$ at $2000\, cm^{-1}$). As a result, Raman spectra are usually taken from the Stokes side of the Rayleigh line as these are generally very much more intense and are not broadened by emissions from 'hot' states.

As was seen in the above discussion, normal Raman spectra are usually weak. In addition, in an NR solution experiment, the scattering comes from species in a c. 2 mm depth of liquid centred on the focused laser beam. Thus, the intensity of a spectrum from a monolayer of an adsorbed sample would be expected to be six orders of magnitude weaker than a NR solution spectrum, rendering NR too insensitive to be of use as an *in situ* probe.

Almost all Raman spectra of adsorbed species have therefore relied on enhancement effects. Thus, if the incident radiation can simultaneously excite an electronic absorption (see Figure 2.52), then the polarisability change associated with this change in the electronic configuration is very large. The resulting resonance Raman (RR) emission is enhanced 10^4–10^6-fold. However, not all of the vibrational modes are enhanced; in particular, there is a high selectivity for those vibrational modes directly associated with the light-absorbing fragment of the molecule (i.e. the chromophore site). Resonance Raman is not a surface-specific effect and depends only on the sample molecule having at least one chromophore site. However, there is a second major enhancement effect, surface enhanced Raman spectroscopy (SERS), though this is only observed at a limited number of metal surfaces. For such SERS-active surfaces a large increase in the intensity of the Raman emission, even at monolayer coverage, of up to six orders of magnitude is found.

The first report of the SERS spectrum of a species adsorbed at the electrode/electrolyte interface was by Fleischman *et al.* (1974) and concerned pyridine on silver. The Raman spectrum of the adsorbed pyridine was only observed after repeated oxidation/reduction cycles of the silver electrode, which resulted in a roughened surface. Initially, it was thought that the 10^6-fold enhancement in emission intensity arose as a result of the substantially increased surface area of the Ag and thus depended simply on the amount of adsorbate. However, Jeanmarie and Van Duyne (1977) and Albrecht and Creighton (1977), independently reported that only a single oxidation/reduction cycle was required to produce an intense Raman spectrum and calculations showed that the increase in surface area could not possibly be sufficient to give the observed enhancement.

Silver was the first metal to be found to be SERS-active and remains the best but gold and copper also work well. For all three metals excitation with red light yields the maximum enhancement, falling off rapidly as the excitation moves into the blue–green. However, green light (Ar^+, 514.5 nm) excitation is commonly used to obtain SERS from silver, as the optical throughput, and detector sensitivity of most Raman spectrometers is higher in the green than in the red. For Au and Cu electrodes red light (Kr^+, 647.6 nm) excitation is essential as the metals can absorb wavelengths below this. Much weaker SERS have been reported for other metals but with only *c.* 7-fold enhancement.

Despite nearly two decades of research, the enhancement mechanism is still not fully understood and remains a source of controversy. What does appear to be generally agreed is that the roughening of the metal surface is an essential requirement to obtain enhancement and that two major mechanisms may operate.

The relative importance of the two mechanisms – the non-local electromagnetic (EM) theory and the local charge transfer (CT) theory – remains a source of considerable discussion. It is generally considered that large-scale rough surfaces, e.g. gratings, islands, metallic spheres etc., favour the EM theory. In contrast, the CT mechanism requires chemisorption of the adsorbate at special atomic scale (e.g. adatom) sites on the metal surface, resulting in a metal/adsorbate CT complex. In addition, considerably enhanced Raman spectra have been obtained from surfaces prepared in such a way as to deliberately exclude one or the other mechanism.

The requirement for atomic-scale sites in the CT mechanism is thought to be due to the formation of adatom–adsorbate complexes. Such complexes require a coordination site on the adsorbate through which the strong interaction can occur, as was demonstrated by the controlled adsorption of isonicotinic acid and benzoic acid on thin island films of silver (see the work of Chen *et al.*, 1980). This clearly showed that SERS spectra could only be obtained from the molecules when coordinating sites were exposed to the Ag film.

The charge transfer model of the SERS enhancement can be explained in terms of Figure 2.53. The incident laser light, of frequency $\bar{v}_0$, excites an electron from the half-filled sp band of the metal to one of the empty continuum states in the band. Providing this occurs very near the surface of the metal, the adsorbate can capture the electron in a empty electronic state (see Figure 2.53(a)). This results in electron-hole separation, a temporarily negatively charged adsorbate molecule and an occupied state that has both metal and adsorbate character. Three options are then open to the state so produced. In the first instance, the electron and hole recombine in a process analogous to Rayleigh scattering and a photon of frequency $\bar{v}_0$ is re-emitted (Figure 2.53(b)). Otherwise, inelastic processes may occur, in which energy is lost to the metal

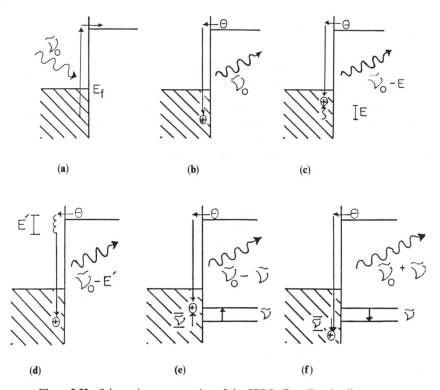

Figure 2.53   Schematic representation of the SERS effect. For details see text.

as phonon excitations (lattice vibrations) via (Figure 2.53(c)) the hole rising towards the surface of the filled band before recombination occurs, with the emission of a photon of lower frequency and the excitation of phonons, and/or (Figure 2.53(d)) the electron cascading down some of the empty continuum states before re-emitting a photon of lower energy. Both of these processes give rise to a continuum background. Finally, if the state lives long enough, electronic reorganisation of the adsorbate molecule occurs, resulting in a large polarisability change during the subsequent electron-hole re-combination (giving a resonance condition). If this adsorbate vibration can couple to phonon excitation, or de-excitation, then the resulting emission is of energy $\bar{v}_0 \pm \bar{v}_i$ (see Figures 2.53(e, f)) where $\bar{v}_i$ is the frequency of the ith vibration. This gives Stokes and anti-Stokes SERS bands. A similar mechanism can be envisaged for an adsorbate to metal electron transfer.

The fact that the CT model is not sufficient to account for all SERS effects is most clearly illustrated by the work of Murray and Allara (1982), who used well-defined polymer films as separators between suitable test molecules and a roughened silver surface.

Figure 2.54 shows a schematic representation of one of the samples used in the experiments. Results from this and similar experiments clearly showed that the SERS enhancement extended as far as 120 Å from the silver film; clear evidence of a long-range, and therefore electromagnetic (EM), mechanism. The enhancement decayed by an order of magnitude for separations of 30 to 50 Å between the active molecule and Ag film. On the basis of this long-range enhancement the authors suggested SERS as a possible means of obtaining vibrational spectra from any interface providing it was within 100 Å of a suitably roughened Ag film.

One source of EM enhancement may be attributed to the excitation of surface plasmons (SP) in the metal. A plasmon is a collective excitation in which all of the conduction electrons in a metal oscillate in phase. In the bulk, there is essentially only *one* allowed fundamental plasmon frequency,

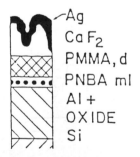

Figure 2.54 The geometry of one of the substrates employed by Murray and Allara in their 'spacer' experiments. PMMA = polymethyl methacrylate spacer of thickness d, PNBA = monolayer of p-nitrobenzoic acid. The mass thicknesses of the other layers are: CaF$_2$ = 800 Å, Al = 2500 Å, Al = oxide c. 30 Å, Ag = 200 Å. From Murray and Allara (1982).

ω_p, whose value is given by:

$$\omega_p = (\rho e_0^2/\varepsilon_0 m^*)^{1/2} \tag{2.65}$$

where ρ is the number of free electrons per unit volume of charge e_0 and effective mass m^*.

The lower symmetry associated with small metal particles gives rise to a range of plasmon frequencies which can couple to incoming electromagnetic radiation. The resultant oscillations in electronic density from one side of the metal particle to the other gives rise to large changes in dipole and this results in the emission of light at the same frequency as the oscillating dipole. The movement of charge across the sphere corresponds to a collective oscillation of the electrons in the metal and is termed a localised dipolar surface plasmon (DSP).

The total enhancement, $\Im$, of the intensity of the Raman emission from the molecule is directly proportional to the total average field enhancement normal to the surface of the sphere and inversely proportional to the distance the molecule is away from the surface of the sphere. For a molecule adsorbed *on* the surface $\Im$ is given by:

$$\Im = (1 + 2g_L)^2(1 + 2g_R)^2/9 \tag{2.66}$$

where g_L relates to enhancement induced by the incident laser beam, and g_R is a function taking into account the enhancement induced by the oscillating dipole on the adsorbate, which is a function of the electric field of the emitted Raman light. The two g functions are given by:

$$g_L = [\hat{\varepsilon}(\omega_L) - \varepsilon_S]/[\hat{\varepsilon}(\omega_L) + 2\varepsilon_S] \tag{2.67}$$

$$g_R = [\hat{\varepsilon}(\omega_R) - \varepsilon_S]/[\hat{\varepsilon}(\omega_R) + 2\varepsilon_S] \tag{2.68}$$

where ε_S is the relative permittivity of water, and $\hat{\varepsilon}(\omega_L)$ and $\hat{\varepsilon}(\omega_R)$ are the (complex) permittivities of the metal at the (radial) frequencies of the irradiating laser light (ω_L) and the emitted Raman light (ω_R), respectively.

As can be seen from equation (2.67), when:

$$\hat{\varepsilon}(\omega_L) = -2\varepsilon_S \tag{2.69}$$

the enhancement becomes infinitely large. In principle, this will arise at a given value of ω_L. Thus the system could be tuned into resonance by changing the excitation wavelength, the resonance condition corresponding to the excitation of the DSP at the plasma frequency ω_p. However, $\hat{\varepsilon}(\omega_L)$ and $\hat{\varepsilon}(\omega_R)$ are complex variables, having the form $\hat{\varepsilon}(\omega) = \varepsilon'(\omega) - i\varepsilon''(\omega)$, while ε_S is real. Thus, in practice, the resonance condition shown in equation (2.69) cannot be met. Instead, the resonance condition becomes:

$$\varepsilon'(\omega_L) = -2\varepsilon_S \tag{2.70}$$

Fulfillment of this condition gives large, but not infinite, enhancement.

For water at optical frequencies, $\varepsilon_S = 1.77$, hence the resonance condition becomes $\varepsilon'(\omega_L) = -3.54$. When this condition is fulfilled the DSP is considered to have been excited by the absorption of the laser radiation. This is analogous to the resonance Raman effect, with the exception that it is the metal substrate that undergoes the resonance absorption resulting ultimately in the enhancement of the Raman spectrum of the adsorbate.

Calculations using the experimentally determined permittivities of silver and gold show, however, that this model is not adequate as it stands: the predicted enhancement effects are too small and are maximal at the wrong wavelength.

This difficulty was addressed successfully by Birke and Lombardi (1988), who showed that if the particles were considered to be elongated not only did the maximum in enhancement shift to higher wavelengths, as observed experimentally, but the extent of the enhancement, particularly near the 'sharp edges', is also much larger. In fact, for a given 'aspect ratio' (i.e. ratio of long to short axis of the particle) resonance takes place at a specific frequency.

In fact, Birke and Lombardi's calculations suggested much *higher* enhancements than those encountered. The enhancements observed in practice would not be expected to be anywhere near as large since:

1. For a given ω_L resonance will only occur at particles having the correct aspect ratio. Molecules located on particles having larger or smaller aspect ratios will not be enhanced.
2. Not all the adsorbed molecules will be located at the tips of the particles and so experience the maximum field enhancement.
3. Maximum enhancement will be obtained by molecules adsorbed in such a way as to maximise vibrational dipole changes perpendicular to the particle surface.
4. The particles will have a distribution of sizes; particles larger than *c.* 10 nm will show a much reduced enhancement.

As a result of the above considerations and the observed experimental findings, the best estimate of the electromagnetic enhancement expected from the DSP-excitation EM model is *c.* 10^2–10^4. Again, this clearly shows the requirement to invoke other enhancement mechanisms.

In addition to its role in enhancing the local electromagnetic field, the surface roughness also plays an important part in helping to couple the incident electromagnetic beam to the surface metal particles. The role of surface roughness in SP excitation has also been elucidated by UV-visible reflectance (electroreflectance, ER) studies. UV-visible reflectance studies can be performed in much the same way that EMIRS spectra are obtained; the potential of a reflective working electrode is modulated and the reflected intensity detected via lock-in amplification as an AC signal. The intensity of the reflected light is a strong function of the optical properties of the electrode surface, and is extremely sensitive to changes in the conduction band electron density, induced by potential modulation. This has been clearly illustrated by studies on silver in which the coupling of a reflected light beam to the surface plasmons on a roughened (111) face has been reported by Kolb (1988). However, if a silver electrode is cycled in Cl^--containing media (and most SERS spectra have been obtained in such media) a strong SERS signal is obtained from an adsorbate yet ER shows that no SP are present. In such media the silver ions do not dissolve and redeposit during potential cycling but precipitate as AgCl on the electrode surface on oxidation. The net result of this is that the Ag surface remains remarkably smooth, in terms of large-scale roughness, though atomic-scale surface features have been shown to be produced and the electromagnetic mechanism plainly fails to account for the SERS effect here.

Turning now to the practical aspects of SERS, the cell is of standard three-electrode design and employs a fused silica or glass window. The fact that SERS is an optical technique using light in the visible region may be expected to remove the necessity for the thin-layer aspect of the design. However, the film of electrolyte between the window and reflective working electrode is usually kept as thin as possible to minimise the absorption of the laser light, particularly if a coloured electrolyte is being employed, and to minimise the bulk solution features in the spectra. For electrodes with strong optical absorption at the wavelength of the exciting light, i.e. carbon coated with roughened silver, a rotating working electrode, or flow cell, must be employed to prevent heating and/or damage.

Two main types of cell configurations have been used for SERS studies: the 90° geometry shown in Figure 2.55(a) and the backscattering geometry shown in Figure 2.55(b). In both configurations great care must be taken to eliminate backscattered laser lines Rayleigh scattered by the sample since these lines are $c.\ 10^3 \times$ more intense than the Stokes and anti-Stokes lines and thus will show up as intense bands in the Raman spectrum. With modern holographic lasers this presents little difficulty.

As well as two common cell configurations there are two common detection arrangements in SERS. The first is the conventional scanning monochromator system as found in a dispersive infrared instrument. As was discussed in the section on infrared spectroscopy, the spectrum is obtained by dispersing the scattered light via a grating. The dispersed light is then scanned across the exit slit of the monochromator linearly with time and the intensity is measured by a photomultiplier tube plotted as a function of wavelength. This can yield good resolution and accurate band positions with a scan rate typically of $\leqslant 45\,cm^{-1}$/min, giving 71 minutes for a single complete mid-IR scan at $2\,cm^{-1}$ resolution; (compared to 500 co-added and averaged FTIR scans in 3 minutes). The second, more modern, method of detection is to use an optical multichannel analyser. The dispersed light from the spectrograph is focused on a multichannel detector plate such that sensors detect the intensity at multiple wavelengths simultaneously. In this multiplex mode a $1000\,cm^{-1}$ spectral range can be recorded in seconds or even down to a few milliseconds.

The oxidation-reduction cycles (ORCs) necessary to roughen the electrode surface are generally performed as potential steps, linear sweeps with a

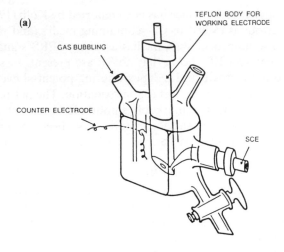

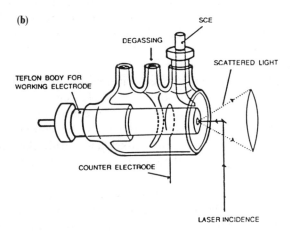

Figure 2.55 Typical electrochemical cells used for SERS measurements. (a) 90° collection geometry, (b) backscattering geometry.

triangular waveform, or a combination of both. A typical cyclic voltammogram resulting from a linear sweep ORC is shown in Figure 2.56 for Ag in 1 M KCl; the SERS scans are collected at potentials > 0.1 V. As with *in situ* infrared spectroscopy, potential modulation is employed in the SERS technique to observe the absorptions of potential-dependent species that would otherwise be masked by the spectrum of the electrolyte; modulation frequencies as high as 20 kHz have been employed. As was discussed in the section on the IR spectroscopies, high modulation frequencies demand fast, fully reversible systems to be studied.

Whilst SERS represents a potentially extremely sensitive technique for the study of adsorbed species in electrochemistry and an enormous number of

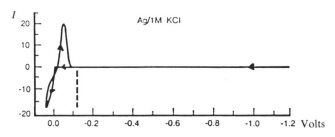

Figure 2.56 Oxidation–reduction cycle for the roughening of an Ag electrode. After R.K. Chang and B.L. Laube in *CRC Critical Reviews in Solid State Materials Science*, Vol. 12, pp. 1–73, CRC Press Inc., Boca Raton, Florida (1984).

papers have appeared in the literature in the years since the discovery of SERS, the technique suffers from the limited number of active substrates available. More importantly, the surface enhancement demands the production of an atypical surface and there is always the awareness that the adsorbates under study are adsorbed at artificially created 'special' sites that may not be representative of the sites found under 'real' conditions. However, the paper by Murray and Allara (1982) represents an extremely important pointer to the full exploitation of SERS in electrochemistry with a wide range of electrodes having untreated ('real') surfaces. Thus, by coating a SERS active substrate with a thin layer of the required electrode material, the SERS effect can still be obtained even from an untreated surface. Seminal work on this approach was performed by Fleischmann and co-workers who realised that the design of SERS active electrodes other than Ag, Cu or Au was of prime importance in the quest to widen the range of possible systems open to study by SERS. More particularly, the ability to obtain Raman spectra from adsorbates on iron would be of great interest with respect to the study of corrosion and corrosion inhibition. To this end, Mengoli *et al.* (1987) reported that enhanced Raman spectra of pyridine adsorbed on Fe could be obtained by electrodepositing thin films of Fe on roughened silver electrodes.

Silver substrates having a very high SERS activity were produced by pulsing the potential to $+0.15\,V$ vs. SCE in a base electrolyte for a time such that the redeposition charge reached $200\,mC\,cm^{-2}$. This was followed by a constant current reduction at $7.5\,mA\,cm^{-2}$ until the potential reached $-0.25\,V$, at which value the potential was then maintained until the iron deposition was performed. The iron was deposited from $0.1\,M\ FeSO_4(NH_4)_2SO_4/0.1\,M$ KCl/HCl at pH 3.5; the potential was held at $+0.2\,V$ for $2\,s$ before cycling the potential to negative values at $50\,mV\,s^{-1}$. On the roughened surface, the charge passed was equivalent to the deposition of 350 monolayers. The same process carried out on a smooth Ag electrode resulted in *c.* 100 monolayers being deposited. The authors interpreted this in terms of the activated Ag surface having a roughness factor (RF) of 3.5, where RF is given by:

$$RF = [\text{observed surface area}]/[\text{geometric area of the electrode}]$$

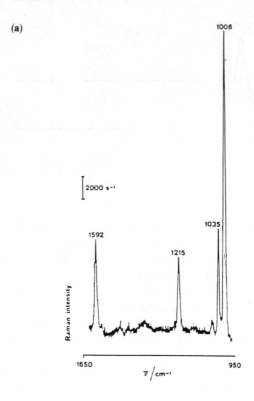

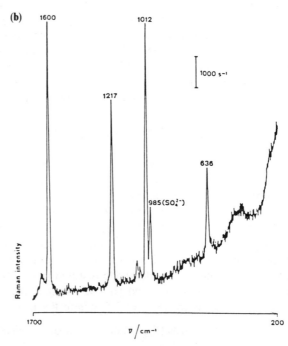

Figure 2.57(a) shows a typical SERS spectrum of pyridine adsorbed at the roughened silver electrode collected at $-0.6\,V$ vs. SCE in pH 4 0.1 M KCl/0.05 M pyridine and Figure 2.57(b) is a typical spectrum of pyridine adsorbed at the Fe surface at $-0.8\,V$ in pH 3.5 0.1 M $FeSO_4(NH_4)_2$/0.1 M KCl/HCl/0.05 M pyridine. From the charge passed 100 monolayers of Fe were deposited on this particular electrode; allowing for the roughness factor, this gives c. 30 monolayers.

The work discussed above is really intended to communicate the fact that SERS spectra can be obtained from Fe. Consequently, the authors were only looking for clearly enhanced spectra, preferably showing differences in peak positions and intensity to differentiate the pyridine adsorbed on iron from that adsorbed on silver. All of these, as can be seen from Figures 2.57(a) and (b), were observed. Raman scattering via electrodeposition on activated silver surfaces has been observed from a wide range of metals clearly indicating that SERS is no longer a restricted technique and can be used as an *in situ* probe for a wide range of electrochemical systems.

2.1.8 Ellipsometry

External reflectance spectroscopy obtains information by investigating the change in intensity of a light ray on reflection from an electrode surface. By considering only the change in *intensity* of reflected light, other information-containing properties of the light ray are being ignored.

Ellipsometry is concerned with the measurement of the changes in polarisation state, as well as light intensity, on reflection since these parameters are highly sensitive probes of the thickness and refractive index, $\hat{n}_f$, of a surface film. A full treatment of the principles involved in ellipsometric measurements can be found in any one of several excellent reviews (see references).

Consider a linearly polarised monochromatic light ray incident on a metal surface. Such a ray can always be resolved into two orthogonal components and if the plane of reflectance at the metal is chosen as the reference, then these components correspond to S- and P-polarised light, as discussed in the previous section.

Figure 2.58 shows the S- and P-components of a ray polarised at 45° to the plane of reflection incident at a metal surface; this will be the polarisation state of the incident light assumed throughout the discussion below. On being reflected both the phase and the intensities of the S- and P-components

---

Figure 2.57 (a) Spectra at $-0.6\,V$ vs. SCE from an Fe electrode covered by Ag growth centres immersed in 0.1 M KCl/0.05 M pyridine at pH 4. (b) Spectrum of pyridine at $-1.0\,V$ vs. SCE adsorbed on a monolayer of Fe deposited on a SERS active Ag substrate from a solution containing 0.1 M $FeSO_4(NH_4)_2SO_4$/0.1 M KCl/0.05 M pyridine at pH 3.5. Reprinted from *Electrochimica Acta*, **32**, G. Mengoli, M. Musiani, M. Fleischmann, B. Mao and Z.Q. Tian, 'Enhanced Raman Scattering From Iron Electrodes', pp. 1239–1245 (1987), with kind permission from Pergamon Press Ltd., Headington Hill Hall, Oxford OX3 0BW, UK.

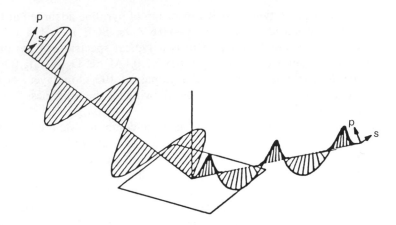

Figure 2.58 Schematic representation of the generation of elliptically polarised light *via* the reflection of plane-polarised light from a reflective surface.

may change. If the light is S-polarised, then the reflection at a metal surface leads to a phase-change, δ_S, of 180°. P-polarised light may suffer a phase change, δ_P, of between 0 and 180°, dependent on the angle of incidence. In both cases, the reflected ray remains linearly polarised.

Complex reflection coefficients can be defined thus:

$$r_S = (|\mathbf{E_r}|/|\mathbf{E_i}|)_S \cdot \exp(i\delta_S) \qquad (2.71)$$

$$r_P = (|\mathbf{E_r}|/|\mathbf{E_i}|)_P \cdot \exp(i\delta_P) \qquad (2.72)$$

where $\mathbf{E_r}$ and $\mathbf{E_i}$ are the electric vectors of the reflected and incident rays, respectively. δ_S and δ_P are the phase changes on reflection: $\delta_S = 180°$, $\delta_P = 0\text{–}180°$. These reflection coefficients define both the change in magnitude of the electric field, $|\mathbf{E}|$, on reflection and the change in phase. The intensity of the light of polarisation ρ at any point I_ρ is simply $|\mathbf{E}_\rho|^2$ and the change in this intensity on reflection is given by the reflectivity, R_ρ:

$$R_\rho = |r_\rho|^2 = (|\mathbf{E_r}|^2/|\mathbf{E_i}|^2)_\rho$$

i.e. $|r_\rho|^2 = r_\rho \times r_\rho^*$, where r_ρ^* is the complex conjugate of r_ρ. R_ρ does not contain any information on the phase change, this can only be obtained directly from the complex reflection coefficients, but the complex variables r_S and r_P are extremely difficult to measure. It is, however, possible to measure the ratio r_P/r_S, where:

$$r_P/r_S = (|r_P|/|r_S|) \cdot \exp(-i[\delta_P - \delta_S]) \qquad (2.73)$$

which we can write as:

$$r_P/r_S = \tan \Psi \cdot \exp(-i\Delta) \qquad (2.74)$$

where $\tan \Psi = (|r_P|/|r_S|)$, and $\Delta = [\delta_P - \delta_S]$.

Ψ, Δ and R_ρ are strong functions of the thickness and optical properties of any film present on the electrode, and the technique of ellipsometry involves the measurement of R_ρ (although some instruments do not have this facility) and the (indirect) measurement of Ψ and Δ, and thence the ratio r_P/r_S. These are then employed to produce values of n_f, k_f and the film thickness via an appropriate model.

The path followed by the tip of the electric vectors of the individual components after reflection can be described by the equations:

$$\mathbf{E_P} = \mathbf{E_P^0} \cdot \cos(\omega t + \delta_P) \tag{2.75}$$

$$\mathbf{E_S} = \mathbf{E_S^0} \cdot \cos(\omega t + \delta_S) \tag{2.76}$$

where $\mathbf{E_P^0}$ and $\mathbf{E_S^0}$ are the maximum values of the electric field, $\omega = 2\pi v$, v is the frequency of the light and t is time; these components must then be convolved to form the reflected wave. When $\Delta = 0$, the ray is linearly polarised and is circularly polarised when $\Delta = \pi/2$. However, the most common state of polarisation has Δ neither 0 nor $\pi/2$; in this case, recombining the S- and P-components after their individual reflection off the surface results in an electric vector whose tip describes a helix in space. The projection of this helix onto a plane normal to the direction of propagation is an ellipse (see Figure 2.59) hence the name of the technique.

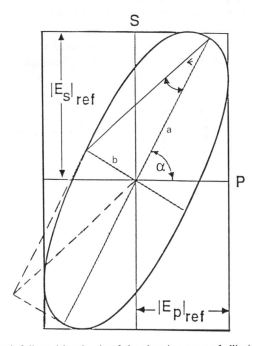

Figure 2.59   The path followed by the tip of the electric vector of elliptically polarised light. $|E_S|_{ref}$ and $|E_P|_{ref}$ show the directions of the planes of the S-polarised and P-polarised light. For details see text.

The ellipse in Figure 2.59 can be defined in terms of two lengths and two angles: the major and minor axis lengths a and b, the angle made by the major axis with respect to the plane of incidence, α, the *azimuthal angle* and the angle ε, defined as $\tan^{-1}(b/a)$. The angles α and ε are the parameters actually measured in an ellipsometry experiment and are related to Δ and Ψ by:

$$\tan 2\alpha = -\tan 2\Psi \cdot \cos \Delta \qquad (2.77)$$

$$\cos 2\Psi = -\cos 2\varepsilon \cdot \cos 2\alpha \qquad (2.78)$$

for light polarised linearly at 45°.

The simplest method of measuring α and ε, and the one that underlines the general principles involved, is schematically represented in Figure 2.60. Monochromatic light polarised at 45° is incident on a reflective electrode immersed in electrolyte solution. An analyser, A, is rotated until a maximum in the intensity of the reflected light is obtained: the angle between this maximum and the plane of incidence is the azimuthal angle. The ratio of the maximum reflectivity, R_a, to the minimum reflectivity, R_b (obtained by further rotation of the analyser by 90°), gives ε as:

$$\varepsilon = \tan^{-1}(\sqrt{R_b}/\sqrt{R_a}) \qquad (2.79)$$

Although such experiments could be carried out, in practice it is usually much more accurate to measure complete extinction rather than ratios of non-zero intensities. This is accomplished by using a compensator (a wave plate) to restore linear polarisation to the reflected ray by the introduction of a known phase difference between the S- and P-components. The restored

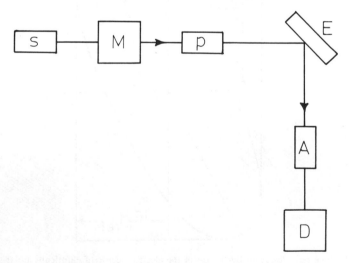

Figure 2.60 Simple schematic representation of an ellipsometer. S = source, M = monochromator, P = polariser, E = reflective electrode, A = rotating analyser, D = detector.

linear polarisation is recognised by the *extinction* with a linear analyser, i.e. the angle at which no light is transmitted is perpendicular to that of the linear polarisation of the ray. Ψ is obtained from this azimuthal angle and Δ from the phase difference imposed by the compensator.

By determining Δ and Ψ, the refractive index $\hat{n}_m = n_m - ik_m$ of an *uncoated* metal electrode can be determined directly from the equations:

$$n_m = n_0[0.5\{[G^2 - E^2 + \sin^2\theta] + ([G^2 - E^2 + \sin^2\theta]^2 + 4G^2E^2)^{0.5}\}]^{0.5}$$

(2.80)

and

$$k_m = n_0[0.5\{-[G^2 - E^2 + \sin^2\theta] + ([G^2 - E^2 + \sin^2\theta]^2 + 4G^2E^2)^{0.5}\}]^{0.5}$$

(2.81)

where n_0 is the refractive of the electrolyte above the metal, θ is the angle of incidence and G and E are given by:

$$G = [\sin\theta \cdot \tan\theta \cdot \cos 2\Psi]/[1 + \sin 2\Psi \cdot \cos\Delta] \qquad (2.82)$$

$$E = [\sin\theta \cdot \tan\theta \cdot \sin 2\Psi \cdot \sin\Delta]/[1 + \sin 2\Psi \cdot \cos\Delta] \qquad (2.83)$$

It is assumed that the electrolyte does not absorb in the wavelength region under discussion (i.e. $k_0 = 0$).

For the case of a thin film on the electrode (see Figure 2.61) the total reflection coefficients of the S- and P-components from the ambient phase/film and film/electrode interfaces, r_P and r_S, are given by:

$$r_P = [r_{1P} + r_{2P}Z]/[1 + r_{1P}r_{2P}Z] \qquad (2.84)$$

$$r_S = [r_{1S} + r_{2S}Z]/[1 + r_{1S}r_{2S}Z] \qquad (2.85)$$

where 1 refers to the ambient/film interface, 2 to the film/electrode interface

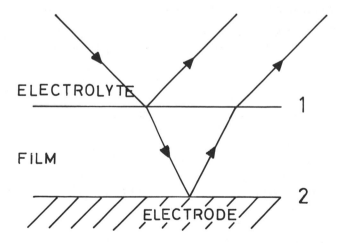

Figure 2.61   Schematic representation of the electrode/film/electrolyte interfaces.

and Z is the delay imposed on the light rays as they pass through the film of refractive index $\hat{n}_f$ given by:

$$Z = \exp([-i \cdot 4\pi L \hat{n}_f \cos \phi_f]/\lambda_0) \tag{2.86}$$

where λ_0 is the vacuum wavelength of the light, L is the thickness of the film and ϕ_f is the complex angle of refraction in the film, which may actually be a complex quantity if the film is absorbing at λ_0.

The individual reflection coefficients can be shown to be given by:

$$r_{1P} = \tan(\theta - \phi_f)/\tan(\theta + \phi_f) \tag{2.87}$$

$$r_{1S} = -[\sin(\theta - \phi_f)]/\sin(\theta + \phi_f) \tag{2.88}$$

$$r_{2P} = \tan(\phi_f - \phi_m)/\tan(\phi_f + \phi_m) \tag{2.89}$$

$$r_{2S} = -[\sin(\phi_f - \phi_m)]/\sin(\phi_f + \phi_m) \tag{2.90}$$

where ϕ_m is the complex angle of refraction into the metal. The various angles can be obtained by applying Snell's law to the two interfaces:

$$\hat{n}_0 \sin \theta = \hat{n}_f \sin \phi_f \tag{2.91}$$

$$\hat{n}_f \sin \phi_f = \hat{n}_m \sin \phi_m. \tag{2.92}$$

The measurement of Ψ and Δ gives r_P/r_S via equation (2.74), for the coated electrode system, with both of these parameters highly sensitive to the properties of the film. However, whereas a knowledge of Δ and Ψ allowed the direct calculations of the optical properties of an uncoated metal via equations (2.80) and (2.81), these equations no longer apply in the case of a coated substrate. At this point, the central problem in ellipsometry is encountered. To characterise the film completely, three parameters are required; n, k and L. However, the above discussion includes only the determination of two parameters, α and ε, giving Ψ and Δ. One of two approaches is commonly employed to solve the equations to give n, k and L.

1. The reflectivity $R = 0.5[|r_S^2| + |r_P^2|]$, can be measured. R is independent of both Δ and Ψ and thus provides a third variable. In order to obtain n_f, k_f and L, values of these parameters are estimated. R, Δ and Ψ are then calculated from equations (2.84) to (2.92) and compared to the experimentally observed values. n_f, k_f and L are altered and the calculations repeated. Regression analysis eventually yields values of the thickness and refractive index of the film that would give rise to the observed R, Ψ and Δ.

2. If the measurement of the intensity of the reflected light is not available, then a second approach is employed. Thus, Δ and Ψ are measured at a large number of wavelengths. The thickness L is estimated, perhaps from electrochemical data, and values of n_f and k_f are then calculated at each of the wavelengths using the measured Δ and Ψ values. The resultant values of the optical parameters are then scrutinised and retained or rejected on the basis of judgements on whether or not they fall within

acceptable ranges. L itself can be altered and the refractive indices again evaluated and there will be a narrow range of thickness over which reasonable values of the optical constants are obtained for all wavelengths. Thus, a solution is found for $2N$ data points, where N is the number of wavelengths investigated, each one yielding values of Δ and Ψ. While to mathematically solve the problem for n_m, k_m and L, $2N + 1$ parameters are needed, in practice, such a solution that fits all $2N$ data points and is also under the constraint that the thickness is *not* a function of the wavelength, is almost unique and highly convincing that the resultant model is correct.

It may be thought that since ellipsometry is most commonly employed as an optical technique its resolution will be limited to the same order of magnitude as the wavelength of the probing light, i.e. several thousand angstroms. However, ellipsometry measures *changes* in phase (Δ) and *changes* in amplitude (Ψ). These changes are measured mechanically in terms of changes in the azimuthal angle α. The resolution on the determination of changes in α can be as high as $0.01°$, i.e. 3×10^{-5} revolutions. In many cases an increase in film thickness equivalent to λ results in a change in α of $360°$. Thus changes in film thickness of down to $3 \times 10^{-5}\lambda$, $c.$ 0.2 Å, can be detected.

As can be seen, while the practical aspects of ellipsometry are (relatively) straightforward, the calculations are prodigious. Hence, it is not surprising that renewed interest in ellipsometry (the technique having been in existence at least since 1888) coincided with the availability of high-speed computers and comprehensive programs in the late sixties. Additional factors were the advent of fully-automated instruments, manual alignment and operation being a potentially tedious and time-consuming business, and the recognition of its potential value in electrochemistry.

The importance of ellipsometry in electrochemistry lies in the following attributes:

1. It is a visible technique and hence there are no problems arising from light absorption by the supernatant electrolyte.
2. The high precision in the determination of film thickness.
3. It is a spectroscopic technique, hence the optical properties of the film can be probed over the entire spectral range of the instrument, typically UV to near-IR.
4. The changes in α and ε can be determined very rapidly, allowing changes in film thickness and optical properties to be determined in real time, hence the technique can be used to gain kinetic information. The speed of data collection is limited only by the response of the instrument and this may be as low as a few tens of microseconds.

A paper that demonstrates the power of the ellipsometric technique is that of Hamnett and Hillman (1988) which investigated the nucleation and growth of polythiophene on an electrode *in situ*. The ellipsometer employed by the

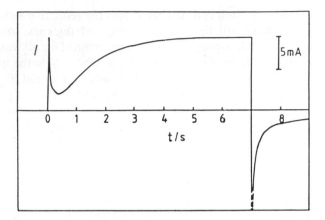

Figure 2.62   The I/t transient obtained during the growth of a polythiophene film on a $6.30\,cm^2$ Pt electrode at 1.80 V vs. SCE in acetonitrile/0.1 M tetraethylammonium tetrafluoroborate electrolyte. At $t = 7\,s$, the potential was switched to 0 V to terminate the growth process. After Hamnett and Hillman (1988).

authors included the measurement of reflected light intensity, hence three-parameter fits of n_f, k_f and L were achieved without recourse to the assumption of a fixed value of one of the parameters.

Figure 2.62 shows a current–time response collected during the electro-polymerisation of thiophene, C_4H_4S, at a Pt electrode. The potential was stepped from 0 V vs. SCE to 1.8 V for 7 s, before stepping the potential back to 0 V. The area under the I/t curve bounded by the minimum, i.e. from $t = 0$ to 0.4 s, apparently corresponded (in terms of the charge passed) to the formation of a monolayer of polythiophene. However, this hypothesis was not in accord with subsequent results obtained with the ellipsometer. Figures 2.63(a) to (c) show the values of Ψ, Δ and intensity obtained during the growth of the film.

As can be seen from Figures 2.63(a) to (c), Δ, Ψ and intensity all lag behind the current response. More particularly, Δ and Ψ lag more than the intensity and this has important implications for the growth mechanism since Δ and Ψ are only effected by the film on the electrode surface, whereas the intensity responds to species both in solution and at the surface. In addition, when the growth was carried out for only 0.4 s, the responses of Δ, Ψ and the intensity during the growth were identical to those shown in Figures 2.63(a) to (c). However, on stepping the potential down to 0 V, where consumption of the monomer ceases, no evidence for a polymer film is observed.

The authors suggested that the charge passed in the first 0.4 s was not due to the formation of a monolayer of adsorbed polythiophene, but the generation of oligomers in solution. These oligomers attain a certain length before solvation can no longer sustain their presence in solution and they deposit on the surface of the electrode. This hypothesis was strongly supported by values of n_f, k_f and L calculated from the data, under the assumption that

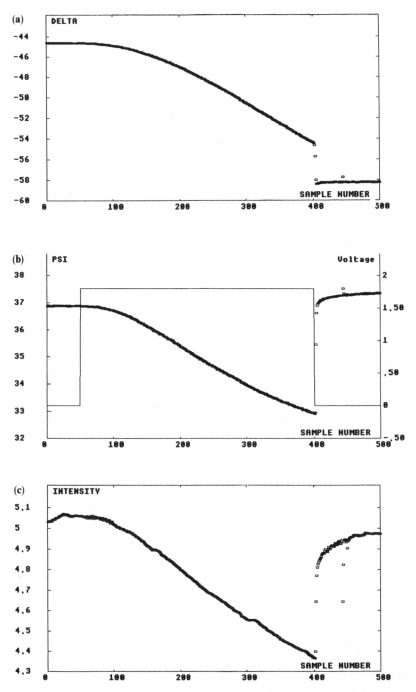

Figure 2.63 Experimental data, Δ, Ψ and intensity, as obtained as a function of time during the growth depicted in Figure 2.62. The data were acquired at 633 nm at 20 ms intervals. After Hamnett and Hillman (1988).

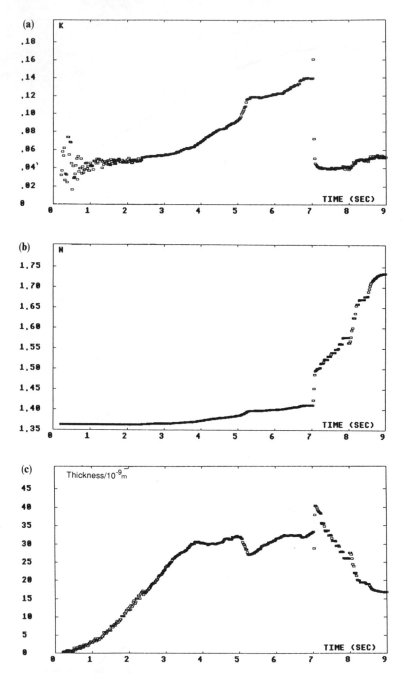

Figure 2.64 The processed data from Figure 2.63. The values of n, k and the thickness L were obtained in a three-parameter fit to Δ, Ψ and the intensity. The abrupt changes after 7 s are due to the potential being stepped back to 0 V. No points are shown for the first 200 ms because the authors were unable to obtain fits to the experimental data due to the fact that no film is present until $t > 200$ ms. After Hamnett and Hillman (1988).

the film grows as a uniform 'slab' of material. Thus Figures 2.64(a) to (c) show the variations in the thickness of the film and the optical constants during the growth shown in Figures 2.62 and 2.63. Clearly, L does not increase until after c. 0.4 s and k_f is undefined until above 1 s after the potential is stepped to 1.8 V as it is calculated from a model which assumes film growth from $t = 0$ s.

Thus, ellipsometry gives direct evidence for a model of the initial stages of polythiophene growth, disproving the conclusions based purely on coulometry. In the same paper, Hamnett and Hillman were able to obtain valuable and complementary information not just on the initial stages of the polymerisation but also on the mechanism of the subsequent nucleation and growth. The unique piece of information that the ellipsometer was able to extract, the changes in film thickness (*in real time*), when combined with coulometric data allowed a wealth of information to be deduced, e.g. with respect to the film composition, and ably showed the power of the technique.

2.1.9 X-ray spectroscopies

No investigation of a solid, such as the electrode in its interface with the electrolyte, can be considered complete without information on the physical *structure* of that solid, i.e. the arrangement of the atoms in the material with respect to each other. STM provides some information of this kind, with respect to the 2-dimensional array of the surface atoms, but what of the 3-dimensional structure of the electrode surface or the structure of a 'thick' layer on an electrode, such as an under-potential deposited (upd) metal? At the beginning of this chapter, electrocapillarity was employed to test and prove the theories of the double layer, a role it fulfilled admirably within its limitations as a somewhat indirect probe. The question arises, is it possible to 'see' the double layer, to determine the location of the ions in solution with respect to the electrode, and to probe the double layer as the techniques above have probed adsorption? Can the crystal structure of a upd metal layer be determined? In essence, a technique is required that is able to investigate long- and short-range order in matter.

The early investigations into the structure of matter at the micron level were carried out using optical spectroscopy. However, in the 19th century, Abbé determined that this approach was essentially wavelength limited; to 'see' matter at the atomic level would require light of wavelength comparable to atomic dimensions. The discovery of X-rays in 1895 thus appeared to be a godsend and the X-ray region of the electromagnetic spectrum is the most generally useful for structural studies, even though the X-ray scattering cross-sections commonly encountered are rather small, giving poor surface sensitivity.

The problem of surface sensitivity could be overcome to a certain extent by using grazing angle incidence, however, the refractive indices of solids in

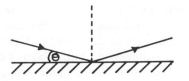

Figure 2.65 Schematic representation of a beam of X-rays incident at a reflecting surface, showing the angle of incidence θ. The convention is different to that for the reflection of UV/visible or IR radiation, where the angle of incidence is that between the incident ray and the surface normal.

the X-ray region are < 1. As a consequence very low angles of incidence, θ, must be employed (see Figure 2.65). (By convention in X-ray studies the angle of incidence is measured between the reflecting surface and the incident ray, rather than between the ray and the surface normal, as is the case in optical and IR techniques.) These low angles of incidence require long solution pathlengths which can result in considerable attenuation of the beam due to scattering. In addition, X-ray absorption methods require a 'white' source of emission, a difficult requirement to fulfil until the recent availability of synchrotron radiation sources. The increasing interest in the possible use of X-ray spectroscopy as a general surface probe, i.e. not just *in situ* in an electrochemical cell, is a direct consequence of the construction around the world of electron storage rings, synchrotrons, dedicated to the generation of high intensity radiation tunable over a wide range of frequencies, from hard (high energy) X-rays to the infrared.

A full description of the technology and principles of X-ray spectroscopy is outside the scope of this book, such a treatment may be found in any one of a number of excellent texts, some examples of which are referred to at the end of this chapter. Briefly, when an element is bombarded with high energy electrons an electron is ejected from one of the core levels. An electron from a higher energy orbital then drops into the vacancy so formed and the excess energy is emitted as a photon. In general, the separation of the inner energy levels is of sufficient energy that the emitted photon has a frequency in the X-ray region. Furthermore, the core orbitals of an atom are highly contracted and do not overlap significantly with neighbouring atoms in the solid. As a result, the corresponding X-ray emission lines are very narrow and are only slightly affected by the formation of chemical bonds. In fact, their energy is characteristic of the element and X-ray analysis is often used to establish elemental composition.

The emission from copper is shown in Figure 2.66. The two prominent lines in the copper emission are termed the copper K_α and K_β lines. The transitions responsible are to the *K shell*, that with principal quantum number one, the 2s and 2p levels are referred to as the L shell, etc. The Greek letters indicate from where the transition originates: the 2p → 1s transition gives the K_α line and the 3p → 1s the K_β line. Sometimes the K_α line is split into a doublet as a result of exchange terms.

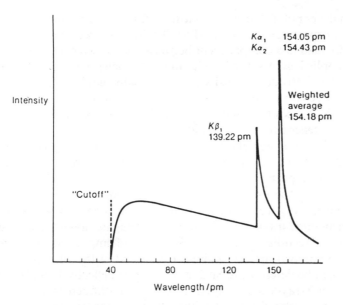

Figure 2.66 The X-ray emission spectrum of copper. From A.R. West, *Solid State Chemistry and its Applications*, John Wiley and Sons, Chichester (1984). Reprinted by permission of John Wiley and Sons, Ltd.

As a result of the atomic nature of the core orbitals, the structure and width of the features in an X-ray emission spectrum reflect the density of states in the valence band from which the transition originates. Also as a result of the atomic nature of the core orbitals, the selection rules governing the X-ray emission are those appropriate to atomic spectroscopy, more especially the orbital angular momentum selection rule $\Delta l = \pm 1$. Thus, transitions to the 1s band are only allowed from bands corresponding to the p orbitals.

The most common way of generating X-rays in the laboratory is via an X-ray tube. A hot tungsten cathode generates electrons *in vacuo* which are accelerated towards the target anode (e.g. copper) in the presence of a large potential difference between the anode and cathode. The X-rays emitted from the anode pass out of the cell through beryllium windows. X-rays from such a source will not only contain several intense lines but will also have a substantial background arising from the electron deceleration process during classical inelastic scattering and must be filtered or monochromated if a single wavelength is required. This substantially reduces the intensity of the X-ray beam and modern synchrotron technology undoubtedly offers a superior, if less convenient, X-ray source of high brilliance, whose intensity is high even after monochromatisation.

Several means are employed to detect X-rays with ion chambers and scintillation detectors sufficient for use in the adsorption and reflection methods where only the intensity of the X-ray beam is of interest. However, increasingly popular detectors used in diffraction experiments are position sensitive proportional detectors (PSPDs). Such devices consist of a central anode wire held in a compartment between one or two cathodes with a large voltage applied across the anode and cathode(s), e.g. 1500 V. The space between the electrodes is filled with an ionisable gas, Ar or Xe, mixed with a quencher (CH_4) in the ratio c. 9:1, at a pressure of a few atmospheres. When an X-ray photon enters the chamber, it ionises a gas molecule, the electron so generated moves rapidly and directly to the anode under the influence of the applied field. This electron can collide with any gas molecules in its path and so cause an avalanche of electrons to reach the anode wire, giving a charge pulse in the wire. The magnitude of the charge is directly proportional to the energy of the incident photon. The resistance of the anode, and the electronics of the detector, are such that when the charge pulse is generated it takes a finite time to propagate down the wire and does so in both directions away from the point of impact. By measuring the difference in arrival times of pulses at each end of the wire, the location of the event can be determined. Such a detector can effectively replace the strip of film in an X-ray 'camera', used for obtaining diffractograms.

X-rays are scattered by electrolyte solutions; one possible way to reduce this problem would be to use high energy, 'hard', X-rays which are scattered much less. However, such X-rays can cause electrode damage and are difficult to detect since the detectors require high pressures of expensive Xe gas, and soon become contaminated. For this reason, soft X-rays are prevalent even though they have pathlengths of only a few tenths of a millimetre in aqueous solution, owing both to scattering and absorption by water. These problems, coupled with the demand for grazing angle incidence if surface sensitivity is required, mean that the design of the cell in electrochemical X-ray experiments is critical.

The cell designs most commonly employed are based around two configurations depending on the systems studied (see Figures 2.67(a) and (b)). In diffraction studies the transmission, or Laue, configuration is used for studying adsorbates (Figure 2.67(a)) while the reflection, or Bragg, configuration is more suited to the study of thin surface films, and is shown in Figure 2.67(b). In absorption or reflection experiments only the Bragg mode is employed. The cell windows are usually Mylar or Melinex plastic, 5–25 μm thick. These materials are strong, stable in common electrolytes and have high transmission coefficients for X-rays.

The three principal X-ray techniques that have been applied to the study of structure at the electrode/electrolyte interface are diffraction, absorption and specular reflection.

2.1.9.1   X-ray diffraction.   The mechanism by which atoms diffract or scatter electromagnetic radiation via the coupling of the electron cloud of the atom to the incident oscillating electric field was discussed in the section on SERS. The X-rays scattered by an atom are the resultant of the waves

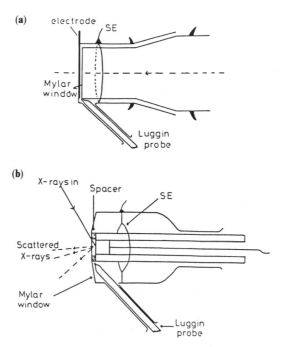

Figure 2.67 Typical cell configurations employed in obtaining *in situ* electrochemical X-ray diffraction measurements: (a) transmission (Laue configuration), (b) reflection (Bragg configuration).

scattered by each electron, hence the 'scattering power', or scattering factor f, of an atom is proportional to its atomic number Z. The intensity of a particular diffracted line is then proportional to f^2.

Thus, a beam of X-rays incident on a solid may be scattered from any of the crystal planes that make up its structure. If θ_i is the angle of incidence of an X-ray beam on a particular set of planes i of separation d_i, then constructive interference will be observed providing the scattering obeys the Bragg equation, $\lambda = 2d_i \sin \theta_i$. Each of the different sets of lattice planes in the solid will give rise to a diffraction maximum at a particular angle θ_i. By tradition, a convenient reference point from which to measure θ is the path of the undefracted beam, as shown in Figure 2.68. From the figure, it can be seen that the diffracted ray makes an angle of 2θ with this beam. If the distances A and B can be measured, then the angle 2θ can be calculated. As a result, X-ray diffraction patterns are represented as plots of intensity vs. 2θ, with peaks at the angles θ_i where constructive interference occurs. Detection could be via an ion chamber or scintillation detector moved stepwise through an arc encompassing the diffracted rays. However, this is extremely time-consuming and prone to the instrumental drift and instability of the detector. Instead, Fleischmann *et al.* employed a PSPD to obtain the diffractogram. This is much less time-consuming (although the experiment still takes several hours) as at any instant information on the whole

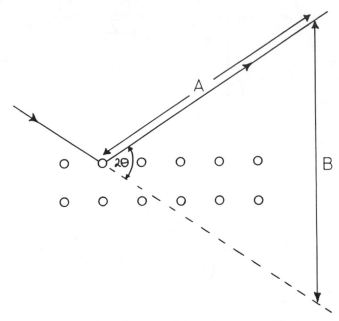

Figure 2.68   The diffraction of X-rays from a crystal lattice.

diffractogram is collected, the time over which the data is taken depending only on the signal to noise ratio required.

As with *in situ* infrared (see section 2.1.6) there is a problem with sensitivity. The X-rays will be scattered in a reflection or transmission diffraction experiment not only by the sample but by the windows and electrolyte. Thus it would be impossible, for example, to identify directly any scattering from a film on an electrode by an examination of the diffractogram obtained at a single potential. The solution is the same as that employed in *in situ* infrared: to use potential modulation. Two approaches can then be envisaged. First, the step-potential method in which an initial diffraction pattern is acquired and a second diffractogram is then collected after film formation has taken place. The first diffraction pattern is then subtracted from the second. However, X-ray emission sources are unstable over such long periods of time. Consequently, a more effective approach is that employed by Fleischmann and colleagues, analogous to the PDIR technique. The potential is modulated between two limits, one before the change or process of interest and the other at which the reaction occurs. The modulation frequency is sufficiently high to avoid experimental drift but sufficiently low to allow the change under investigation, and its reversal, to go to completion. Typical frequencies are 10^{-1} to 10^{-3} Hz. Diffractograms are collected at each potential flip and stored separately. At the end of the experiment the diffractograms at each potential are co-added. Subtraction of the total diffractograms for each potential then give the final difference diffraction pattern. The time for a complete experiment, up to 100 hours, is such that a means of monitoring the stability of the electrochemical system must be incorporated into the experiment. This was achieved by measuring the current at a fixed time after each pulse.

The electrochemical cell used by Fleischmann and co-workers (1986) employing the Bragg configuration is shown in Figure 2.67(b). The source is a copper anode X-ray tube employing a Ni filter to select out the Cu K_α line; the detector is a PSPD.

One of the systems investigated by the authors using the Bragg configuration was the structural changes taking place during the potential cycling, and ageing, of electrodeposited nickel hydroxide on nickel. The *in situ* study of the structural changes taking place during the charge–discharge cycles of battery systems, as well as the effects of ageing on storage capacity, are of great interest, particularly with respect to the importance of the nickel/cadmium battery.

It is known that α-Ni(OH)$_2$ films electrodeposited on Ni become more crystalline and dehydrate slowly to form β-Ni(OH)$_2$ when immersed in 8 M KOH. The study of this process would thus be a reasonable test of the technique. This ageing process can be seen in Figure 2.69 which shows cyclic voltammograms of a 7500 Å thick film of nickel hydroxide on a nickel electrode immersed in 8 M KOH. In order to slow the ageing process down, and so reduce the possibility of structural changes taking place during the collection of the diffractograms, the X-ray experiments were performed in 1 M KOH solution after the film had been aged for various times in the 8 M

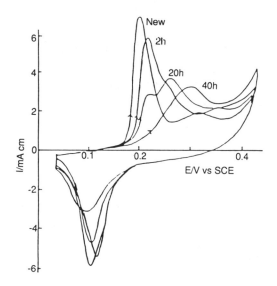

Figure 2.69 Cyclic voltammograms recorded for a 7500 Å-thick Ni(OH)$_2$ film on a nickel electrode in 8 M KOH. Scan rate = 1 mV s^{-1}. The periods of ageing are shown in hours. Reprinted from *Electrochimica Acta*, **31**, M. Fleischmann, A. Oliver and J. Robinson, *In situ* X-ray diffraction studies of the electrode solution interface, pp. 899–906 (1986), with kind permission from Pergamon Press Ltd, Headington Hill Hall, Oxford OX3 OBW, UK.

electrolyte. The α-Ni(OH)$_2$ films, c. 1500 Å thick, were cathodically deposited from Ni(NO$_3$)$_2$ solution onto smooth nickel substrates. The diffractograms were collected at $+100$ mV and $+400$ mV at 5×10^{-3} Hz for 6 hours. As was discussed above, the diffractograms at each potential were co-added separately and the sum diffractogram at 400 mV subtracted from that at 100 mV.

Figure 2.70(a) shows the sum diffractogram of the fresh film collected at 100 mV, before the subtraction. The broad peak in the middle of the figure is due to scattering from the water while the narrow peaks can be attributed

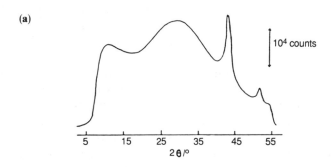

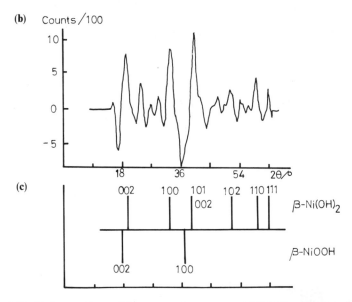

Figure 2.70. (a) *In situ* X-ray diffractogram for an unaged Ni(OH)$_2$ coated-electrode held at $+0.1$ V vs. SCE in 1 M KOH. The data required 3 hours' collection time using Cu K_α radiation. (b) Difference diffractogram for the Ni(OH)$_2$/NiOOH system in 1 M KOH for an unaged electrode. The total data acquisition time was 6 hours at a modulation frequency of 5×10^{-3} Hz. The data are displayed as the diffractogram at $+100$ mV minus that at $+400$ mV. (c) The assignment to reflections of α-Ni(OH)$_2$ and γ-NiOOH. Reprinted from *Electrochimica Acta*, **31**, M. Fleischmann, A. Oliver and J. Robinson, *In situ* X-ray diffraction studies of the electrode solution interface, pp. 899–906 (1986), with kind permission from Pergamon Press Ltd, Headington Hill Hall, Oxford OX3 OBW, UK.

to the [111] and [200] diffractions of the bulk nickel substrate. This clearly shows that features due to the coating on the electrode cannot be discerned from the unsubtracted sum diffractogram, as was stated above. Figure 2.70(b) shows the difference diffractogram and Figure 2.70(c) the corresponding assignments: bands pointing up, to +ve intensity in Figure 2.70(b), can be attributed mainly to the β-Ni(OH)$_2$ present at 400 mV; the peaks pointing down, to −ve intensity, can be assigned to β-NiOOH. Figure 2.71(a) and (b) show the corresponding diffractogram and assignments for the unaged film; the potential step induces the change from α-Ni(OH)$_2$ at 400 mV to γ-NiOOH at 100 mV. The absence of any data for the underlying Ni substrate in Figures 2.70(b) and 2.71(a) shows that the bulk metal suffers no potential-induced structural changes. Data from the incompletely aged films was reported to show that a mixture of phases was present.

Thus this elegant experiment showed that X-rays could be used to probe potential- and time-induced structural changes in a film having reasonably good long-range order. For more crystalline samples the authors did not need to employ potential modulation although they did state the need for reducing the noise on the diffractograms.

On considering more amorphous systems, having only short-range order, X-ray diffraction is not a suitable probe. Instead, a form of X-ray absorption can provide invaluable information on the local structure around the atoms in a surface layer.

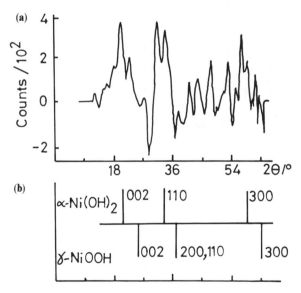

Figure 2.71 As for Figure 2.70(b) and (c) except for the aged electrode. (a) Experimental data, (b) assignment to reflections of β-Ni(OH)$_2$ and β-NiOOH. Reprinted from *Electrochimica Acta*, **31**, M. Fleischmann, A. Oliver and J. Robinson, 'In-situ X-ray diffraction studies of the electrode solution interface', pp. 899–906 (1986), with kind permission from Pergamon Press Ltd, Headington Hill Hall, Oxford OX3 OBW, UK.

2.1.9.2 Surface extended X-ray absorption fine structure (SEXAFS). In order to understand the SEXAFS approach it may first be useful to give a brief recap of X-ray absorption. In addition to the scattering process described above, an X-ray photon may instead be absorbed. X-ray absorption results in the excitation of a core electron into one of the continuum of empty levels in the lowest unoccupied band of the sample; this is shown for copper in Figure 2.72. To a first approximation, the attenuation of the beam is given by a form of the Beer–Lambert expression:

$$I = I_0 e^{-\mu\tau} \tag{2.93}$$

where I_0 is the incident intensity, I is the reflected or transmitted intensity, μ is the linear absorption coefficient and τ is the pathlength through the absorbing material. The variation of μ with λ is approximately:

$$\mu \approx k\lambda^3 = kc^3/\bar{v}^3 \tag{2.94}$$

where k is a constant: A plot of μ against the energy of the incident light should show μ decreasing in a smooth curve with increasing energy. However, there are discontinuities in μ at energies just sufficient to knock an electron out of a core orbital. The absorption coefficient undergoes an abrupt increase as the energy of the light is increased to this (threshold) value. Above this

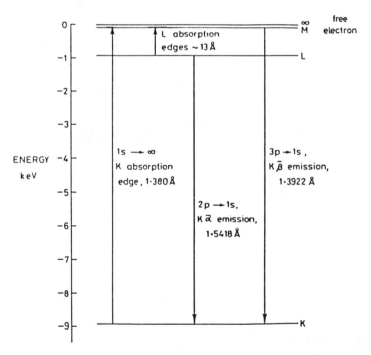

Figure 2.72 Schematic representation of the transitions involved in the X-ray absorption and emission spectra of copper.

value the cross-section again decreases rapidly until the next threshold is encountered.

Figure 2.73(a) shows the X-ray absorption spectrum of copper. The K edge is the minimum energy required to ionise an electron from the 1s orbital;

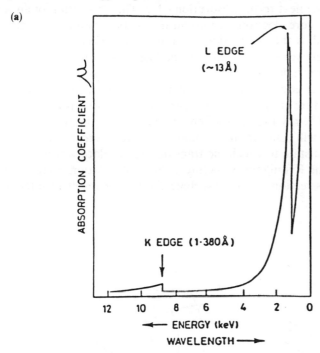

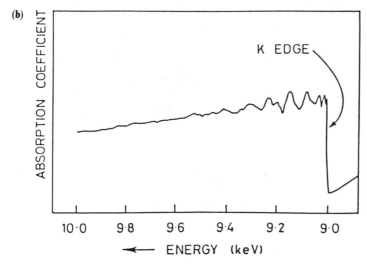

Figure 2.73   (a) The X-ray absorption spectrum of copper showing the K- and L-absorption edges. (b) The K-edge in more detail. From A.R. West, *Solid State Chemistry and its Applications*, John Wiley and Sons, Chichester (1984). Reprinted by permission of John Wiley and Sons, Ltd.

the L edge is actually three closely-spaced edges due to ionisation from 2s and 2p. Again, due to the atomic nature of the core levels, the wavelength at which an absorption edge occurs is characteristic of each element.

Quantitative data on local structure can be obtained via an analysis of the decaying slope next to the absorption edge. The absorption of an X-ray photon boosts a core electron up into an unoccupied band of the material which, in a metal, is the conduction band above the Fermi level. Electrons in such a band behave as if nearly free and no fine structure would be expected on the absorption 'tail'. However, fine structure is observed up to 500 to 1000 eV above the edge (see Figure 2.73(b)). The ripples are known as the Kronig fine structure or extended X-ray absorption fine structure (EXAFS).

Absorption of X-ray radiation of energy well above the threshold for an X-ray transition will result in the ejection of a photoelectron since the *initial* unoccupied band state to which the transition takes place will be above the vacuum level. The Kronig fine structure is due to oscillations induced in the absorption cross-section of the absorbing atom as a result of interference

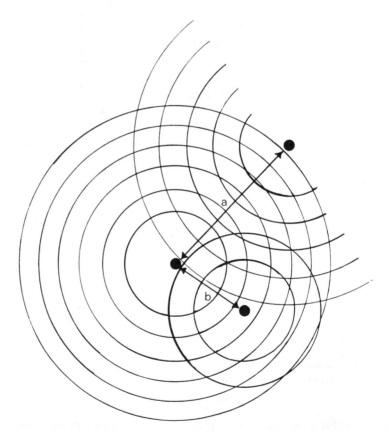

Figure 2.74 Schematic representation of the electron wave interference effects giving rise to the Kronig fine structure on X-ray absorption edges (see text).

effects caused by this photoelectron wave as shown in Figure 2.74. Thus the emitted electron wave is reflected from the nearest and next nearest neighbouring atoms back towards the absorbing atom. These backscattered waves from atoms of differing distances (a, b, etc., see Figure 2.74) can interact with the outgoing wave to give constructive and destructive interference. Constructive interference increases the absorption cross-section of the central atom and hence the probability of absorption of incoming X-ray photons. Destructive interference reduces the absorption cross-section. The degree of interference depends on the wavelength of the electron wave, and thus the wavelength of the incident photon, and on the local structure around the absorbing atom, including interatomic distances.

Thus the Kronig fine structure is effectively a *convolution* of all of the interference produced as a result of backscattering from each of the neighbours around the target atom. The application of Fourier transform techniques then allows the deconvolution of these components to produce a plot resembling a radial distribution function (RDF), i.e. a graph showing the probability of finding an atom as a function of the distance from the absorbing atom. Unfortunately, as a result of the fact that the backscattering produces an unknown phase shift in the returning electron wave, the peaks in the transform (RDF) plot are not exactly at positions that reflect the local structure. The phase shift is generally determined by using EXAFS data from model compounds containing the same atoms as the sample, for which structural data are sought. This approach has two important advantages: first, by selecting the absorption edge of each element present in the material of interest, the partial RDFs so obtained can be used to construct a total structural description of the sample; second, the technique does not require the presence of long-range order but finds its major application in the study of amorphous systems having only short-range order. Since electrochemical systems are frequently of this type, it was recognised that the technique would be an ideal structural tool if it could be applied *in situ*.

The problems faced by electrochemists wishing to use EXAFS are exactly the same as those discussed above: the scattering or absorption of X-rays by the solvent and the requirement for surface sensitivity. *In situ* surface extended edge X-ray fine structure (SEXAFS) seeks to overcome these two problems by employing grazing-angle incidence (giving total external reflectance from the underlying metal substrate) but minimising the pathlength through the electrolyte as much as possible.

The ideal experimental arrangement to run a SEXAFS experiment is shown in Figure 2.75. Synchrotron radiation is rendered monochromatic, and the wavelength selected, via a double crystal monochromator. Two ionisation chambers, one before the sample and one after, provide the measurement of I and I_0, and thus the absorption as a function of the incident X-ray wavelength. The grazing-angle Bragg configuration inevitably results in relatively long pathlength cells and thus requires X-rays of high energy. This, in turn, limits the technique to probing higher-energy absorption edges, i.e.

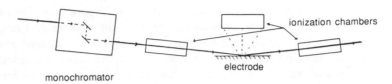

Figure 2.75   A schematic representation of the detection system for obtaining surface EXAFS
data.

those of heavier elements. In addition, grazing incidence requires reflection
but the technique measures absorption, thus it finds application in the
investigation of layers or coatings on electrodes.

The use of a central facility, such as a synchrotron, means that the user
has very little control over the equipment. Thus it may not be possible to
obtain grazing-angle incidence; the angles of incidence may be restricted to
a value greater than the critical angle θ_C. At angles appreciably greater than
θ_C, no reflected beam may be detected. In this case, advantage may be taken
of the fact that the absorption of X-rays is usually accompanied by the re-
emission of X-ray photons as fluorescence. The intensity of this fluorescence
is a direct measure of the X-ray absorption cross-section of the target atom,
and a plot of fluorescence intensity vs. the wavelength of the incident light
can be used to obtain the absorption spectrum. The fluorescence is detected
via a series of detectors mounted around the sample. This configuration is
generally less desirable than the grazing incidence arrangement, as the
fluorescence generally contains information on the underlying substrate. This
can be turned to some advantage, however, if the angle of incidence can be
varied; obtaining SEXAFS spectra as a function of θ for thick layers is a
means of depth profiling.

An elegant example of the early application of SEXAFS *in situ* in electrochemistry is provided by the work of Blum *et al.* (1986) who studied the upd
of copper on gold in $1 M\ H_2SO_4/2 \times 10^{-5}\ M\ CuSO_4$. The electrode was the
gold [111] surface held at 0.13 V vs. an Ag/AgCl reference electrode. A
monolayer of copper was allowed to deposit, as determined from stripping
voltammetry, prior to the acquisition of the SEXAFS spectra. The absorption
spectra were obtained at grazing angle incidence via the detection of the Cu
K_α fluorescence.

Figure 2.76(a) shows the raw data from the upd copper monolayer and,
for comparison, Figures 2.76(b) and (c) show the corresponding spectra from
aqueous $CuSO_4$, and a 2% Cu–Au alloy. As was stated above, the position
of the edge provides a clear indication of the oxidation state of the target
atom; the K edge of Cu^{2+} is shifted by 4 eV relative to that of copper metal
(see Figures 2.76(b) and (c)), a phenomenon employed by the authors to
check the potential control of the working electrode.

The spectrum of the upd monolayer (see Figure 2.76(a)) showed the near-
edge peak to be somewhat different from the doublet observed for the Cu–Au

alloy. In contrast, the absorption attributable to the copper ions in solution is as expected (see Figure 2.7.6(b)). The Kronig fine structure is strongly damped, indicating a loosely-bound solvation shell.

Fourier analysis of the raw data in Figure 2.76(a) yielded reasonable results only if it was assumed that the upd copper was bound to the gold substrate atoms *and to an oxygen atom*, with a Cu–O bond length of *c.* 2.1 Å, i.e. the copper is coordinated to water or SO_4^{2-} via their oxygen atoms. However, this does not agree with the near-edge structure which corresponds more closely to Cu^0. The authors resolved this problem by postulating that the copper was in the reduced state on the gold substrate, with strongly bound oxygen atoms above. Two models fitted the data, and these are depicted in Figures 2.77(a) and (b). In both cases, the best fit was obtained with a Cu–Au bond length of 2.08 Å, the 'best' Cu–O bond lengths being 2.05 Å and 2.08 Å, as shown.

Interestingly, the work of Blum *et al.* (1986) showed that a surface selection rule operates in X-ray reflection absorption. Thus the synchrotron radiation employed in their experiments was polarised in the plane of reflection and the authors noted that bonds perpendicular to the plane of reflection do not contribute to the SEXAFS (cf. the infrared SSR, discussed above).

2.1.9.3 Long period X-ray standing waves (XSW). Grazing-angle X-ray diffraction relies on scattering from surface atoms and SEXAFS is concerned with X-ray reflection absorption from atoms on the reflective surface. In contrast, there are two important X-ray reflection techniques that do not rely either on scattering or on absorption of the X-rays to obtain information on the structure of species on or near the electrode surface. The first approach is that of long-period X-ray standing waves (XSW) as reported by Bedzyk *et al.* (1990). This technique uses the standing wave that is set up on reflection of X-rays at a suitable mirror surface to modify the X-ray fluorescence of target species in the diffuse double layer. Measurement of the fluorescence can then yield information on the structure of the double layer in terms of the Debye length and occupancy of the inner Helmholtz plane.

Thus, when an X-ray beam impinges on the boundary separating two materials that have different refractive indices, the wave suffers both refraction and reflection, the division of the incident energy depending on a number of parameters including the angle of incidence, θ. If the angle of incidence is $\leqslant \theta_C$, the critical angle, then the beam is totally *externally* reflected. This arises as a result of the fact that solids show refractive indices of less than 1 in the X-ray region. Under these conditions, interference between the incident and reflected rays produces a standing wave above the mirror surface (see Figure 2.78). As in the case of total internal reflection of IR radiation, an evanescent wave exists below the surface of the crystal.

The standing wave has nodal and antinodal planes of intensity parallel to the mirror surface, the period of which is a function of the angle of incidence, θ. On increasing θ from 0 to θ_C, the period between the antinodes decreases

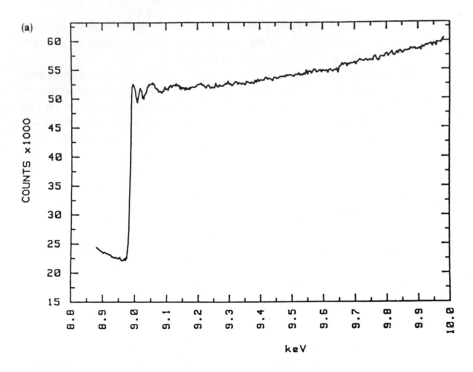

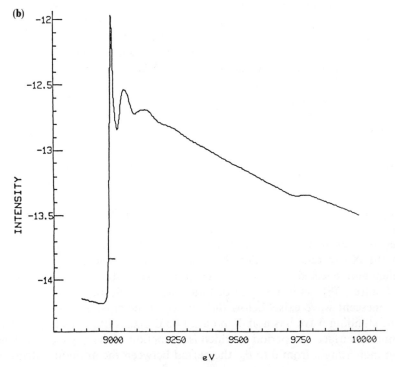

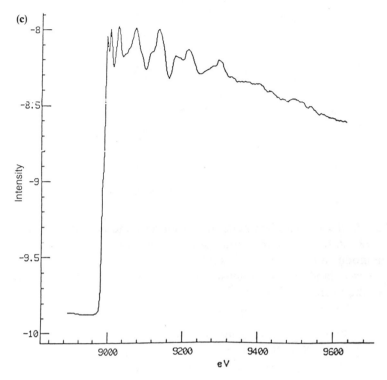

Figure 2.76 (a) Raw SEXAFS spectra of an upd Cu monolayer on Au(111). (b) Raw EXAFS spectra of aqueous $CuSO_4$. (c) Raw SEXAFS spectra of a 2% Cu–Au alloy. After Blum *et al.* (1986)

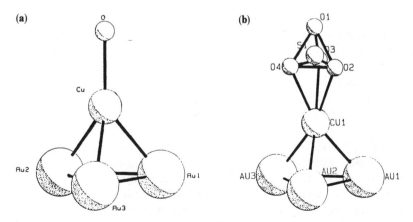

Figure 2.77 Possible geometric arrangement of (a) the Cu and O atoms on the Au(111) surface and (b) the Cu and SO_4^{2-} ions on the Au(111) surface. After Blum *et al.* (1986).

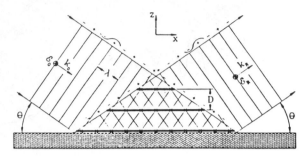

Figure 2.78   Illustration of the X-ray standing wave field formed by the interference between the incident and reflected plane waves above a mirror surface (see text for details). After Bedzyk *et al.* (1990). Copyright 1990 by the AAAS.

from ∞. At $\theta = 0$, there is a node at the surface, and an antinode at ∞. As θ is increased, the nodes and antinodes move in from ∞ until, at $\theta = \theta_C$, the first antinode coincides with the surface.

The X-ray standing wave intensity at an angle of incidence θ and distance z above the mirror surface is:

$$I(\theta, z) = |E_r + E_i|^2 \tag{2.95}$$

where E_i and E_r are the magnitude of the incident and reflected electric vectors. This can be expressed as:

$$I(\theta, z) = |E_i|^2 [1 + R + 2\sqrt{R} \cdot \cos(\delta - 2\pi Q z)] \tag{2.96}$$

where R is the reflectivity (given by $R = [|E_r|/|E_i|]^2$, see above), δ is the phase change on reflection, Q is the wavevector transfer for the reflection, i.e. effectively the momentum transfer, and is equal to $(2 \sin \theta)/\lambda = 1/D$. When θ is equal to the critical angle, $D = D_C$.

For total reflection in a non-absorbing medium, $R = 1$ and $I(\theta, z)$ at the mirror surface, $z = 0$, is then given by:

$$I(\theta, z = 0) = |E_i|^2 [2 + 2 \cos \delta] \tag{2.97}$$

Thus, the intensity of the standing wave at the surface increases smoothly from 0 to $4|E_i|^2$ as θ is increased from 0 to θ_C, as shown in Figure 2.79. At distances $> 0, I(\theta, z)$ modulates as a cosine wave between 0 and $4|E_i|^2$, the number of modulations being directly proportional to z. Thus, at a distance $N D_C$ above the mirror surface there will be $(N + 1/2)$ modulations between $\theta = 0$ and θ_C. The case for $z = 2D_C$ is also shown in Figure 2.79.

The essence of the XSW technique now lies in the effect these modulations have on the photoelectric cross-section of a target atom a distance z above the mirror surface. The incident X-rays can eject a core electron from the atom so generating a vacancy and resulting in the emission of a fluorescent X-ray photon. The probability of an incident photon ejecting the core electron, the photoelectric cross-section, is directly proportional to the electric field experienced by the atom. Hence, the fluorescence yield, $Y(\theta, z)$, for an atom or ion distribution $N(z)$ a distance z above the mirror surface can be written

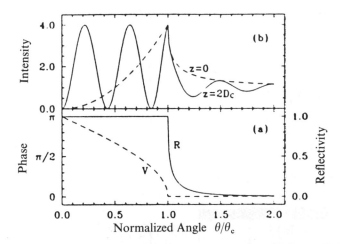

Figure 2.79 For $\beta = 0$. (a) The angular dependence of the reflectivity R and relative phase V of the reflected plane wave. (b) The angular dependence of the electric field intensity at $z = 0$ and $z = 2D_C$ for $|E_i| = 1$. After Bedzyk *et al.* (1990) and M. J. Bedzyk, *Synchrotron Radiation News*, 3 (1990) 25. Copyright 1990 Gordon and Breach Science Publishers, S.A.

as in equation (2.98):

$$Y(\theta, z) = \int_0^\infty N(z) \cdot I(\theta, z) \, dz \qquad (2.98)$$

As θ is varied, the fluorescent yield from target species a distance z above the mirror will be modulated according to equations (2.97) and (2.98) as a result of the movement inwards of the nodes and antinodes of the XSW. The phase and amplitude of this modulation are a measure of the mean position z and the width of the atom distribution.

While the diffraction and SEXAFS techniques discussed above are ideal for probing the atomic-scale structure of solids, they are somewhat restricted in their application. Thus, the Å-scale wavelength of X-rays renders them unsuitable for probing larger-scale structure, such as that of the diffuse double-layer. More importantly with respect to this particular application, SEXAFS requires at least short-range order in the system under study and diffraction demands long-range order. Consequently, the two techniques are aptly suited to the study of solid coatings on electrodes but cannot probe the structure of the double-layer. However, the long-period XSW approach does provide a means of deducing the ion distribution in the diffuse double layer and so supplies the first spectroscopic means of supporting or refuting the fundamental conclusions of electrocapillarity measurements.

Bedzyk and co-workers used the XSW technique to probe the ion distribution in the electrolyte above a charged cross-linked phospholipid membrane adsorbed onto a silicon–tungsten layered synthetic microstructure (LSM) as shown in Figure 2.80(a). The grazing-angle incidence experimental set-up

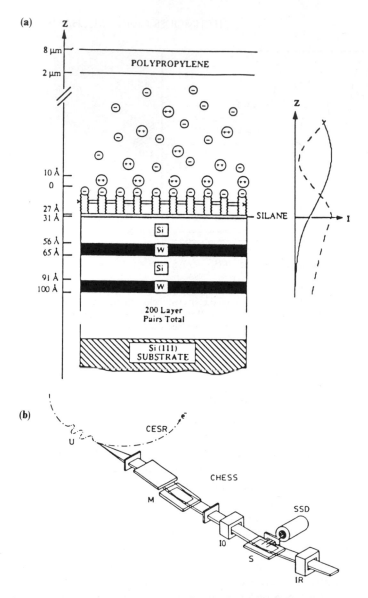

Figure 2.80 (a) Schematic showing a cross-linked phospholipid membrane deposited on a silanated Si/W layered synthetic microstructure (LSM). The top of the lipid contains a PO_4^- headgroup which attracts Zn^{2+} ions in the 0.1 mM aqueous solution. The encapsulating 6 μm thick polypropylene film traps a 2 μm thick water layer by capillary action between the film and the lipid. Along the right side is shown the electric field intensity at $E_\gamma = 9.8$ keV for $\theta = 2.5$ milliradians (solid curve) and $\theta = \theta_C = 4.5$ milliradians (dashed curve). When $\theta \leqslant \theta_C$, an X-ray standing wave exists above the reflecting surface and an evanescent wave below. (b) Experimental system for collecting X-ray standing wave data. The X-ray radiation from the undulator (U) passes through a Si(111) monochromator (M) and is then reflected by the sample (S). As θ is scanned, X-ray fluorescence spectra are simultaneously collected with a solid state detector (SSD) and reflectivity is measured with ion chambers to measure the incident (IO) and reflected (IR) beam intensities. CHESS stands for the Cornell High Energy Synchrotron Source, and CESR is the Cornell Electron Storage Ring. After Bedzyk *et al.* (1990). Copyright 1990 by the AAAS.

is shown in Figure 2.80(b). The electrolyte was aqueous $ZnCl_2$ (0.1 µM) adjusted to the required pH (6.8, 4.4 and 2.0) with HCl. The paper was not strictly concerned with probing the double-layer as the interaction of the Zn^{2+} with the phospholipid PO_4^- head groups involved the formation of a complex with most of the negative charge of the phospholipid neutralised by the ('condensed') layer of complexed Zn^{2+}. However, the remaining negative charge is compensated by Zn^{2+} in the Gouy–Chapman diffuse layer and the XSW technique showed itself to be sensitive to subtle changes in the diffuse layer, as well as confirming the presence of the condensed layer of Zn^{2+}.

Thus, the long-period XSW technique allows the direct measurement of the ion distribution above a charged surface, a measurement hitherto not possible. The approach does have its limitations, however: it requires perfect crystalline mirrors and can only probe ions having X-ray fluorescence at high enough energies to penetrate the electrolyte and cell window. In addition, it requires the target species to have a much larger concentration in the region of interest compared to the bulk of the solution. This prevents the fluorescence of the solution species swamping that of the target atoms. It is not inconceivable that some form of subtraction approach would rectify this last problem.

Whereas the XSW technique takes advantage of the standing wave established on the total reflection of X-rays from a mirror surface, a conceptually more straightforward approach is that of simply specularly reflecting an X-ray beam from an electrode coated with the film of interest, measuring the ratio of the intensities of the incident and reflected rays, and fitting the data, using the Fresnel equations, to a suitable model; an approach similar to optical ellipsometry.

2.1.9.4 Specular X-ray reflection   While the specular X-ray reflection (SXR) approach is again a means of probing solid coatings at the atomic level, it does have the advantage that the model obtained at the surface can include information on the atomic-scale roughness of the 'buried', i.e. usually inaccessible, electrode/surface layer interface.

Melendres *et al.* (1991) reported the *in-situ* study of the electrode/oxide and oxide/electrolyte interfaces for a copper electrode in pH 8.4 borate buffer under potential control. The grazing-angle incidence arrangement employed by the authors is shown in Figure 2.81(a) and a cyclic voltammogram of the Cu electrode in the buffer is shown in Figure 2.81(b).

Waves I and II in Figure 2.81(b) are due to the formation of Cu(I) and Cu(II) surface oxides. Subsequent reduction of these films occurs during the cathodic sweep to give waves III and IV. The points A to D represent the potentials at which reflectivity data were collected during the voltammetric scan. The potential was ramped at 10 mV/s until one of these potentials was reached, at which the scan was stopped for the duration of the data acquisition. The spectrum collected at A represents the condition of the electrode surface

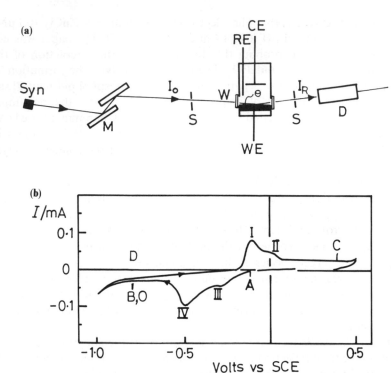

Figure 2.81 (a) Schematic of the system for *in situ* X-ray reflectivity measurements. Syn = synchrotron source; M = monochromator; S = slit; I_0, I_R = incident and reflected X-rays beams, respectively; θ = angle of incidence; W = teflon windows; WE = working electrode; RE = reference electrode; CE = counter electrode; D = scintillation detector. (b) Cyclic voltammogram of Cu-on-Si electrode in borate buffer solution (pH 8.4), scan rate = $10\,\text{mV}\,\text{s}^{-1}$. From Melendres *et al.* (1991).

at the open circuit voltage of 0.12 V, at which corrosion takes place, the reduction of the film so formed at -0.8 V (B), the subsequent oxidation at 0.4 V (C), and the re-reduction at -0.8 V (D). The results are shown in Figure 2.82(a) as plots of I/I_0 (where I_0 is the intensity of the incident ray and I the intensity of the reflected ray), against $2\pi Q$, where Q is the wavevector transfer and is equal to $2\sin\theta/\lambda$ (see previous section). As λ is fixed, the plot is essentially I/I_0 vs. $\sin\theta$.

The authors assumed a model of the various interfaces as shown in Figure 2.82(b). The Si was the underlying substrate on which the copper electrode was deposited. Such a structure will give rise to the reflections as shown, each one characterised by a wavevector transfer for both sides of the interface at which the reflection occurs. By applying the Fresnel equations to the reflections in the model in Figure 2.82(b), the reflectivity R can be derived.

Using a least-squares fit program, the copper film thickness D and the thickness of the Cu_2O film, Δd, densities ρ_n and σ (roughness) parameters can be derived from the data.

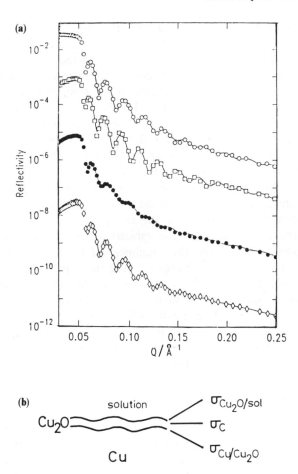

Figure 2.82 (a) Reflectivity of Cu-on-Si electrode at various potentials in borate buffer solution (pH 8.4). A, B, C and D correspond to potentials indicated in the cyclic voltammogram of Figure 2.81(b). Solid lines represent calculated curves while symbols correspond to experimental data. Open circles, A, -0.12 V; open squares, B, -0.80 V; both y-axes are reflectivity $\times 10^{-1}$. Filled circles, C, 0.40 V, reflectivity $\times 10^{-3}$; open diamonds, D, -0.80 V, reflectivity $\times 10^{-5}$. (b) Schematic of multi-layer model for Cu-on-Si electrode (not to scale). The oxide film is represented as Cu_2O. From Melendres *et al.* (1991).

The oscillations in the reflectivity curves arise from interference between the X-rays reflected from the various interfaces. The frequency of the oscillations is proportional to layer thickness and the amplitude depends on the interface roughness.

The authors found that measurements on a clean copper electrode *ex situ* under nitrogen gave a value of D of 283 Å, consistent with the value of c.

250 Å as measured by the quartz-crystal thickness monitor employed to follow the thickness of the copper during the vapour deposition process. At open circuit 24 Å of oxide forms at the expense of the underlying copper as would be expected for a corrosion process. The total $Cu + Cu_2O$ thickness increases due to the Cu_2O being less dense than the metal. Cathodic reduction of the oxide at -0.8 V resulted in the apparent complete recovery of the copper metal from the oxide. Subsequent oxidation of the film at 0.4 V considerably reduced the thickness of the copper to 250 Å, the thickness of the oxide film still only 26 Å, indicating that electrochemical corrosion results in a mixture of oxides, Cu_2O, $Cu(OH)_2$ and CuO (note the different film density) as well as some dissolution of the copper into solution. Again, reduction of the film at -0.8 V resulted in the recovery of the copper from the oxides, with a small loss due to the dissolution.

The roughness parameters of the various interfaces were also obtained from the model showing that the roughness of the Cu/Cu_2O and $Cu_2O/$ solution interfaces were little changed during the experiment.

Thus, the SRX technique allowed the state of the electrode surface *during corrosion* and the subsequent reduction to be followed *in situ*. In addition, information on the microstructure of the buried Cu/Cu_2O interface was also obtained by the authors, as well as the exposed $Cu_2O/$solution interface. The technique has potential application in a wide range of research concerned with film formation and corrosion.

2.1.10   AC techniques

One of the most powerful methods for the investigation of electrochemical reactions is the use of *alternating current* (AC) impedance techniques. In order to understand the basis of this approach, however, it is helpful first to consider some simple electronic analogues which mimic closely the systems of electrochemical interest.

In this section the nomenclature employed is that specific to the technique of AC impedance.

If we consider first a simple resistor, then applying a steady potential difference V_0 across this resistor will cause a current to flow, with a magnitude of $I_0 = V_0/R$, where R is the resistance of the resistor. Suppose that we now apply an alternating voltage $V_1 \sin \omega t$ across the resistor, where the functional form 'sin ωt' ensures that the voltage cycles from zero through a maximum of V_1, again through zero to a minimum of $-V_1$ and then back to zero with a frequency $f = \omega/2\pi$. The alternating form of the voltage is found to lead to a similarly alternating current $I_1 \sin \omega t$, where $I_1 = V_1/R$, and it is also clear that if a composite voltage $V_0 + V_1 \sin \omega t$ is applied the resultant current will take the form $I_0 + I_1 \sin \omega t$. The current and voltage are shown in Figure 2.83(a) and it is clear that the alternation in both V and I are *in phase* (i.e. I increases at the same time as V).

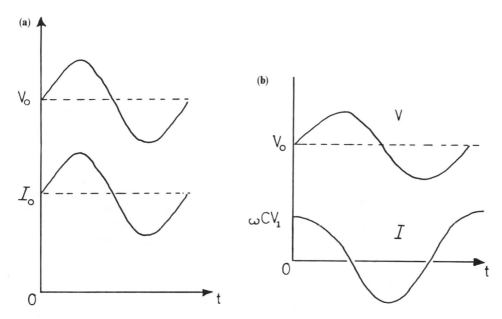

Figure 2.83   (a) The variation of the resulting voltage, V, and current, I, with time when a composite voltage $V_0 + V_1 \sin \omega t$ is applied across a resistance R. (b) The variation of the resulting voltage, V, and current, I, with time when a composite voltage $V_0 + V_1 \sin \omega t$ is applied across a capacitance C.

Let us now turn our attention to a capacitor which can be envisaged as two thin conducting plates which are *not* in electrical contact directly but are separated by a thin non-electronically conducting membrane. Simple capacitors are essentially charge-storage devices and, as was discussed in the section on electrocapillarity, the charge stored, Q, is related to the voltage V applied across the capacitor by the expression $Q = CV$, where C depends only on the geometrical arrangement of the capacitor plates and has units of CV^{-1} or *farads*. The current passing at any point is simply $I = dQ/dt = CdV/dt$ and it is evident that a capacitor will only pass a current if V is time-dependent. If we apply a potential $V = V_0 + V_1 \sin \omega t$, then the current passed will be $I = \omega C V_1 \cos \omega t$ and a plot of this is shown in Figure 2.83(b). This plot shows clearly that the current is now *out of phase* with the applied potential. In fact the current appears to have been moved 90° forward with respect to the voltage and is said to *lead the voltage* by 90°. It is also evident that the current is purely alternating: it has no DC (*direct current*) component.

It should be evident that if we had a black box that we knew contained either a resistor or a capacitor, we would immediately be able to tell both the nature and magnitude of the component by applying a potential of the form $V_1 \sin \omega t$ and measuring the resultant current. Real electrochemical systems are, unfortunately, rather more complex than this and tend to behave as mixtures of components arranged to form more complex *circuits*. A simple example is the series circuit:

which consists of a capacitor (represented as $-||-$) and a resistor (represented as $-\bigwedge\bigwedge-$ connected together *in series*.

To analyse the response of this circuit to an alternating voltage, it turns out to be rather easier to replace the simple sinusoidal form of the voltage used above by $V = V_1 e^{i\omega t}$, where the complex number $e^{i\omega t} = \cos \omega t' + i \sin \omega t$ and $i = \sqrt{-1}$. Any component of a circuit such as that shown can be defined as having an *impedance Z*, which can also be thought of as a complex number, containing both phase and magnitude information. For a resistor, Z is entirely real and simply equal to the resistance R, but for a capacitor:

$$Z = V/I = V_1 e^{i\omega t}/(i\omega C V_1 e^{i\omega t}) = 1/(i\omega C)$$

The impedance of a series circuit of the type shown is simply the algebraic sum of the impedances of the individual elements:

$$Z_{tot} = Z_R + Z_C \equiv R + 1/(i\omega C)$$

The inverse of the impedance of a circuit is its *admittance*, $Y_{tot} \equiv 1/Z_{tot}$ and clearly:

$$I = V Y_{tot} = V_1 e^{i\omega t} \cdot \{i\omega C/(1 + i\omega CR)\}$$

By manipulating the complex variables, it can then be shown that we may write:

$$I \equiv (V/|Z|) \cdot e^{i(\omega t + \alpha)}$$

where $|Z| = (1 + \omega^2 C^2 R^2)^{1/2}/\omega C$ and $\alpha = \tan^{-1}(1/\omega CR)$.

It is evident that the current still leads the voltage but that the 'phase angle', α, will vary from close to $90°$ at low frequencies to close to $0°$ at high frequencies. Also, at low frequency $|Z| \to 1/\omega C$ and at high frequency $|Z| \to R$. In other words, at low frequencies, the circuit behaves like a pure capacitor but at high frequencies it behaves like a pure resistor. Moreover, by fitting the observed current data as a function of frequency to calculated values of $|Z|$ and α, an accurate estimate of both R and C can be made.

This is the essence of AC impedance techniques: the cell is replaced by a suitable model system, in which the properties of the interface and the electrolyte are represented by appropriate electrical analogues and the impedance of the cell is then measured over as wide a frequency range as possible. By comparing the measured results with values calculated from the model system, we can evaluate both the suitability of our model and the values of the parameters required.

In order to show how this works in more detail, we will consider a very simple example in which we work out the impedance of an electrode in contact with an electrochemically active redox couple of the form:

$$O + e^- \rightleftharpoons R$$

The current density that flows, J, will be a function of the potential drop E across the interface and the concentrations of O and R at the electrode surface. Mathematically:

$$J = J(E, C_O(x = 0), C_R(x = 0))$$

where we must bear in mind that J is the positive current associated with a positive potential E, and x is the distance from the electrode surface. Small changes in J can be related to changes in the other variables; provided all changes are *small*:

$$\delta J \approx (\partial J/\partial E)_{C_O, C_R} \delta E + (\partial J/\partial C_O)_{E, C_R} \delta C_O + (\partial J/\partial C_R)_{E, C_O} \delta C_R \quad (2.99)$$

We can progress from here provided that we can find expressions for the partial derivatives of equation (2.99). Provided that the concentration of supporting electrolyte is sufficiently high that all the potential difference across the interface is accommodated within the Helmholtz layer, then transport of O and R near the electrode will only take place via diffusion (i.e. we can neglect *migration*). The equation of motion for *either* O or R is given by the differential form of Fick's equation, as discussed in chapter 1:

$$\partial C/\partial t = D\partial^2 C/\partial x^2 \quad (2.100a)$$

where x is the distance from the electrode surface (which is at $x = 0$) and the flux of C in the *positive* x-direction is $\boldsymbol{J} = -D(\partial C/\partial x)$.

If the concentrations of O and R can be written:

$$C_O = C_{O0} + C_{O1} e^{i\omega t}$$
$$C_R = C_{R0} + C_{O1} e^{i\omega t} \quad (2.100b)$$

which is equivalent to imagining that a steady state is established at some potential E_0. If we then superimpose a small AC potential modulation $E_1 e^{i\omega t}$, remembering that C_{O0}, C_{O1} and C_{R0}, C_{R1} and *not* functions of time and inserting (2.100b) into (2.100a), the diffusion equation becomes:

$$i\omega C_{O1} e^{i\omega t} = D_O(d^2 C_{O0}/dx^2 + d^2 C_{O1}/dx^2 \cdot e^{i\omega t}) \quad (2.100)$$

where D_O is the diffusion coefficient for species O. A similar expression can be derived for C_R.

Now we come to an important simplification: we are measuring *only* the AC components in our cell, i.e. we are measuring only those quantities that multiply $e^{i\omega t}$. This allows us to neglect the time independent term $d^2 C_{O0}/dx^2$ in equation (2.100), finally giving us:

$$D_O(d^2 C_{O1}/dx^2) = i\omega C_{O1} \quad (2.101)$$

This can be solved immediately under the conditions $C_{O1} \to 0$ as $x \to \infty$ and $C_{O1} = C_{O1}^s$ at $x = 0$ to give:

$$C_{O1} = C_{O1}^s \exp[-(i\omega/D_O)^{1/2}x] \qquad (2.102)$$

and similarly

$$C_{R1} = C_{R1}^s \exp[-(i\omega/D_R)^{1/2}x] \qquad (2.103)$$

The *flux* of C_{O1} at the surface must be related to the current, since it is the passage of the current that leads to the flux of components of the redox couple. We need to be careful here since there are subtleties of sign but for an increase in *positive* current density, δJ, there will be an increase in *positive* flux of C_O. The change in *flux* of C_O at the surface ($x = 0$) is given by Fick's first law as

$$\delta J = -D_O(\partial C_{O1}/\partial x)_{x=0}$$

which, from (2.102), can be evaluated immediately as

$$\delta J = +D_O(i\omega/D_O)^{1/2}C_{O1}^s$$

The change in current is simply the product of the molar flux and the Faraday, so:

$$\delta J = +F\cdot(i\omega D_O)^{1/2}C_{O1}^s \qquad (2.104)$$

and similarly for R:

$$\delta J = +D_R F(\partial C_{R1}/\partial x)_{x=0} = -F\cdot(i\omega D_R)^{1/2}C_{R1}^s \qquad (2.105)$$

From (2.104) and (2.105), C_{O1}^s and C_{R1}^s can be expressed in terms of δJ and replacing the δC_O and δC_R in (2.99) by the expressions for C_{O1}^s and C_{R1}^s, we obtain:

$$\delta J\{1 - (1/F(i\omega D_O)^{1/2})\cdot(\partial J/\partial C_O)_{E,C_R} + (1/F(i\omega D_R)^{1/2})\cdot(\partial J/\partial C_R)_{E,C_O}\}$$
$$= \delta E\cdot(\partial J/\partial E)_{C_O,C_R} \qquad (2.106)$$

The partial derivative on the right-hand side represents, in essence, the electrochemical rate constant; the larger the change in current with potential the easier charge transfer is and we can define R_{CT}, the charge-transfer resistance, as $1/(\partial J/\partial E)_{C_O,C_R}$. Given that $(1/i)^{1/2} = (1 - i)/\sqrt{2}$ and *defining* the transport part of (2.106):

$$\sigma = (1/F)\cdot\{(1/(2D_R)^{1/2})\cdot(\partial J/\partial C_R)_{E,C_O} - 1/(2D_O)^{1/2})\cdot(\partial J/\partial C_O)_{E,C_R}\} \qquad (2.107)$$

then, since the impedance of the interface is $\delta E/\delta J = Z$:

$$Z = R_{CT} + (1 - i)\cdot R_{CT}\cdot(\sigma/\omega^{1/2}) \qquad (2.108)$$

Evidently, the impedance of the interface consists of two components: a charge-transfer resistance R_{CT}, which will depend on the electrochemical rate constants, and a more unusual element arising from the diffusion of the redox couple components to and from the interface. The magnitude of this element *decreases* with increasing frequency as might be expected, since the faster the potential is alternated, the less far the solution couple can diffuse before the potential is reversed. The most characteristic aspects of this part of the impedance are the dependence of its magnitude on the inverse square root of the frequency and the presence of the $(1 - i)$ term which will give rise to a fixed phase angle of $45°$, *independent of frequency*. These characteristics were first verified by Warburg in the last century and the second term of equation (2.108) is termed a 'Warburg' impedance, denoted by —W—.

Our equivalent circuit now looks like:

though there are two further physical contributions that we have not discussed.

1. There is a displacement contribution of δJ arising from the charging of the double layer. This can simply be represented by a *capacitance, C_{dl}*, that is *in parallel* with R_{CT} and —W—.
2. The electrolyte will have a finite resistance, R_{so}, which will be in *series* with all the interfacial terms.

There is also a term representing the impedance of the second electrode in the cell and a term representing the geometrical capacitance of the whole cell. These latter two can, however, be minimised by proper choice of cell geometry, but we cannot eliminate the first two in any practical measurement, with the result that our final equivalent circuit for the cell looks like:

This circuit is usually referred to as the 'Randles' circuit and its analysis has been a major feature of AC impedance studies in the last fifty years. In principle, we can measure the impedance of our cell as a function of frequency and then obtain the best values of the parameters R_{CT}, σ, C_{dl} and R_{so} by a least squares algorithm. The advent of fast micro-computers makes this the normal method nowadays but it is often extremely helpful to represent the AC data graphically since the suitability of a simple model, such as the Randles model, can usually be immediately assessed. The most common graphical representation is the 'impedance plot' in which the real part of the measured impedance (i.e. that in phase with the impressed cell voltage) is plotted against the $90°$ out-of-phase 'quadrature' or *imaginary* part of the impedance.

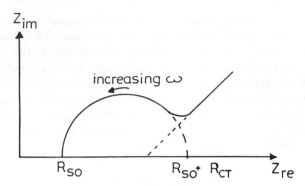

Figure 2.84   A plot of the real part of the measured impedance, Z_{im}, vs. the imaginary part of the impedance, Z_{re}.

For the Randles circuit this plot, for reasonable values of the components, is shown in Figure 2.84. It can be seen that there is a rising spur on the right-hand side of the figure whose slope will allow us to work out $R_{CT}\sigma$. The first intercept on the real axis gives $R_{CT} + R_{so}$ and the second, high frequency, intercept gives R_{so} alone. Finally, the frequency corresponding to the top of the semicircle will be $1/C_{dl}R_{CT}$ approximately, allowing us to read off the approximate values of our parameters immediately.

The values of the parameters derived from the best fit can be related to the fundamental physical constants, such as the electrochemical rate constants, by explicit calculation. From the Butler–Volmer equation,

$$J = F \cdot k^0 \cdot \{ C_R(0) \, e^{(1-\beta)\eta F/RT} - C_O(0) \, e^{-\beta\eta F/RT} \}$$

If the DC potential is adjusted to zero overpotential, then by direct differentiation we have:

$$(\partial J/\partial E)_{C_O, C_R} = (k^0 F^2/RT) \cdot C_O^{1-\beta} C_R^{\beta} = 1/R_{CT} \tag{2.109}$$

and by calculating the derivates with respect to C_O and C_R we find

$$R_{CT}\sigma = (RT/F^2\sqrt{2}) \cdot \{ 1/(C_R^* D_R^{1/2}) + 1/(C_O^* D_O^{1/2}) \} \tag{2.110}$$

where the asterisk means concentrations in the bulk of the solution. The equations (2.109) and (2.110) constitute Randles' model and it is evident that the electrochemical rate constant, k^0, can be obtained immediately from the AC impedance measurements.

It is also the case that this simple approach can be extended to more complex mechanisms including those in which surface adsorbed intermediates are involved in the reaction. For example, consider the mechanism:

$$A \longrightarrow B_{ads} + e^- \quad J_1 \tag{2.111a}$$

$$B_{ads} \longrightarrow P + e^- \quad J_2 \tag{2.111b}$$

If θ is the coverage of adsorbed species, B_{ads}, then since J_1 is a function of

E, C_A and θ alone, we can write

$$\delta J_1 = (\partial J_1/\partial E)_{C_A,\theta} \cdot \delta E + (\partial J_1/\partial C_A)_{E,\theta} \cdot \delta C_A + (\partial J_1/\partial \theta)_{E,C_A} \cdot \delta \theta \quad (2.112)$$

and, as before, (2.104):

$$\delta J_1 = F(i\omega D_A)^{1/2}\delta C_A,$$

so replacing δC_A in (2.112) we have, as for (2.106):

$$\alpha_1 \delta J_1 = (\partial J_1/\partial E)_{C_A,\theta} \cdot \delta E + (\partial J_1/\partial \theta)_{E,C_A} \cdot \delta \theta \quad (2.113)$$

where

$$\alpha_1 = 1 - (1/F(i\omega D_A)^{1/2}) \cdot (\partial J_1/\partial C_A)_{E,\theta} \quad (2.114)$$

Similarly:

$$\alpha_2 \delta J_2 = (\partial J_2/\partial E)_{C_P,\theta} \cdot \delta E + (\partial J_2/\partial \theta)_{E,C_P} \cdot \delta \theta \quad (2.115)$$

where

$$\alpha_2 = 1 + (1/F(i\omega D_P)^{1/2}) \cdot (\partial J_2/\partial C_P)_{E,\theta} \quad (2.116)$$

Now, it is clear that:

$$d\theta/dt = \delta J_1 - \delta J_2 \quad (2.116a)$$

and if we can write $\theta = \theta^0 + \delta\theta \cdot e^{i\omega t}$, then from (2.116a) using (2.113) and

$$i\omega\delta\theta = \{(1/\alpha_1) \cdot (\partial J_1/\partial E)_{C_A,\theta} - (1/\alpha_2) \cdot (\partial J_2/\partial E)_{C_P,\theta}\} \cdot \delta E$$
$$+ \{(1/\alpha_1) \cdot (\partial J_1/\partial \theta)_{E,C_A} - (1/\alpha_2) \cdot (\partial J_2/\partial \theta)_{E,C_P}\} \cdot \delta\theta \quad (2.117)$$

The total current density, $J = J_1 + J_2$, and a rather complicated formula may be obtained by straightforward algebra. In fact, for many practical purposes, the value of ωD is sufficiently large for us to neglect the effects of diffusion. This is equivalent to putting $\alpha_1 = \alpha_2 = 1$ in the formulae above, which then simplify to:

$$\delta J = (1/R_{CT}) \cdot \delta E + \delta E/\{R_0(1 + i\omega\tau)\} \quad \text{or} \quad (2.118)$$

$$Y_f = 1/R_{CT} + 1/R_0(1 + i\omega\tau) \quad (2.118a)$$

where:

$$R_{CT} = \{(\partial J_1/\partial E)_\theta + (\partial J_2/\partial E)_\theta\}^{-1} \quad (2.119)$$

$$R_0^{-1} = \tau \cdot \{(\partial J_1/\partial \theta)_E + (\partial J_2/\partial \theta)_E\} \cdot \{(\partial J_1/\partial E)_\theta - (\partial J_2/\partial E)_\theta\} \quad (2.120)$$

$$\tau = F\{(\partial J_2/\partial \theta)_E - (\partial J_1/\partial \theta)_E\}^{-1} \quad (2.121)$$

The corresponding equivalent circuit has the form:

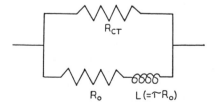

R_{CT}

R_0 $L(=\tau \cdot R_0)$

and contains an *inductive* component L which has an impedance $i\omega L$. Provided the model is correct, then an analysis of the AC data will allow us to estimate the rate constants for the two steps, as well as information about the adsorbed species B.

Experimentally, the main difficulty with AC impedance studies is the need for very great care in setting up the measurement. The potentiostat used must have high quality operational amplifiers which possess a very wide frequency response. All leads to the cell should be as short as possible, to minimise stray impedances, and considerable attention must be paid to cell design to ensure that the measured impedance actually reflects *only* that of the working electrode interface under study. Modern day equipment is, in essence, fully automated, using phase-sensitive detection to explore a dynamic range between 10^{-4} Hz and 10^4 Hz. Even with this equipment, however, measurements above 10^4 Hz are likely to reflect the presence of artifacts unless enormous care is taken and below 10^{-4} Hz the long-term stability and drift in the cell will prove exceedingly difficult to counter.

To summarise, AC methods have proved most successful where the system is straightforward and can be modelled analytically. By measurement over a wide range of frequencies the constants for the reaction steps constituting the model can be established and, particularly if adsorbed species are involved, AC methods have proved very powerful indeed, with a major area of application being in the study of metal passivation, as discussed in detail elsewhere in the book. An example of this behaviour in practice is provided by the work of Conway's and Hillman's groups on chlorine evolution at platinum. Several mechanisms for this reaction have been proposed, including both Volmer and Heyrovsky types:

$$\text{mechanism (1)}$$
$$Cl^- \longleftrightarrow Cl^{\bullet}_{ads} + e^-$$
$$2Cl^{\bullet}_{ads} \longrightarrow Cl_2$$

and:

$$\text{mechanism (2)}$$
$$Cl^- \longleftrightarrow Cl^{\bullet}_{ads} + e^-$$
$$Cl^{\bullet}_{ads} + Cl^- \longrightarrow Cl_2$$

and mechanisms involving either Cl^+_{ads} or oxychloro intermediates:

$$\text{mechanism (3)}$$
$$Cl^- \longleftrightarrow Cl^{\bullet}_{ads} + e^-$$
$$Cl^{\bullet}_{ads} \longrightarrow Cl^+_{ads} + e^-$$
$$Cl^+_{ads} + Cl^- \longleftrightarrow Cl_2$$

and

mechanism (4)

$$H_2O \longleftrightarrow OH_{ads} + H^+ + e^-$$

$$OH_{ads} + Cl^- \longrightarrow HOCl + e^-$$

$$HOCl + HCl \longleftrightarrow Cl_2 + H_2O$$

On the basis of Tafel slope studies and potential relaxation transients, mechanism (1) has been suggested for Cl_2 evolution at Pt, whereas chlorine evolution at the technically hydrated RuO_2/TiO_2 electrode appears to proceed by (3) or, possibly, (4).

The AC data of Hillman and co-workers (Li *et al.*, 1992) showed a marked dependence on both geometry and convection effects. By using a thin *ring* electrode of thickness 0.025 cm and diameter 0.8 cm, results showing both simple Warburg behaviour and, in rotation, the underlying slow surface kinetics can be obtained.

Data for the stationary electrode are given in Figure 2.85(a) which shows a simple Randles response. The diffusional process giving rise to the Warburg region at lower frequencies ($< 40\,Hz$) is caused by diffusion of dissolved Cl_2 away from the electrode. If the ring is rotated, the Warburg section disappears and, over a wide range of conditions (potential, rotation speed and electrolyte composition) *three* successive semicircles are observed, as shown in Figure 2.85(b). The first two semicircles were treated using the model depicted as equations (2.111(a)) and (2.111(b)) above, modified to include the effects of finite electrolyte resistance, R_Ω, and the double layer capacitance, C_{dl}. The model gives the equivalent circuit:

At very high frequencies, C_{dl} will have a low impedance (since $Z \approx 1/\omega C_{dl}$) and will effectively short out R_{CT} and $R_0 + i\omega L$; the impedance at high frequencies thus collapses to R_Ω. At lower frequencies the impedance of C_{dl} will be high but if the inductance still has a high impedance, the overall impedance will reduce to $R_\Omega + R_{CT}$, as marked in Figure 2.85(b). At lower frequencies still, $|\omega L| \ll |R_0|$ and the equivalent circuit will be:

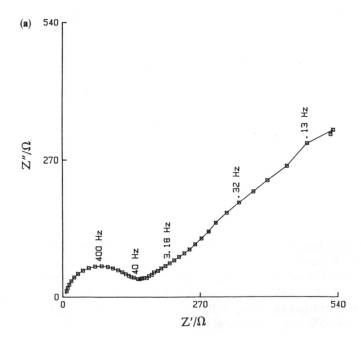

with $Z = R_\Omega + R_{dc}$, where $1/R_{dc} = 1/R_{CT} + 1/R_O$, from the appropriate formula for adding resistors in parallel. In this way, the two higher frequency semicircles can be accounted for. As might be expected, analysis at potentials between 0.98 V and 1.05 V vs. SCE show that R_{CT} decreases strongly with potential. In this case, $R_O < 0$, though its magnitude again decreases strongly with E and τ increases only weakly with E. Hillman and co-workers also studied the variation of these parameters with Cl^- concentration and showed that increasing the chloride concentration closely mimicked the effect of increasing the potential. They were also able to show that C_{dl} values were close to those expected for Pt, a result of some importance in validating the coupling of C_{dl} and Y_f leading to the operation circuit above.

The behaviour of Y_f with respect to potential and concentration can be used to distinguish the mechanisms (1)–(4) above. In the highest frequency semicircle, the impedance $Y_f \approx 1/R_{CT} + i\omega C_{dl}$. Extraction of R_{CT} and its dependence on E and $[Cl^-]$ is now straightforward. Theoretically, for

(b)

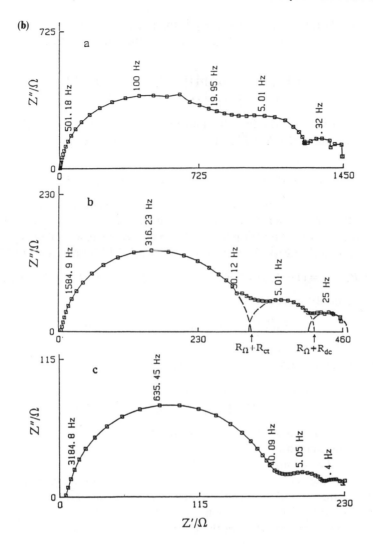

Figure 2.85 (a) Impedance spectra for the chlorine evolution reaction in the presence of a mass transport effect. Pt ring electrode. Electrolyte: 1 M $HClO_4$ + 1.25 M NaCl in water. $E = 1.060$ V vs. SCE. Pre-conditioning time: 2 min, $\omega = 0$ Hz, quiescent solution. Squares represent the experimental data points. (b) Impedance spectra for the chlorine evolution reaction on a rotating thin ring electrode in the absence of a mass transport effect. Electrolyte: 4.0 M NaCl + 1 M HCl in water, $\omega = 7$ Hz. Electrode potential: (a) 1.010 V vs. SCE; (b) 1.035 V; (c) 1.051 V. Pre-conditioning time is 1 min. The broken lines in (b) illustrate the empirical best fitting of the semicircles. Reprinted from *Electrochimica Acta*, **37**, B. Li, A.R. Hillman and S.D. Lubetkin, pp. 2715–2723 (1992), with kind permission from Pergamon Press Ltd., Headington Hill Hall, Oxford OX3 0BW, UK.

mechanism (1), $(\partial v_d/\partial E)_p = 0$ (where p is the partial pressure of chlorine, and v_d the rate of desorption of Cl_{ads}) and since the rate of adsorption, v_a, is given by:

$$v_a = \{k_1'[Cl^-](1 - \theta_{Cl^-})\exp[(1 - \beta_a)F(E - E^0)/RT]\}$$
$$- \{k_{-1}'\theta_{Cl^-}\exp[-\beta_a F(E - E^0)/RT]\} \qquad (2.122)$$

where k_1' and k_{-1}' are the forward and reverse rate constants for the initial adsorption step when $E = E^0$, and θ_{Cl^-} is given by the quasi-equilibrium condition:

$$\theta_{Cl^-}/(1 - \theta_{Cl^-}) = (k_1'/k_{-1}')[Cl^-]\exp[F(E - E^0)/RT] \qquad (2.123)$$

(see section on H_2 evolution reaction in chapter 1). Remembering that V_a is related to the current density J_1 as defined in (2.111a), save that we must now consider the equivalent reaction to (2.111a) to be *reversible*, we obtain $J_1 = FV_a$, and $R_{CT(1)} = \{(\partial J_1/\partial E)_\theta\}^{-1}$ or, by differentiating (2.122) and using (2.123):

$$(1/R_{CT})_{(1)} = (1 - \theta_{Cl^-})(F^2/RT)\cdot[Cl^-]k_1'\exp[(1 - \beta_a)(E - E^0)/RT]$$

If mechanism (2) operates,

$$V_d = k_d'[Cl^-]\theta_{Cl^-}\exp[(1 - \beta_d)F(E - E^0)/RT]$$

where k_d' is the rate constant for the desorption step when $E = E^0$. Clearly, V_d *is* potential dependent. We have now:

$$(1/R_{CT})_{(2)} = (1/R_{CT})_{(1)} + ((1 - \beta_d)F^2/RT)(k_1 k_d'/k_{-1}')$$
$$\cdot[Cl^-]^2\exp[(2 - \beta_d)F(E - E^0)/RT].$$

In fact, $1/R_{CT}$ experimentally is *first* order with respect to $[Cl^-]$, pointing strongly to mechanism (1) as being the most likely and the weak dependence of τ on E is also strongly suggestive that $\theta_{Cl^-} \ll 1$.

Analysis of the second semicircle can also be carried out. In essence, this depends on the second step of the reaction mechanism and a second order dependence on $[Cl^-]$ is found which rules out mechanisms (3) and (4) above.

There remains the problem of the lowest frequency semicircle. In fact, this is associated with $Pt-O_x$. Under high Cl^- concentrations, $Pt-O$ formation is strongly suppressed, coverage is low and can be maintained reasonably constant within the timescale of an AC impedance run. However, if the chloride concentration is reduced, or if the Pt is not 'pre-conditioned' by holding it at the measured potential for several minutes to stabilise a steady state, then this third semicircle changes drastically.

2.2
In situ probes of the near-electrode region

The characterisation and monitoring of the phenomena that occur in the near-electrode region form an extremely important aspect of the study of electrochemical processes. In contrast to the study of the electrochemistry of adsorbates, the electrochemical reactions of solution species commence with their diffusion to the electrode. Thus, the term 'near electrode region' can be defined in two principle ways:

1. That volume bounded by the distance away from the electrode over which a redox-active species can diffuse to the electrode surface, within the timescale of the experiment being undertaken. From considerations of random walk in one dimension it can be shown that the distance l which a species moves in a time t is given by:

$$l = \sqrt{[2Dt]} \tag{2.124}$$

   where D is the diffusion coefficient of the species. Equation (2.124) will be valid only for a short time, after which thermal currents, concentration gradients, etc. will disrupt the diffusion process. In aqueous solution, a typical value of D is $5 \times 10^{-6}\,\mathrm{cm^2\,s^{-1}}$; for a cyclic voltammetry experiment t may be of the order of a few seconds, giving $l = 45\,\mu\mathrm{m}$. In contrast, in a typical FTIR experiment $t = 30\,\mathrm{s}$, giving $l = 170\,\mu\mathrm{m}$; and in an AC experiment, $t \approx 10^{-3}\,\mathrm{s}$, giving $l \approx 1\,\mu\mathrm{m}$.

2. That volume within which the ions having charge opposite to that on the electrode have a concentration higher than those in the bulk of the solution (in the absence of specific adsorption). Under the conditions typically employed in electrochemical measurements, i.e. high ionic strength, this would correspond simply to a volume bounded by the outer Helmholtz plane, a few angstroms (see section on electrocapillarity).

The first of these definitions seems to be the more reasonable and is used throughout the remainder of this book. Equation (2.124) is a simple representation of a 'diffusion layer' that takes no account of the concentration gradient that is set up on discharge of an electroactive species at the electrode or the existence of convective effects. Later in this chapter we will develop this idea to derive the concept of the diffusion layer extending a distance δ out from the electrode.

Having defined our 'near electrode region', we turn now to consider the various techniques that can be employed in the *in situ* investigation of the reactions that occur within it. The various methods that can be employed will each provide different types of information on the processes occurring there. As has already been discussed, cyclic voltammetry is the most common technique first employed in the investigation of a new electrochemical system. However, in contrast to the LSV and CV of adsorbed species, the voltammetry of electroactive species in solution is complicated by the presence of an additional factor in the rate, the mass transport of species to the electrode. Thus, it may be more useful to consider first the conceptually more simple chronoamperometry and chronocoulometry techniques, in order to gain an initial picture of the role of mass transport.

2.2.1 *Chronoamperometry, chronocoulometry and the Butler–Volmer equation*

Quite generally, we can represent an electrochemical reaction as a series of steps, most simply expressed as:

1. Mass transport to the electrode.
2. Electron transfer step(s)
3. Mass transport away from the electrode.

Mass transport can be by migration, convection or diffusion. As discussed in chapter 1, in the presence of strong electrolyte migration can be neglected, as can convection if the solution is unstirred, at a uniform temperature and the timescale of the experiment is short (i.e. a few seconds). Thus, we can make the first distinction between electrode reactions that are dominated by step 1, diffusion-controlled, and those for which steps 1 and 2 contribute to the overall observed rate.

Consider first the diffusion-limited regime. The simplest experiment to perform is a chronoamperometric measurement, i.e. to monitor the current after a potential step to a value where an electroactive species will undergo electron transfer. This effectively allows us to monitor the rate of reaction, v, as a function of time, through the relationship:

$$v = I/FA \tag{2.125}$$

where, as above, A is the area of the electrode.

Consider the reaction:

$$O + ne^- \longrightarrow R \tag{2.126}$$

Fick's second law of diffusion is:

$$\partial[O]_0/\partial t = -D_0 \partial^2[O]_0/\partial x^2 \tag{2.127}$$

where x is the distance as measured from the electrode surface and $[O]_0$ is the concentration of O at the surface. This can be integrated under the assumption of very fast electron transfer to give the Cottrell equation:

$$I_d = -(nFAD_O^{1/2}[O^*])/\pi^{1/2}t^{1/2} \tag{2.128}$$

where $[O^*]$ is the bulk concentration of O and I_d is the diffusion-limited (cathodic) current for the reduction of O and the equation indicates that the current is determined entirely by diffusion. Thus, a plot of I_d vs. $t^{-1/2}$ will yield a straight line if the reaction is diffusion-limited. The slope of the line gives the product $nD_O^{1/2}$ for the reacting species.

If the I/t curve is integrated from $t = 0$, since:

$$Q = \int I \, dt$$

then:

$$\int_0^Q dQ = \int_0^t -(nFAD_O^{1/2}[O^*])/\pi^{1/2}t^{1/2} \cdot dt$$

$$Q_d = -(2nFAD_O^{1/2}[O^*]t^{1/2})/\pi^{1/2} \tag{2.129}$$

Thus, if the reaction is diffusion-controlled, a plot of the charge passed vs. $t^{1/2}$ (Anson's plot) is a straight line that should pass through the origin; however, it is generally found that the line has a positive intercept. This is due to the fact that double-layer charging adds to the observed charge, as does the presence of any adsorbed O. Thus, the charge Q obtained from Anson's plot is given by:

$$Q = Q_d + Q_{DL} + Q_{ads}$$

where Q_{DL} is the contribution from double-layer charging and Q_{ads} the contribution from the reaction of any of the adsorbed electroactive species. The intercept on a Q vs. $t^{1/2}$ plot is thus $Q_{DL} + Q_{ads}$. By performing the chronocoulometric experiment in the presence and absence of electroactive species, the presence of adsorption can be detected, and Q_{DL} and Q_{ads} determined.

We can obtain an additional expression for the diffusion current by considering Fick's first law of diffusion, first introduced in chapter 1, equation (1.34). If $\mathbf{J}$ is the flux of species *to the electrode*, it will be related to the observed current, I, by:

$$\mathbf{J} = I/nFA,$$

where n is the number of electrons involved in the reduction reaction $O + ne^- \to R$. From Fick's law, assuming that the concentration of O varies uniformly over a distance δ from the electrode surface, we can write:

$$d[O]_0/dx \approx \Delta[O]_0/\Delta x \approx ([O]_0 - [O^*])/\delta,$$

where $[O]_0$ is the concentration of O at the electrode surface and $[O^*]$ is that in the bulk. Hence:

$$I/nFA = D_O([O]_0 - [O^*])/\delta \tag{2.130}$$

From equation (2.128) we can write:

$$D_O([O]_0 - [O^*])/\delta = -D_O^{1/2}[O^*])/\pi^{1/2}t^{1/2}$$

Rearranging gives:

$$\delta = D_O^{1/2}\pi^{1/2}t^{1/2}(1 - [O]_0/[O^*]) \tag{2.131}$$

Equation (2.131) is extremely important as it clearly shows that for a simple electrode geometry of the sort considered above, the diffusion layer thickness, δ, increases with time at a fixed potential. This is depicted in Figure 2.86.

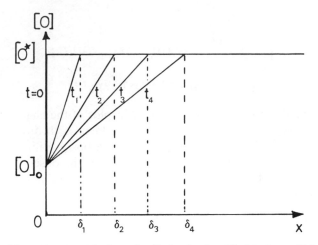

Figure 2.86 Schematic representation of variation in the diffusion layer thickness, δ, as a function of time for the reduction of an oxidant O at a fixed planar electrode. [O*] is the concentration of O in the bulk of the solution, $[O]_0$, is the concentration at the electrode surface, x is the distance into the solution from the electrode surface, and δ_1, etc. are the diffusion layer thickness at various times t_1, etc.

The diffusion current has a maximum, or limiting, value when the kinetics at the surface are so fast that $[O]_0 = 0$. This limiting current is then:

$$I_1 = -D_O nFA[O^*]/\delta_1 \qquad (2.132)$$

and δ_1 is given by:

$$\delta_1 = D_O^{1/2}\pi^{1/2}t^{1/2} \qquad (2.133)$$

Providing that the potential is sufficiently negative, the kinetics of the reduction reaction in equation (2.125) can usually be rendered fast enough to tip the system into the diffusion-controlled regime, as was shown in the discussion of the Butler–Volmer equation in chapter 1.

2.2.2 Voltammetry

The cyclic voltammogram of a surface-bound species has already been considered in section 2.1 and we now turn to the situation where the redox species R/O is in solution. Consider the cyclic voltammogram in Figure 2.87. The figure is typical of a reversible electrochemical system, i.e.:

$$O + e^- \longleftrightarrow R \qquad (2.134)$$

The shape of the voltammogram can be qualitatively explained as follows: on increasing the potential from the region where R is stable, anodic current starts to flow as the potential approaches E^0 for the O/R couple. The current continues to increase with increasing potential until the rate of consumption

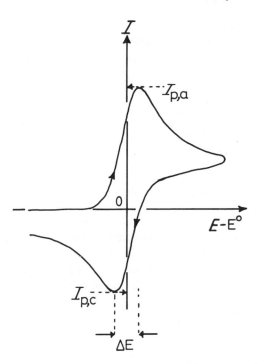

Figure 2.87 Schematic of the cyclic voltammogram expected from a reversible electrochemical redox system $O + e^- \rightleftharpoons R$ having a standard reduction potential E^0. E is the potential of the working electrode, and I the current.

of R at the electrode becomes such that transport from the bulk of the solution is not fast enough to maintain $[R^*] = [R]_0$ and the concentration of R in the near-electrode region drops in order to enhance the flux of R to the surface. As the potential moves past E^0, the near-electrode concentration of R species falls almost to zero, the mass transfer of R reaches a maximum rate, in unstirred solution, this rate then declines as depletion of R further and further from the electrode takes place. Thus, the current passes through a maximum before dropping again. On reversing the potential scan, the above sequence of events is repeated for the reduction of the electrochemically generated O that now predominates in the near-electrode region.

If, as is normal, the solution is not stirred, then the conditions of laminar (uniform) diffusion characterising the above description will hold only for a short time. For longer periods, thermal and concentration gradients induce random convection processes and the resultant currents show sizeable fluctuations.

If the electrochemical kinetics of the process are facile then the overall process will be dominated entirely by mass transport. Kinetic parameters such as the exchange current cannot, therefore, be obtained from such a system by analysis of the cyclic voltammetric response. Systems which satisfy this condition are normally referred to as 'reversible'. This is slightly unfortunate terminology since reversible is also used more generally to imply that O can be re-reduced to R. If the electrochemical kinetics are slow but the redox process still takes place, the system is termed 'quasi-reversible' but care should be taken especially if earlier papers are being read.

The shape of the anodic and cathodic sweeps of the I/E curves in Figure 2.87 can be derived, though the derivations are difficult as the boundary conditions are time-dependent and are beyond the scope of this book. The interested reader is referred to any one of the excellent texts listed at the end of this chapter. The result of the derivation is to state that the anodic current is given by:

$$I = FA[\text{R*}](\pi D_R Fv/RT)^{1/2}\chi(\sigma t) \tag{2.135}$$

where A is the area of the electrode, D_R is the diffusion coefficient of R, v is the scan rate and $\chi(\sigma t)$ is a tabulated number that is a function of the electrode potential and contains the variation of the potential with time.

Of particular interest is the relationship between the peak current and the concentration of the reacting species for the anodic and cathodic peak current, $I_{p,a}$ and $I_{p,c}$:

$$I_{p,a} = 0.4463FA[\text{R*}](F/RT)^{1/2}v^{1/2}D_R^{1/2} \tag{2.136}$$

$$I_{p,c} = -0.4463FA[\text{O*}](F/RT)^{1/2}v^{1/2}D_O^{1/2} \tag{2.137}$$

the dependence of the peak current on $v^{1/2}$ is indicative of diffusion control.

Useful experimental parameters in cyclic voltammetry are (i) the value of the separation of the potentials at which the anodic and cathodic peak currents occur, $\Delta E = E_{p,a} - E_{p,c}$, and (ii) the half wave potential, $E_{1/2}$, the potential mid-way between the peak potentials. A value of ΔE of c. 0.057 V at 25°C is diagnostic of a Nernstian response, such as that shown in Figure 2.87. More generally, if n electrons are transferred from R, then the separation will be $0.057/n$ V. It should be noted that the expected value for ΔE of $0.57/n$ V has no relationship to the usual Nernstian slope of $RT/nF = 0.059/n$ V at 25°C.

The half wave potential is related to E^0 by:

$$E^0 = E_{1/2} + (RT/nF)\ln\left[(D_O/D_R)^{1/2}\right] \tag{2.138}$$

thus, $E^0 = E_{1/2}$ when $D_O = D_R$ and this is a common approximation employed.

For a reversible system, ΔE should be independent of scan rate, however, it is found in practice that ΔE generally increases with v. This is due to the presence of a finite solution resistance between the reference and working electrodes. Thus, if this resistance is R_s, and is large enough such that $I_{p,a}R_s$ and $I_{p,c}R_s$ are appreciable with respect to the accuracy of the potential measurement, then the true potential, E_{true} of the working electrode will differ from that dictated by the potentiostat, E_{appl}. Instead, the potential will be greater or less than E_{appl} by the 'IR drop', i.e. $E_{true} = E_{appl} + I_pR_s$, with the IR term negative for $I_{p,a}$ and positive for $I_{p,c}$. Since the peak current is

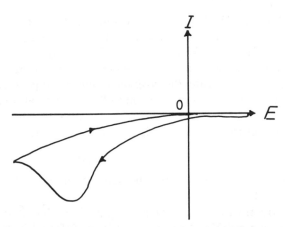

Figure 2.88 Schematic of the cyclic voltammogram expected from an irreversible process of the form $O + e^- \rightarrow R$. E is the potential of the working electrode and I is the current.

proportional to $v^{1/2}$, this IR drop increases with scan rate and the peak potentials move apart.

At the other kinetic extreme, a voltammogram such as that shown in Figure 2.88 is indicative of a completely irreversible reaction. For example, $O + e^- \rightarrow R$ where R cannot be re-oxidised to O (or anything else). In such a case, the (cathodic) current is given by:

$$I = -FAk_c[O]_0 \qquad (2.139)$$

From the previous section, we can write:

$$I = -FA[O]_0 k^0 \exp[-\beta f(E - E^0)] \qquad (2.140)$$

with E the potential at which the reaction is being driven, and:

$$k_c = k^0 \exp[-\beta f(E - E^0)] \qquad (2.141)$$

Introducing the scan rate, such that $E = E_i - vt$, where E_i is the starting potential of the scan (see section on the voltammetry of adsorbed species), gives:

$$I = -FA[O]_0 k^0 \exp[bt] \cdot \exp[-\beta f(E_i - E^0)] \qquad (2.142)$$

where $b = \beta f v$. This equation can be solved to give the current in terms of the more easily measured bulk concentration of O and the diffusion coefficient D_O:

$$I = -FA[O^*]D_O^{1/2} v^{1/2}[\beta F/RT]^{1/2}\pi^{1/2}\xi(bt) \qquad (2.143)$$

where $\xi(bt)$ is another tabulated function. The peak current is given by:

$$I_{p,c} = -2.99 \times 10^5 \beta^{1/2} A[O^*]D_O^{1/2}v^{1/2} \qquad (2.144)$$

and the potential at which this peak current occurs is:

$$E_p = E^0 - (RT/\beta F)[0.78 + \log_e(D_O^{1/2}/k^0) + \log_e(\beta Fv/Rt)^{1/2}] \qquad (2.145)$$

Equations (2.144) and (2.145) are extremely interesting as they show that the *magnitude* of the peak current is independent of the kinetics of the reaction. This is not surprising since from equations (2.139) and (2.141) it can be seen that k_c can be increased to any value simply by increasing the potential. However, at a given scan rate, the *position* at which the peak current occurs is related to the kinetics; E_p occurs beyond E^0 by an activation overpotential related to k^0.

For such an irreversible reaction, it can be shown that:

$$|E_p - E_{p/2}| = 1.857RT/\beta F \tag{2.146}$$

and β can thus be obtained from the separation of peak potential and half-peak potential.

The above treatment is somewhat simplified in that it only considers single-step electron transfer. In practice, it is often found in both reversible and irreversible systems that the electrons are transferred in more than one step. Fortunately, these systems commonly fall into one of two categories:

1. All the steps are facile (in both forward and reverse reactions) and the reaction behaves reversibly; in which case the Nernst equation holds, as does the treatment given above.
2. One step is very much slower than the rest and so is the rate determining step (rds); in such a reaction, kinetic treatments can be applied that are built around the rate of this rds, as explained in chapter 1.

In the case of an irreversible reaction, therefore, two parameters are now defined with respect to the number of electrons transferred in the reaction: n now refers to the electrons transferred overall, while n_a indicates the number of electrons participating in the rds. Thus, for example, we can rewrite equations (2.144) and (2.146) as:

$$I_{p,c} = 2.99 \times 10^5 n(\beta n_a)^{1/2} A[O^*]D_O^{1/2} v^{1/2} \tag{2.144}$$

$$|E_p - E_{p/2}| = 1.857RT/\beta n_a F \tag{2.146}$$

Normally we expect n_a to be 1 but, as was shown in the first chapter, β can now take on values in excess of unity. Determination of β in this way is certainly possible, though in fact it is customary to measure I vs. E plots under stationary conditions, for reasons that will become apparent later.

Thus, cyclic or linear sweep voltammetry can be used to indicate whether a reaction occurs, at what potential and may indicate, for reversible processes, the number of electrons taking part overall. In addition, for an irreversible reaction, the kinetic parameters n_a and β can be obtained. However, LSV and CV are dynamic techniques and cannot give any information about the kinetics of a typical static electrochemical reaction at a given potential. This is possible in chronoamperometry and chronocoulometry over short periods by applying the Butler–Volmer equations, i.e. while the reaction is still under diffusion control. However, after a very short time such factors as thermal convection render kinetic treatments founded on simple diffusion untenable. In order to quantify both the kinetics of an electrode reaction and the role of mass transport, a technique must be employed that allows the measurement of the current under mass-transport conditions that are both controllable and quantifiable. Such a method is termed a controlled-convection technique.

2.2.3   The controlled-convection techniques: the rotating disc and rotating ring-disc electrodes

There are many controlled-convection techniques available but we will restrict our discussion to the two most commonly employed by the electrochemist; the rotating disc electrode (RDE) and the rotating ring disc electrode (RRDE).

Up until the mid-1940s, most physical electrochemistry was based around the dropping mercury electrode. However, in 1942, Levich showed that rotating a disc-shaped electrode in a liquid renders it uniformly accessible to diffusion, yet the hydrodynamics of the liquid flow are soluble and the kinetic equations relatively simple. In addition, in contrast to the case of a stationary planar electrode, the current at an RDE rapidly attains a steady-state value.

Assuming laminar flow to the rotating electrode surface, i.e. the rotation rate is not sufficiently fast to cause turbulence, Karmàn and Cochran solved the pattern of flow and showed that the streamlines are as in Figure 2.89. The rotating disc sucks solution up towards it from the bulk and flings it out centrifugally. The layer of liquid immediately adjacent to the surface is

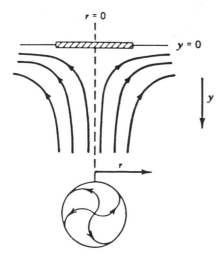

Figure 2.89   The pattern of flow to a rotating disc electrode and across its surface, assuming laminar flow.

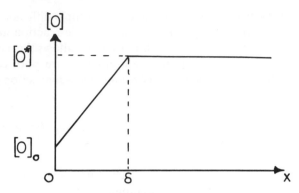

Figure 2.90   The variation of the concentration of a species O with distance x from the electrode surface, for the reduction process $O + e^- \rightarrow R$. $[O^*]$ is the concentration of O in the bulk of the solution and $[O]_0$ is the concentration at the electrode surface. δ is the diffusion layer thickness.

stagnant with respect to movement towards the disc, and rotates with it at the same angular velocity. As a result of the rapid flow of the electroactive species from the bulk of the solution to the edge of this boundary layer, the electroactive species is found at its bulk concentration at all distances greater than δ from the electrode surface. Thus, δ effectively divides the solution into a well-stirred region, where mass transport is via convection and the stagnant diffusion layer (see Figure 2.90). Solution of the convection–diffusion equations gives δ as:

$$\delta = [1.61v^{1/6}D^{1/3}]\omega^{-1/2} \tag{2.147}$$

where ω is the rotation speed in radians s^{-1} ($\omega = 2\pi f$, where f is the rotation speed in revolutions s^{-1}) and v is the kinematic viscosity ($v = $ viscosity/density). From the equation, it can be seen that the thickness of the diffusion-layer can be controlled very precisely simply by altering the rotation speed of the RDE.

A typical RDE response for a reversible system,

$$O + e^- \longleftrightarrow R$$

is shown in Figure 2.91. We can qualitatively describe the events resulting in the observed I/E response by considering the current to be *determined by a combination of the transport of molecules to the electrode and the kinetics of its conversion at the electrode.* We can express this in terms of the rate equations appropriate to the Butler–Volmer treatment. We saw above that the current I can be expressed as:

$$I/FA = k_a[R]_0 - k_c[O]_0 \tag{2.148}$$

where (see chapter 1):

$$k_a = k^0 \exp[(1 - \beta)f(E - E^0)] \tag{2.149}$$

$$k_c = k^0 \exp[-\beta f(E - E^0)] \tag{2.150}$$

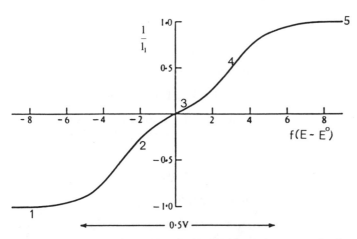

Figure 2.91 Schematic representation of the rotating disc electrode response for the reduction and oxidation of a reversible couple. $f = F/RT$, I_1 is the limiting current, I is the current, E is the potential of the electrode and E^0 is the standard reduction potential of the couple.

and E is the potential on the electrode. Under the steady-state conditions prevalent at an RDE, we can write:

Rate of loss of O at electrode = flux of O at electrode surface

and:

Rate of loss of R at electrode = flux of R at electrode surface

i.e.:

$$-I_c/FA = k_c[O]_0 - k_a[R]_0 = D_0([O^*] - [O]_0)/\delta \qquad (2.151)$$

$$I_a/FA = k_a[R]_0 - k_c[O]_0 = D_0([R^*] - [R]_0)/\delta \qquad (2.152)$$

The discussion can only now proceed if one of the reactions can be taken as being very slow, e.g. if we are in regions 1 or 2 in Figure 2.91, such that k_a is very small in equation (2.151), then equation (2.151) becomes:

$$-I_c/FA = k_c[O]_0 = D_0([O^*] - [O]_0)/\delta \qquad (2.153)$$

and we can write:

$$k_c[O]_0 = k_{m,c}([O^*] - [O]_0) \qquad (2.154)$$

where, for convenience:

$$k_{m,c} = D_0/\delta \qquad (2.155)$$

is the mass transfer constant for the cathodic reaction. Rearranging equation (2.154) we obtain:

$$[O]_0 = k_{m,c}[O^*]/(k_{m,c} + k_c) \qquad (2.156)$$

since $-I_c/FA = k_c[O]_0$ then, from equation (2.154):

$$-I_c/FA = k_c k_{m,c}[O^*]/(k_c + k_{m,c})$$

or:

$$-1/I_c = 1/(FAk_{m,c}[O^*]) + 1/(FAk_c[O^*]) \tag{2.157}$$

From equations (2.153) and (2.154), we have:

$$-I_c/FA = k_{m,c}([O^*] - [O]_0) \tag{2.158}$$

The limiting cathodic current, region 1 in Figure 2.91, corresponds to values of k_c so large that $[O]_0 \ll [O^*]$. Hence, the cathodic limiting current, $I_{1,c}$, is given by:

$$-I_{1,c} = FAk_{m,c}[O^*] \tag{2.159}$$

and we can rewrite equation (2.157) as:

$$-1/I_c = -1/I_{1,c} + 1/(FAk_c[O^*]) \tag{2.160}$$

Replacing $FAk_c(O^*)$ by $-I_k$, the kinetic contribution to the overall (observed) current, then equation (2.160) becomes:

$$1/I_c = 1/I_{1,c} + 1/I_k \tag{2.161}$$

Equation (2.161) expresses the relative contributions of mass transport and kinetics to the observed current and is one expression of the Koutecky–Levich equation.

In order to turn equation (2.161) back into a useable form we replace the I terms by the relevant expressions, $k_{m,c}$ by D_0/δ and δ by equation (2.147), to obtain:

$$-1/I_c = 1/(0.62FAD_0^{2/3}\omega^{1/2}v^{-1/6}[O^*]) + 1/(FAk_c[O^*]) \tag{2.162}$$

Thus, at potentials in the range where mass transfer *and* electrokinetic effects are significant (region 2 of Figure 2.91), a plot of $1/I_c$ vs. $1/\omega^{1/2}$ should be linear, D_0 available from the slope and k_c available from the intercept. As a result of the exponential dependence of the rate constant on potential, the actual value of the intercept is strongly dependent on potential.

$$k_c = k^0 \exp[-\beta f(E - E^0)] \tag{2.150}$$

A plot of $\log_e k_c$ (obtained from the intercept) vs. the potential at which the measurements were obtained is a form of Tafel plot; the plot should be linear, with slope $\beta F/RT$ and intercept k^0.

If k_c is very large such that we are in the mass transport-limited region of the current and the second term in equation (2.162) $\rightarrow 0$, then the observed current is its limiting value:

$$I_c = I_{1,c} = -0.62FAD_0^{2/3}\omega^{1/2}v^{-1/6}[O^*] \tag{2.163}$$

which is the Levich equation. A plot of current vs. $\omega^{1/2}$ that is linear and passes through the origin is *an excellent test that I is diffusion controlled*. The slope gives the diffusion coefficient.

In contrast to voltammetry, reversal techniques are not available with the

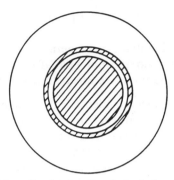

Figure 2.92 Schematic of a ring-disc electrode (see text for details).

RDE as the products of the electrode reaction are swept away from the disc. However, it is possible to intercept solution-borne intermediates or products in their flight away from the disc, by using the rotating ring-disc electrode (RRDE), see Figure 2.92. Information equivalent to that obtained by reversal techniques at static electrodes can be deduced by the addition of an independently-controlled ring electrode surrounding the disc. For instance, by maintaining the potential of the ring at some value where any intermediates or products (produced at the disc) undergo electron transfer, some knowledge of the events at the disc electrode surface can be obtained.

The essentials of RRDE operation are effectively as for the RDE with the addition of one extremely important parameter: the collection efficiency, N, defined for the reversible reaction:

$$O + e^- \longleftrightarrow R$$

as:

$$N = -I_R/I_D \qquad (2.164)$$

where I_R and I_D are the ring- and disc-currents and the negative sign indicates that the currents are opposed.

If O is being reduced at the disc, even if the ring is set at a potential at which the limiting current for the re-oxidation of R occurs, not all of the reduced form produced at the disc will reach the ring surface. For a given system, the fraction of R collected will depend on the geometry of the disc and ring only. The value of N is actually given by the Albery–Bruckenstein equation (see Albery and Hitchman (1971), p. 22), though, in practice, for a particular RRDE cell system, N is determined using a standard redox couple, such as $Fe(CN)_6^{4-/3-}$ or Br_2/Br^-. A value of ≈ 0.3 is common for N.

There are three main approaches employed in RRDE experiments:

1. The disc is held at a potential where the reaction of interest takes place, and an I/E curve is then recorded at the ring. This allows the identification of any solution-free intermediates and/or products.
2. An I/E curve is recorded at the disc while the ring potential is held at a constant value where the intermediates or products are reduced. This

allows the identification of the exact potential range over which they are formed.

3. The disc is held at a potential where intermediates or products are formed and the ring is maintained at a potential at which they undergo electron transfer. This allows quantitative kinetic measurements to be obtained.

The rotation speed provides an independent variable; increasing ω decreases the time taken for a species to travel between the disc and the ring and hence increases the likelihood of detecting a short-lived intermediate. All of the above methods critically depend on the species generated at the electrode being active towards further electron transfer at the ring.

An informative example of the application of the RRDE concerns the study of the reduction of O_2 at a platinum disc electrode and the detection of the intermediates and/or products at the Pt ring. The technological importance of oxygen reduction in such devices as fuel cells, metal–air batteries, etc. has led to an extensive investigation of the reduction mechanism and particularly of the role of hydrogen peroxide as an intermediate. The analysis of kinetic data has largely been concerned with this aspect of oxygen reduction rather than with any derivation of *molecular* mechanisms, the latter being still at an early stage of investigation. For an extensive review see Paliteiro (1985).

Any description of the reduction of oxygen must be concerned with the following processes:

(a) O_2 transport of the electrode with some rate constant $k_{m,c}$.
(b) O_2 adsorption on the surface and subsequent reduction to:
   (i) water directly by a 4-electron process with rate constant k_1 and/or
   (ii) hydrogen peroxide with rate constant k_2.
(c) Desorption of hydrogen peroxide from the surface with rate constant k_5 and subsequent adsorption with rate constant k_6.
(d) Transport of hydrogen peroxide away from the surface with rate $k'_{m,c}$ and/or
(e) Reduction of hydrogen peroxide to water with rate constant k_3 and/or
(f) Disproportionation of hydrogen peroxide with rate constant k_4 to give O_2 and water, this latter reaction being catalysed by the electrode surface.

These processes can be summarised as:

$$
\begin{array}{ccccccc}
 & & & & [H_2O_2]_0 & \xrightarrow{k'_{m,c}} & \text{ring} \\
 & & & & \uparrow\downarrow{\scriptstyle k_5\;k_6} & & \\
 & & \overset{k_1}{\overbrace{\hspace{3cm}}} & & & & \\
[O_2]^* & \underset{k_{m,c}}{\rightleftharpoons} & [O_2]_0 & \underset{k_{2b}}{\overset{k_2}{\rightleftharpoons}} & [H_2O_2]_{ads} & \xrightarrow{k_3} & H_2O \\
 & & & & \underset{k_4}{\underbrace{\hspace{3cm}}} & &
\end{array}
$$

Distinguishing between these processes has primarily been achieved with the rotating-ring-disk electrode. In O_2-saturated solutions, a positive potential, high enough to oxidise any H_2O_2 formed at the disc, is applied to the ring and the concentration of H_2O_2 in the solution bulk is kept to a minimum. The potential on the disc is then scanned cathodically and the ring and disc currents simultaneously monitored. A second type of experiment can also be carried out in O_2-free solutions containing H_2O_2. Now the ring potential is fixed so that all O_2 molecules arriving at the ring are *reduced*. Thus, any O_2 formed at the disc by disproportionation will be reduced at the ring and a current seen.

From the discussion above (equation (2.150) onwards), the flux of O_2 to the disc electrode is given by:

$$\mathbf{J}_{O_2} = D_{O_2}([O_2^*] - [O_2]_0)/\delta \equiv k_{m,c}([O_2^*] - [O_2]_0)$$

Clearly, in the limit of fast electrochemical reaction, $[O_2]_0 \to 0$, and the transport rate will become potential independent, having the value $k_{m,c}[O_2^*]$. Transport of H_2O_2 to the ring from the disc will take place according to the model above. If the collection efficiency is N and the concentration of H_2O_2 in solution is zero, the *flux* of H_2O_2 away from the surface will be $k'_{m,c}[H_2O_2]_0$, where $k'_{m,c}$ is the mass transfer constant for H_2O_2 and is different from that for O_2 through a different diffusion coefficient. The total amount of H_2O_2 transported away from the disc per unit time is then $A k'_{m,c}[H_2O_2]_0$, where A is the disc area, and of this a fraction N is detected at the ring. If the ring current is I_R, then this is related to the concentration of H_2O_2 at the surface by the equation

$$I_R = 2FANk'_{m,c}[H_2O_2]_0$$

To see how this approach works, consider the scheme above. If a steady state can be established, then we can analyse the various components of the kinetic scheme by assuming in each case that the rate of transport to or from the electrode must equal the nett sum of all the possible reactions.

Consider first the concentration of O_2 at the electrode surface. Then assuming that the rate of adsorption of O_2 is always fast:

rate of arrival of O_2 + rate of re-oxidation of adsorbed H_2O_2

+ rate of production of O_2 through disproportionation

= total rate of reduction of O_2

or:

$$k_{m,c}([O_2^*] - [O_2]_0) + k_{2b}[H_2O_2]_{ads} + \tfrac{1}{2}k_4[H_2O_2]_{ads} = (k_1 + k_2)[O_2]_0 \quad (1)$$

where the factor of 1/2 arises from the fact that each disproportionation step will yield only a half a molecule of O_2, as:

$$H_2O_2 \longrightarrow H_2O + \tfrac{1}{2}O_2$$

Similarly, we can write down equations for $[H_2O_2]_{ads}$:

$$k_2[O_2]_0 = (k_{2b} + k_3 + k_4 + k_5) \cdot [H_2O_2]_{ads} - k_6[H_2O_2]_0 \qquad (2)$$

and finally, for the concentration of H_2O_2 at the electrode surface,

$$k_5[H_2O_2]_{ads} = k_6[H_2O_2]_0 + k'_{m,c}[H_2O_2]_0 \qquad (3)$$

Now, the disc current actually measured for the model above will be

$$I_D = 4FAk_1[O_2]_0 + 2FAk_2[O_2]_0 + 2FA(k_3 - k_{2b})[H_2O_2]_{ads} \qquad (4)$$

where the first two terms refer to the reduction of oxygen and the last term to the nett current for reduction and re-oxidation of adsorbed hydrogen peroxide.

The ring current will be related to the surface concentration of H_2O_2 by the equation

$$I_R = 2FANk'_{m,c}[H_2O_2]_0 \qquad (5)$$

as above, and combining equations (1) to (5) we obtain

$$NI_D/I_R = \{1 + 2k_1/k_2 + X\} + \{k_6 X/Z'\} \cdot \omega^{-1/2} \equiv Y_1 + S_1 \cdot \omega^{-1/2} \qquad (6)$$

where

$$X = 2k_1(k_{2b} + k_3 + k_4)/k_2 k_5 + (2k_3 + k_4)/k_5 \qquad (7)$$

and we have written $k'_{m,c} \equiv Z'\omega^{1/2}$ to expressly indicate the way in which $k'_{m,c}$ varies with rotation speed.

Clearly a plot of NI_D/I_R vs. $\omega^{-1/2}$ will allow us to make some deductions. If, for example, $k_1 = 0$; i.e. there is no direct four-electron reduction pathway for O_2 at the electrode, and we obtain from equation 6:

$$(1 + X) + \{k_6 X/Z'\} \cdot \omega^{-1/2} \equiv Y_1 + S_1 \cdot \omega^{-1/2}. \qquad (8)$$

Comparing the two sides of this equation, it is clear that

$$Y_1 = 1 + X$$

and:

$$S_1 = \{k_6 X/Z'\}$$

hence

$$Y_1 = 1 + Z'S_1/k_6$$

Given that we expect k_6 to vary little with potential, a plot of Y_1 vs. S_1 measured at different potentials should be linear with intercept of unity. However, if $k_1 \neq 0$, then a plot of Y_1 vs. S_1 will still be linear if the potential dependence of k_1 and k_2 happens to be the same, though the intercept will certainly differ from unity. If the potential dependence of k_1 and k_2 differs, then the plot of Y_1 vs. S_1 will be non-linear.

To summarise, for the scheme above, analysis of the ring and disc currents allows us unequivocally to establish whether or not a four-electron process is in operation. The analysis does *not* tell us anything about the molecular species involved in the reduction scheme, with the exception of those species such as H_2O_2 that are sufficiently stable to escape from the disc and reach the ring. More complex schemes can be considered with the same basic approach, though it will be appreciated that there is an important limit to the amount of unambiguous information that can be obtained for more complex models.

2.2.4   *Electron paramagnetic resonance*

Electron paramagnetic resonance (epr) spectroscopic methods are used for the detection and identification of species that have a nett electronic spin: radicals, radical ions, etc. It is extremely sensitive, capable of detecting species down to concentration levels of 1×10^{-12} moles dm^{-3}, and produces spectra that are distinctive and generally easily interpreted. Consequently, the technique has found extensive application in electrochemistry since the late 1950s. In order to understand epr, it may be helpful to review some fundamental concepts.

The origins of epr can be traced back to the realisation that a spinning electron acts as a magnet, and may not take up an arbitrary orientation in an applied magnetic field. An elegant demonstration of these facts was provided by the Stern–Gerlach experiment in 1921 in which a low-energy beam of silver atoms (having a lone electron in their outermost 5s orbital) was fired between the poles of an inhomogeneous magnet. In classical theory, spin angular momentum is unquantised and a spinning magnet can assume any orientation in a magnetic field. Consequently, the atoms emerging from the magnet should strike the photographic strip used as a detector in a disordered broad band. In contrast, the observed result was two discrete bands corresponding to only two discrete orientations in the field, a phenomenon called *space quantisation*.

In the absence of an applied magnetic field, the spin vector of the electron *can* be in any direction. The *magnitude* of the spin vector is:

$$S = [1/2(1/2 + 1)]^{1/2} \qquad (2.165)$$

In the presence of a magnetic field, the energy of interaction between the magnetic moment, μ, of the electron and the field **B** is:

$$E = -\mu\mathbf{B} \qquad (2.166)$$

The magnetic moment of the electron can be expressed as:

$$\mu = -ge_0h\mathbf{S}/4\pi m_0 c \qquad (2.167)$$

where g is a dimensionless constant known as the g-factor. The g-factor depends on the molecule under study; for a free electron $g = 2.0023$ and for many radicals g lies in the range 1.9 to 2.1. **S** represents the direction and

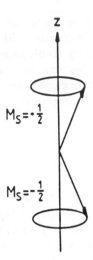

Figure 2.93 The precession of the spin vector S around a magnetic field. The component of the vector in the z direction, the direction of the field, is M_S.

magnitude of the spin angular momentum. Equation (2.167) is more usually expressed in terms of the Bohr magneton, μ_B:

$$\mu = - g\mu_B S \qquad (2.168)$$

with $\mu_B = - e_0 \hbar/2 m_0 c$.

In the presence of the applied field, **B**, the spin vector precesses around the direction of the field, taken by convention as the z-direction (see Figure 2.93). This precession is quantised such that the component of S in the z-direction, S_Z, can only take one of two values, $+1/2$ or $-1/2$, and the quantum number M_S is used to label the allowed values of S_Z. Thus:

$$S_Z = M_S = \pm 1/2 \qquad (2.169)$$

Under these conditions, the energy of interaction of the field with the spin, from equations (2.166) to (2.168), is given by the scalar product:

$$E = M_S g\mu_B B \qquad (2.170)$$

where B is the magnitude of the magnetic field. Or:

$$E = \pm g\mu_B B/2 \qquad (2.171)$$

Thus the energy of interaction is confined to two values, depending on whether the spin is oriented with its magnetic moment parallel to the field, $M_S = -1/2$, or antiparallel with $M_S = +1/2$. These orientations of the spin are commonly labelled as α and β spins, respectively. At zero field, these two spins are degenerate, however, once the field is switched on, they separate in energy as shown in Figure 2.94 with increasing field strength. The separation

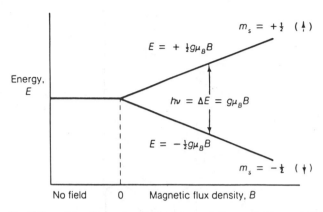

Figure 2.94 The lifting of the degeneracy of the $M_S = -1/2$, α, and $M_S = +1/2$, β, spin states on the application of a magnetic field of flux density B. g is the dimensionless g-factor, $\mu_B = -e_0\hbar/2m_0c$.

in energy, ΔE, of the α and β levels is given by:

$$\Delta E = g\mu_B B/2 - (-g\mu_B B/2)$$

$$\Delta E = g\mu_B B \tag{2.172}$$

If the sample is bathed in electromagnetic radiation of frequency v, then a transition between the two spin states, corresponding to the electron changing its component of spin, can take place providing the resonance condition:

$$hv = g\mu_B B \tag{2.173}$$

is fulfilled.

As was stated above, a typical value of g is c. 2, the Bohr magneton has a value of $9.02732 \times 10^{-24} \, m^2 \, A$, and a typical applied field is 0.34 T. This gives a frequency of 9.4×10^9 Hz, i.e. in the microwave region of the spectrum, with a wavelength of 3 cm.

From equation (2.173) it can be seen that there are two possible methods of obtaining epr spectra: to monitor the absorbed light intensity as a function of its frequency at fixed field strength, or as a function of the field strength at fixed microwave frequency. In practice, for technical reasons the latter approach is always employed.

The basic features of an epr spectrometer are shown in Figure 2.95. The microwave source is a Klystron tube that emits radiation of frequency determined by the voltage across the tube. Magnetic fields of 0.1–1 T can be routinely obtained without complicated equipment and are generated by an electromagnet. The field is usually modulated at a frequency of 100 kHz and the corresponding in-phase component of the absorption monitored via a phase-sensitive lock-in detector. This minimises noise and enhances the sensitivity of the technique. It is responsible for the distinctive derivative nature of epr spectra. Thus, the spectrum is obtained as a plot of dA/dB vs.

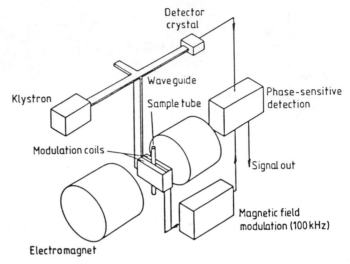

Figure 2.95   Basic features of an epr spectrometer. From *Comprehensive Chemical Kinetics*, **29** (1989), 297.

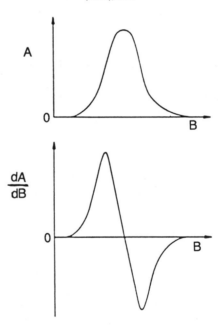

Figure 2.96   The form of the absorption spectrum (top) and first derivative spectrum (bottom) observed in epr.

B, where A is the absorbance as shown in Figure 2.96. The microwaves emitted by the Klystron tube travel down a waveguide to the epr cavity, in which is positioned the electrochemical cell. The characteristics of the waveform set up in the cavity are a function of its geometry. For the flat, rectangular cavity commonly used, the standing wave has its electric and magnetic components spatially separated. The magnetic intensity is at a maximum and the electric intensity at a minimum in the centre of the cavity. This has the great advantage that aqueous samples can be investigated if the electrochemical cell is carefully designed and accurately positioned. If this is the case, then considerable absorption of the microwave radiation by water can be minimised. The distribution of the magnetic field within the cavity is such that the sensitivity varies with the position of the absorbing species, the maximum sensitivity occurring at the cavity centre.

The electrochemical cells employed in epr measurements are generally fabricated out of silica since glass contains paramagnetic impurities that can interfere with the signal. The need to minimise the interaction of the sample with the electric component of the radiation in the cavity, and the geometry and size of the cavity itself, have generally led to electrochemists employing flat cells that are $c. < 0.5$ mm thick.

The information that can be obtained from an epr spectrum can be divided into three forms: (i) the g-factor, (ii) the hyperfine splitting of the spectra due to the interaction of the spin with magnetic nuclei in the radical and (iii) the shape of the observed bands.

2.2.4.1 The g-factor.   The g-factor takes into account the fact that the local magnetic field experienced by a particular atom in a molecule may not be the same as the applicd field owing to the existence of local field effects. In the absence of such effects, g for any particular radical would simply have the same value as that of the free electron, 2.0023, and all radicals would come into resonance at the same applied field for a given microwave frequency. We can thus express the resonance condition (equation 2.173) as:

$$hv = 2.0023\mu_B B_{local} \tag{2.174}$$

where B_{local} is the local field experienced by the lone electron and can be given in the form:

$$B_{local} = B - \sigma B$$

where σ is a constant dependent on the molecular structure and constitution. Thus:

$$hv = 2.0023(1 - \sigma)\mu_B B$$

where $g = 2.0023(1 - \sigma)$.

The g-factor also depends on the orientation of the molecule in the field. Fortunately, in solution, rapid tumbling gives rise to an average g factor and therefore to sharp epr signals. It also depends on the electronic structure of the molecule and the measurement of g can give some such information. However, it is more commonly employed as an aid to the identification of the radical species.

2.2.4.2 Hyperfine splitting. As was discussed above, one consequence of placing a free electron onto a molecule is to alter its *g*-value. Another is that the electron spin comes under the influence of any magnetic *nuclei* present in the radical, with the result that the spectrum is split into a number of lines centred on the position of the single resonance expected for the simple β–α transition discussed above. This hyperfine structure is the most useful characteristic of epr spectra in the identification of an unknown radical species.

Hyperfine structure arises through the interaction of the electron spin with a nuclear spin. Consider first the interaction of the electron spin with a single magnetic nucleus of spin *I*. In an applied magnetic field the nuclear spin angular momentum vector, of magnitude $(I(I+1))^{1/2}$, precesses around the direction of the field in an exactly analogous way to that of the electron spin. The orientations that the nuclear spin can take up are those for which the spin in the z-direction, M_I, has components of:

$$M_I = I, (I-1), (I-2)\ldots\cdots - I\cdot\mu_B$$

Hence, the local field experienced by the lone electron now depends not only on the magnitude of the applied magnetic field but also on the magnitude and direction of the magnetic moment of the magnetic nucleus.

The energy levels for a one-electron one-magnetic nucleus system are given by:

$$E = M_S[g\mu_B B + aM_I]$$

$$E = \pm(1/2)\cdot[g\mu_B B + aM_I] \tag{2.175}$$

where *a* is the hyperfine coupling constant for the magnetic nucleus. Thus, for $I = 3/2$, the energies of the various possible levels are:

$$E = \pm(1/2)\cdot[g\mu_B B + a/2] \quad \text{and} \quad \pm(1/2)\cdot[g\mu_B B + 3a/2] \tag{2.176}$$

as shown in Figure 2.97. Providing that when the electron changes its spin the nuclear spin is unaffected (i.e. that the transitions are first-order), then the selection rules for the allowed transitions are $\Delta M_I = 0$ and $\Delta M_S = 1$. These selection rules give rise to the transitions shown in the figure.

The separation in energy of the levels involved in the allowed transitions can be obtained from equation (2.175), remembering that $\Delta M_I = 0$ and $\Delta M_S = 1$, thus:

$$E_\alpha - E_\beta = [g\mu_B B + aM_I]/2 - \{-[g\mu_B B + aM_I]/2\}$$

$$\Delta E = h\nu = [g\mu_B B + aM_I] \tag{2.177}$$

rearranging gives:

$$B = [h\nu - aM_I]/g\mu_B$$

the transitions thus occur (see Figure 2.97) at:

$$B = [h\nu - 3a/2]/g\mu_B, [h\nu - a/2]/g\mu_B, [h\nu + a/2]/g\mu_B \quad \text{and} \quad [h\nu + 3a/2]/g\mu_B.$$

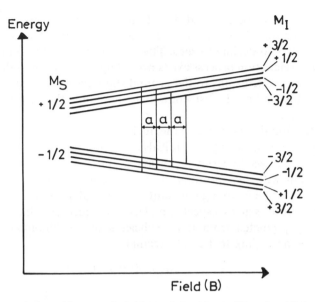

Figure 2.97   The variation with magnetic field, B, of the $M_S = +1/2$ and $-1/2$ levels in the presence of a single magnetic nucleus having $I = 3/2$. The allowed transitions have $\Delta M_I = 0$.

In the centre of the hyperfine structure:

$$B = h\nu/g\mu_B$$

from which the g-factor can also be calculated.

When there are several magnetic nuclei present in the radical, the extension of the argument is straightforward. Each nucleus contributes to the splitting of the spectrum and the α and β levels of the unpaired electron are accordingly split by all the nuclei, with the energies of the resultant levels given by:

$$E = M_S[g\mu_B B + a_1 M_{I,1} + a_2 M_{I,2} +, a_3 M_{I,3} + \cdots] \qquad (2.178)$$

where a_i is the hyperfine coupling constant of nucleus i, and $M_{I,i}$ its corresponding nuclear spin in the direction of the field. Thus, for an electron spin interacting with two nuclei, of spin $I = 1/2$ and 1, the hyperfine structure is composed of a triplet of doublets. The triplet arises from hyperfine interaction with the $I = 1$ nucleus and the doublet is due to the interaction with the $I = 1/2$ nucleus.

At this point, some mention of the intensity of the lines observed in epr spectra should be made. In the case of a radical having no magnetic nuclei the difference in the energy of the α and β levels is of the order of $0.3\,\text{cm}^{-1}$. At room temperature, kT is $c.\ 200\,\text{cm}^{-1}$ and the Boltzmann equation thus gives the ratio of the number of radicals in the lower state to those in the upper, N_β/N_α, as 1.0015. The absorption intensity is proportional to this difference and as a consequence epr absorptions are very weak. The phase-sensitive detection employed in epr spectrometers is thus a necessity if these weak absorptions are to be detected. In addition, it may be expected that such a small population difference would soon be equalised and thus the net absorption would decline to zero. The reason that absorption is maintained at all that a relaxation process exists providing a radiationless decay pathway to return the excited spins to the ground state: the excess energy is then lost as heat. This process is fast enough to maintain the original population distribution.

It is often found that several of the nuclei in an organic radical have identical coupling constants, e.g. organic radicals containing equivalent protons ($I = 1/2$). Figure 2.98 shows hyperfine splitting patterns for the interaction of an electron spin with two protons of (a) different hyperfine coupling and (b) identical hyperfine coupling constants. In general, a radical that contains N equivalent protons gives a spectrum of $(N + 1)$ equivalent lines whose intensities can be predicted from the coefficients of the binomial expansion of $(1 + x)^N$, i.e. according to Pascal's triangle:

N	Intensity distribution
0	1
1	1 1
2	1 2 1
3	1 3 3 1
4	1 4 6 4 1

etc.

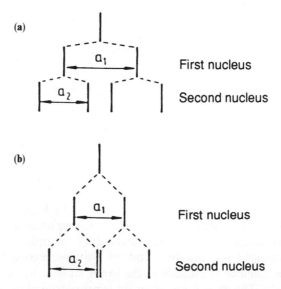

Figure 2.98   The hyperfine splitting patterns resulting from the interaction of an electron spin with two protons with (a) different hyperfine coupling constants and (b) the same hyperfine coupling constant.

It follows that the hyperfine structure is characteristic of the molecular symmetry and of great value in identifying the parent radical species.

2.2.4.3 Lineshape. The two most frequently encountered lineshapes are Lorentzian and Gaussian; the latter generally arises from a large number of unresolved hyperfine splittings. In the absence of such complications, epr lines in solution almost always show a Lorentzian lineshape which is defined in the form:

$$\Upsilon = (\Gamma/\pi) \cdot (\Gamma^2 + [B - B_r]^2)^{-1} \qquad (2.179)$$

where Υ is the shape of the line, B_r is the magnitude of the magnetic field at resonance, and Γ is given by:

$$\Gamma = (-2m_0/e_0)([1/2\tau_1] + [1/\tau_2]) \qquad (2.180)$$

where τ_1 and τ_2 are relaxation times:

- τ_1 is the spin–lattice (or longitudinal) relaxation time. Spin–lattice relaxation is a true energy-relaxation process providing a means by which the excited spins can return to the ground state, and so prevent the epr signal declining to zero (see above). This relaxation is a result of time-dependent fluctuations in the magnetic or electric fields at the unpaired electron, arising from random thermal motions in solution. These random field fluctuations can stimulate the electron to flip to the lower state and transfer the excess energy to the lattice.
- τ_2 is the spin–spin (or transverse) relaxation time and unlike the longitudinal process, is not a true energy relaxation; it does not return spins to the ground state. It is termed a relaxation process by virtue of the fact that it leads to a broadening in the epr line. Spin–lattice relaxation broadens the epr line as a result of shortening the lifetime of the excited spin levels. This leads to the energy of these levels being rendered more imprecise and *lifetime broadening* results. The energy spread of a state is given by:

$$\Delta E = h/2\pi\tau \qquad (2.181)$$

where τ is the lifetime of the level. If τ decreases, ΔE clearly then increases. Spin-spin relaxation, on the other hand, produces broadening as a result of the small differences in local field experienced by a rotating molecule. This leads to resonance over a spread of fields and broadening of the line.

At high concentrations of parent molecule or radical, a third form of broadening may be observed as a result of the exchange of spin between colliding species:

$$X_a + X_b^{\cdot} \longrightarrow X_a^{\cdot} + X_b$$
$$X_a(M_S = +1/2) + X_b(M_S = -1/2) \longrightarrow X_a(M_S = -1/2) + X_b(M_S = +1/2)$$

The real potential of epr studies in electrochemistry was demonstrated by the work of Maki and Geske (1959) who employed a two-electrode cell based around a 3 mm diameter epr capillary sample tube. The counter electrode was separated from the working electrode by a sinter, the latter being inside, and the former outside, the epr cavity. The authors studied the radicals formed from the oxidation of ClO_4^- and the reduction of nitrobenzenes and dinitro-benzenes in non-aqueous media.

An optimised version of the Maiki and Geske cell was developed by Piette, Ludwig and Adams (1962) and versions of this cell are still widely used. The cell employed the standard three-electrode configuration allowing control by potentiostatic means; large numbers of radicals could be produced via electrolysis at a large surface area Pt gauze working electrode. It is the ease and versatility with which radicals can be produced by electrochemical means that has ensured the success of the Adams cell. However, while the Adams cell is ideal for the spectroscopist who wishes only to identify the radical species, it is of very little use in obtaining kinetic and/or mechanistic information on the pathways by which it subsequently decays. This is the result of several factors, including the presence of thermal convection in the cell which causes swirling motions and convection patterns leading to ill-defined mass transport to and from the electrode surface. In addition, the thinness of the cell gives rise to a large solution resistance in the cell, leading to distortion and uncertainty in the current/potential curves.

The above factors impose a severe limitation on the use of *in situ* electrochemical epr as a possible means of establishing the kinetics and mechanism of radical decay. As a consequence, a great deal of effort has been expended in trying to improve the electrochemical behaviour of the epr cell and to design a system that allows the lifetimes and kinetic modes of radical decay to be determined, as well as the identity of the radical. Up until recently these objectives appeared mutually exclusive and led to two alternative methodologies:

1. *Ex-situ generation.* In this approach the radicals are generated away from the epr cavity under well-defined voltammetric conditions and then transferred into the epr cell (usually via a flow system). Such a method imposes a restriction on the radical lifetimes observable due to the dead time involved between generation and detection.
2. *In situ generation.* In this technique, well-defined electrochemistry is sacrificed in favour of generating the radicals *in situ* within the cavity. The advantage is that more shorter-lived radicals can be observed than is possible with the *ex situ* methods.

In the mid-1980s *in situ* cells were produced that showed satisfactory electrochemical behaviour, so-called simultaneous electrochemical and electron spin resonance (SEESR) techniques, and the *ex situ/in situ* division disappeared. A division that has remained, however, is the separation of epr approaches into those dedicated to the detection and identification of short-lived radicals and those that are capable of studying the kinetics of the decay of radicals, as well as obtaining their identity. The latter techniques are not capable of studying radicals having the very short lifetimes that can be studied by the former. It is not within the scope of this book to give an exhaustive history of the development of electrochemical epr; the reader is referred to any one of several excellent reviews for such a treatment (see the further reading section at end of this chapter). Here, we concentrate on the modern methods that have been successfully applied *in situ*.

The most sensitive *in situ* epr approaches for the detection of short-lived radicals are those developed by Allendoerfer and colleagues (1973). The cells were designed to fit into epr spectrometers having a cylindrical cavity and were based around a quartz tube of internal diameter 6 mm. Into this is inserted a platinum wire wound into the form of a shallow-pitched helix, giving a large working electrode surface area of 20–30 cm^2; this large area results in large amounts of radicals being generated and thus increases the sensitivity. The helical working electrode fits snugly into the quartz cell trapping a thin layer of electrolyte between it and the inner surface of the tube (see Figure 2.99). Electrolyte also penetrates into the inside of the helix, however, the microwaves only see the portion of the solution between the helix and the quartz tube wall, the solution inside the helix is effectively invisible to the probing radiation. By placing the reference electrode and a platinum rod counter electrode within this dead volume, excellent electro-

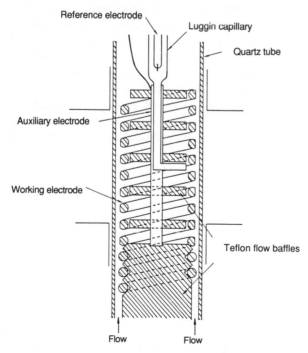

Figure 2.99 The Allendoerfer cell as modified by Carroll for the capability of solution flow over the electrode. From *Comprehensive Chemical Kinetics*, **29** (1989), 297.

chemical characteristics are found, with cyclic voltammograms indistinguish-able from those run under normal optimal conditions.

In order to observe very short-lived radicals, some means of installing hydrodynamic flow is required such that a steady supply of electroactive material to the working electrode can be sustained. Carroll (1983) adapted the cell design above in order to include such a flow, by employing baffles to prevent solution flowing into the inactive volume. Using such a cell, it has been suggested that it should be possible to detect and identify radicals having lifetimes as low as 10^{-5} s.

Whilst the Allendoerfer cell is an excellent means of detecting and identifying short-lived radicals, it is impossible mathematically to model the flow of electrolyte inside the cell. Consequently, it cannot be used in determining the kinetics and mechanisms of radical decay. Several cells have been designed with this aim is mind, one of which is the Compton–Coles cell. Compton and colleagues (1986) have developed a channel electrode cell for use in *in situ* epr (see Figure 2.100). The silica cell was found not to impair the sensitivity of the epr spectrometer when placed near the centre of a rectangular cavity. Compton and colleagues calculated the well-defined pattern of solution flow past the flat electrode located in the cell and so could calculate the concentra-tion profiles of the radicals. This allowed the steady-state epr signal to be related to the electrolysis current, solution flow rate and electrode geometry.

One example of the application of *in situ* electrochemical epr concerns the study of the Kolbe reaction. As was discussed in section 1.3, the Kolbe reaction involves some extremely complex processes and considerable effort has been expended in the search for the identities of the radical intermediates. Evidence for such intermediates remains sparse but one system that has provided such evidence is the electro-oxidation of triphenyl acetic acid (TPA) at a platinum electrode in acetonitrile (Waller and Compton, 1989). The case history of epr in the study of this system is a very good example of the application of the technique to provide details of a reaction mechanism. In

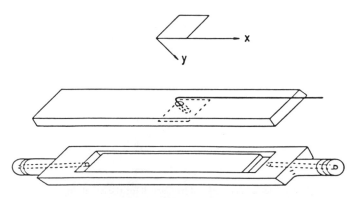

Figure 2.100 The Compton–Coles *in situ* flow cell.

addition, it clearly shows that the results obtained by *in situ* epr and their interpretation, are critically dependent on the cell design.

The first studies reported on the electro-oxidation of TPA were by Kondrikov *et al.* (1972) using an *in situ* cell adapted from the Adams design. A major fault with the design turned out to be that the counter and working electrodes were in close proximity in the sensitive part of the epr cavity. On oxidation a spectrum was obtained which was attributed to the triphenylacetoxy radical, $Ph_3CCO^•$, and was interpreted in terms of hyperfine interaction with four *ortho*-protons and two *para*-protons. The authors quoted coupling constants of 2.2×10^{-4} T and 3.5×10^{-4} T, respectively. These coupling constants were not consistent with the published spectrum which appeared to be a 1:4:6:4:1 quintet with a coupling constant of *c.* 2×10^{-4} T. Moreover, there is evidence to suggest that the triphenylacetoxy radical undergoes rapid decarboxylation.

Goodwin *et al.* (1975) employed another variant on the Adams cell design but obtained rather different results to those of Kondrikov and colleagues. Direct oxidation of the TPA at 2.0 V vs. an Ag pseudo reference electrode produced no epr spectrum. However, if the potential was then stepped to potentials less than -0.35 V, an epr spectrum was observed, that could be assigned to the triphenylmethyl radical. This was interpreted by the authors in terms of the following mechanism:

$$Ph_3CCO_2H \longrightarrow Ph_3C^+ + CO_2 + H^+ + 2e^- \quad (2.0\,V)$$
$$Ph_3C^+ + e^- \longrightarrow Ph_3C^• \quad (\leqslant 0.35\,V)$$

The clear disagreement between the two studies lead Compton *et al.* (1976) to study the system using the Compton–Coles cell. The authors showed that they could reproduce the spectra of Kondrikov and colleagues in wet acetonitrile by using a similar oxidation/reduction sequence to that employed by Goodwin and colleagues. If the potential was stepped to $+2.025$ V vs. SCE and then stepped to -1.8 V, then a spectrum was observed that consisted of a 1:4:6:4:1 quintet, with $a = 2.41 \times 10^{-4}$ T (see Figure 2.101(a)). This spectrum closely resembled that of the benzoquinone radical anion produced by the direct reduction of the parent molecule in acetonitrile and the g-values and coupling constants of the two spectra were identical. The authors thus concluded that the epr signals observed by Kondrikov and co-workers were due to the formation of benzoquinone radicals through the reduction of working electrode products at the counter electrode; a source of interference excluded from the Compton–Coles cell as a result of the counter electrode being located outside the epr cavity and downstream of the working electrode. The authors repeated the experiment in dry acetonitrile and obtained a different spectrum (see Figure 2.101(b)). The spectrum was shown to be a result of the superposition of two spectra due to the benzoquinone radical anion and the benzophenone radical. The mechanism by which these two radicals were formed was rationalised by the authors in terms of the following

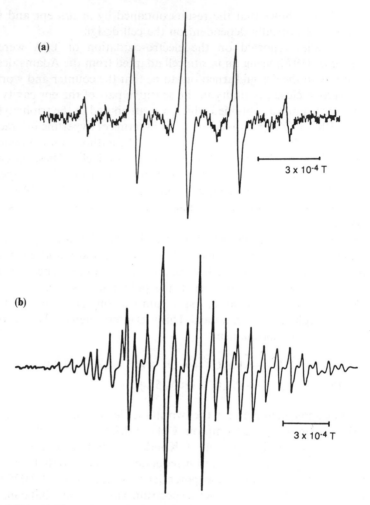

Figure 2.101 (a) The spectrum obtained by oxidation of triphenylacetic acid at $+2.0\,V$ before stepping to $-1.8\,V$ in wet acetonitrile. (b) Spectrum obtained by the oxidation of triphenylacetic acid at $+2.0\,V$ before stepping to $-1.9\,V$ in slightly moist acetonitrile. From Compton *et al.* (1986). Reprinted from *Comprehensive Chemical Kinetics*, **29** (1989), 297.

mechanism:

Oxidation

$$Ph_3C–CO_2H \longrightarrow Ph_3C^+ + CO_2 + H^+ + 2\,e$$

$$H_2O \longrightarrow 2H^+ + \tfrac{1}{2}O_2 + 2\,e$$

Reduction

$$Ph_3C^+ + e \longrightarrow Ph_3C^{\bullet}$$

$$Ph_3C^{\bullet} + O_2 \longrightarrow Ph_3C–O–O^{\bullet}$$

$$Ph_3C–O–O^{\bullet} + 2H^+ + e \longrightarrow Ph_3C–O–\overset{+}{O}–H_2$$

$$Ph_3C-O-\overset{+}{O}H_2 \longrightarrow Ph_3CO^+ + H_2O$$

$$Ph_3CO^+ \longrightarrow Ph_2\overset{+}{C}OPh$$

$$Ph_2\overset{+}{C}OPh + H_2O \longrightarrow Ph_2C(OH)-O-Ph + H^+$$

$$Ph_2C(OH)-O-Ph \longrightarrow Ph_2C=O + Ph-OH$$

$$Ph-OH \xrightarrow[+H_2, -H^+]{\text{'ox'}} O=Ph=O$$

$$O=Ph=O \xrightarrow[+e]{E_{1/2}=-0.55\,V} \text{benzoquinone radical anion}$$

$$Ph_2C=O \xrightarrow[+e]{E_{1/2}=-1.75\,V} \text{benzoquinone radical anion}$$

2.2.5   UV-visible spectroelectrochemistry

This section is primarily concerned with *in situ* UV-visible spectroscopy and, for the various reasons discussed in the section on electroreflectance, we shall confine the discussion to transmittance techniques.

The simplest, and hence first, most widely employed approach to obtaining the UV-visible spectra of electrogenerated species in solution involves passing the probe light beam directly through the working electrode. In this way the change in the absorbed light intensity as a result of the depletion, or generation, of chromophores can be measured directly. Such an approach relies on the existence of optically transparent electrodes (OTEs) which fall essentially into three groups:

1. Metal minigrids
2. Thin coatings of semiconductors or metals on glass
3. Reticulated vitreous carbon.

Minigrids of platinum, gold, silver, etc. can comprise several hundred wires/cm and allow up to 80% transmittance of the light incident on them. They can be used in both UV-visible and IR cells and behave as a planar electrode providing that the time of the spectral data collection is sufficient to allow the diffusion layers around each wire to overlap (see Figure 2.102).

Thin layers of metals such as Pt, Au, Ag or even carbon, do give planar surfaces but they do have the disadvantage of balancing good conductivity with high enough transmittance, as was discussed in the section on IR-ATR. Thin layers of certain oxides such as In/SnO_2 (ITO) on glass substrates are metallic, yet almost completely transparent in the visible. At first sight, this seems a contradiction in terms: a transparent metal. However, ITO is metallic by virtue of its extensive doping but transparent because the bandgap of the material is in the UV, and the free carrier absorption is primarily in the IR. Coated-glass OTEs allow simple cell designs, with the OTE forming one of the cell windows, as depicted in Figure 2.103.

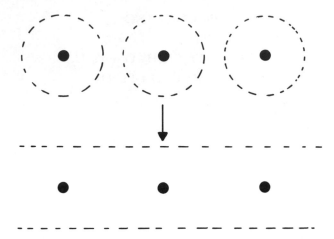

Figure 2.102   The development of the diffusion layer around a wire grid electrode to give effectively planar diffusion.

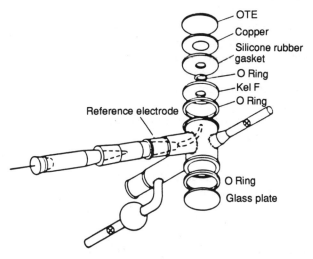

Figure 2.103   A spectroelectrochemical cell based on a coated-glass optically transparent electrode.

Reticulated vitreous carbon (RVC), is a sponge-like form of glassy carbon, having a very high free void volume (i.e. up to 97%) and a correspondingly high surface area (up to 66 cm^2/cm^3). The transmittance of a 0.12 cm-thick slice of RVC is *c*. 24%.

Spectroelectrochemical cells for use in the UV-visible region are not, of course, constrained by solvent absorption and can thus be of a reasonable size to give acceptable electrochemical behaviour. However, as with all the *in situ* techniques discussed in this book, a thin-layer approach is one of the methods employed.

Both the identification of a species and the determination of the kinetics of its formation or decay can be achieved with longer pathlength cells, such as that depicted in Figure 2.103. In kinetic experiments, however, there is the proviso that the experiment can be performed before natural convection currents interfere with the measurements; i.e. the operator must be certain that the removal of a chromophore from the optical path is due to reaction and not due to convection currents. It should be noted that the strength of UV-visible spectroscopy does not lie primarily in the identification of unknown species as the information it provides is not of a molecularly specific nature.

In order to observe a short-lived species it may be necessary to employ a rapid-scanning spectrometer, such as a diode-array instrument (5 ms for a 240 nm–800 nm spectrum). In addition, the absorbances of electrogenerated species can be very small and signal-averaging or phase-sensitive detection may be necessary to achieve the required signal-to-noise ratio (cf. EMIRS and FTIR).

One important advantage that UV-visible spectrometers (and dispersive IR spectrometers) possess over FTIR instruments is the ability to monitor the absorbance at a particular wavelength or over a wavelength range as a function of time. This allows kinetic measurements to be made on relatively fast timescales with signal averaging being used to increase the signal-to-noise if necessary. A typical experiment in an OTE cell would generally involve stepping the potential from a value where the reactant is stable to a value where reaction occurs at the diffusion-limited rate: the absorbance at a wavelength where either the reactant or product absorbs is then followed as a function of time. This is essentially directly analogous to the chronocoulo-metric experiments discussed in section 2.2.1, where linear diffusion of the electroactive species to the electrode is assumed.

Thus, for the reaction:

$$O + n\,e^- \longrightarrow R$$

we have seen that the dependence of the concentration of R on distance from the electrode will be as in Figure 2.104. If R is the only chromophore at the probing wavelength, then the total absorbance, A_t, at any time t will be the sum of all of the absorbances, dA, over the small distances, dx, in the optical path. Thus, the total absorbance will be given by:

$$dA = [R]_{x,t}\varepsilon\,dx$$

where $[R]_{x,t}$ is the concentration of R in the segment dx at a time t and ε is the molar decadic extinction coefficient at the wavelength of the probing light. Hence, A_t is given by:

$$A_t = \varepsilon \int_0^\infty [R]_{x,t}\,dx \qquad (2.182)$$

The integral is simply the total amount of R present per unit area at time t; hence the integral is equal to Q_d/nFA, where Q_d is the charge passed after

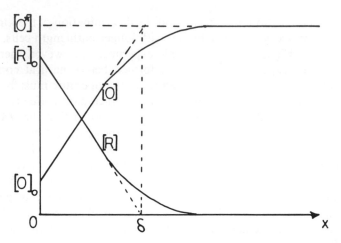

Figure 2.104   The variation in the concentrations of O and R with distance, x, from the electrode surface for the electrochemical reduction $O + e^- \rightarrow R$. $[O]_0$ and $[R]_0$ are the concentrations of the species at the electrode surface and δ is the diffusion layer thickness.

t seconds of electrolysis. Q_d is given by the integrated form of the Cottrell equation (see section 2.2.1) and equation (2.182) becomes:

$$A_t = -(2\varepsilon D_O^{1/2}[O^*]t^{1/2})/\pi^{1/2} \qquad (2.183)$$

The other popular approach to *in situ* spectroelectrochemistry is based on the use of an OTE electrode in a thin-layer, optically transparent thin layer electrode (OTTLE), cell. A schematic representation of one design of OTTLE cell is shown in Figure 2.105.

OTTLE cells have a considerable *IR* drop and are thus unsuitable for kinetic measurements. However, bulk electrolysis can be achieved in a few seconds and the whole solution can thus reach quasi-equilibrium with the electrode potential. The cells are generally designed around the use of Teflon spacers, the minimum thickness of which is around 50 μm. This renders the OTTLE cell capable of obtaining both UV-visible and IR spectra of solution species providing that, in the latter case, the solvent employed is not water.

The major applications of the OTTLE cell are (i) to obtain spectra of electrogenerated species and thence to obtain the extinction coefficients of its major absorption bands, and (ii) to determine the standard redox potential of a reversible couple. The latter experiment relies on the thin-layer electrochemical characteristics of the cell. Thus, for the couple:

$$O + n e^- \longleftrightarrow R$$

a spectrum would be collected at a potential where O is stable and then the potential stepped down to successively lower values with spectra taken at each step after the current has declined to zero. Since the solution has

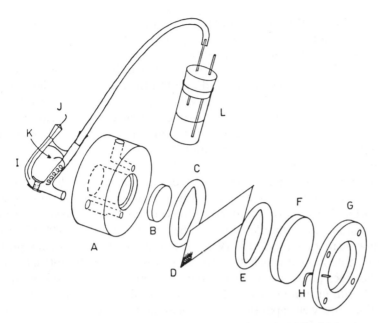

Figure 2.105 Optically transparent thin layer electrochemical (OTTLE) cell. A = PTFE cell body, B = 13 × 2 mm window, (C and E) = PTFE spacers, D = gold minigrid electrode, F = 25 mm window, G = pressure plate, H = gold working electrode contact, I = reference electrode compartment, J = silver wire, K = auxiliary electrode and L = solution presaturator. From Ranjith *et al.* (1990).

equilibrated with the potential, E, at the electrode, we can write:

$$E = E^0 + (RT/nF)\log_e([O^*]/[R^*])$$

If R is the only absorbing species at the probing wavelength and $[O^*]_i$ is the initial concentration of O, then:

$$[R^*] = A\varepsilon l$$

and:

$$[O^*] = [O^*]_i - A\varepsilon l$$

Hence, a plot of the applied potential vs. $\ln([O^*]/[R^*])$ gives n from the slope and E^0 from the intercept.

For an OTTLE cell to retain the conditions for thin-layer electrochemistry, a pathlength of c. 0.02 cm is generally accepted as the upper limit. Such a pathlength poses very real problems of sensitivity when studying systems with very low extinction coefficients. A method of increasing the pathlength but retaining the thin-layer electrochemical characteristics is to employ reticulated vitreous carbon as the OTE. RVC behaves as a thin layer electrode since the average diffusion length of the electroactive species to the electrode surface is sufficiently short that all the solution in its porous structure is electrolysed. In addition, RVC has been coated with a range of metals, including Au, Pt, Rh, Rh/Pt and Hg. In the case of gold, the coating process reduces the transmittance of a 0.12 cm-thick slice of RVC to *c*. 21%.

The use of electrochemical transmittance spectroscopy in both the UV-visible and IR regions of the spectrum is elegantly shown by the work of Ranjith *et al.* (1990) who employed an OTTLE cell to study the reduction of benzoquinone, BQ. The authors were the first to report the UV-visible spectrum of BQ^{2-} and to demonstrate the quantitative aspects of the technique by reporting extinction coefficients for the major bands of $BQ^{\overline{\cdot}}$ and BQ^{2-} in both the UV-visible and IR.

The cell design employed by the authors is shown in Figure 2.105 and employed a 200 μm pathlength. The poor electrochemical characteristics of the cell can be seen in Figure 2.106 which shows a cyclic voltammogram of the BQ/DMSO solution obtained in the cell. The peak-to-peak separations of the anodic and cathodic waves in the two couples are >400 mV even at the slow scan rate used, 0.5 mV/s. However, *IR* drop poses no problem, since the potentials at which the cathodic waves occur can be noted. Nevertheless, this does point to the great importance of obtaining cyclic voltammograms *in situ* in thin-layer based electrochemical cells. The two one-electron reductions are well-separated which facilitated the complete conversion of the BQ, first to $BQ^{\overline{\cdot}}$ and then BQ^{2-}.

Figure 2.107(a) shows UV-visible spectra of the $BQ^{\overline{\cdot}}$ (obtained at −1.0 V) and BQ^{2-} (obtained at −1.8 V) after the potential-step and the re-establishment of equilibrium. The spectrum of $BQ^{\overline{\cdot}}$ is consistent with that obtained

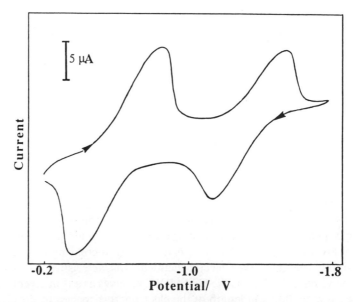

Figure 2.106 OTTLE cell cyclic voltammogram of 0.01 M 1,4-benzoquinone in dimethyl-sulphoxide solution containing 0.5 M tetraethylammonium perchlorate. Scan rate 0.5 mV s^{-1}. From Ranjith *et al.* (1990).

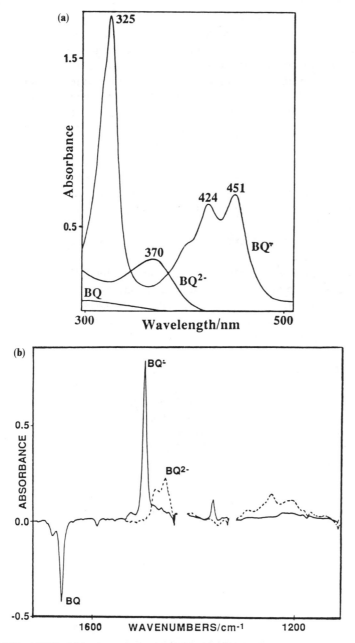

Figure 2.107 (a) UV-visible absorption spectra of 5×10^{-3} M 1,4-benzoquinone (BQ), its radical anion (BQ$^{\cdot-}$) and dianion (BQ^{2-}) in dimethylsulphoxide solution containing 0.5 M tetra-ethylammonium perchlorate. (b) FTIR absorbance difference spectra of 0.02 M 1,4-benzoquinone in dimethylsulphoxide solution containing 0.5 M tetraethylammonium perchlorate. Positive absorbances are due to the 1,4-benzoquinone radical anion (BQ$^{\cdot-}$) and dianion (BQ^{2-}) recorded at -1.00 V and -1.80 V respectively. Negative absorbances are due to 1,4-benzoquinone (BQ) present at the reference potential $+0.1$ V. From Ranjith *et al.* (1990).

in a variety of media and is attributable to $\pi \rightarrow \pi^*$ transitions. From the figure it can be seen that the lowest energy absorption in the $BQ^{\overline{\cdot}}$ spectrum contains a vibrational progression with maxima at 451 nm, 424 nm and 400 nm. This corresponds to an excited state vibrational mode of c. 1410 cm^{-1} which was attributed to the analogue of the ground state a_g C=O symmetric stretch near 1438 cm^{-1}. This suggests that the lower energy electronic absorption in the radical is localised in the vicinity of the C=O groups.

Figure 2.107(b) shows the corresponding FTIR spectra of BQ, $BQ^{\overline{\cdot}}$ and BQ^{2-}. The reference spectrum was collected at $+0.1$ V, with the subsequent spectra collected at the values given above after complete conversion had occurred and ratioed to the reference. The blank areas in the spectrum correspond to regions of strong solvent absorption (and hence large noise). Extinction coefficients were obtained for all the major features in Figures 2.107(a) and (b).

The OTTLE cell therefore:

1. Allows an electrochemical system to be probed by both UV-visible and IR techniques (providing a non-aqueous electrolyte is employed) and so allows information on the excited state to be obtained.
2. Permits straightforward measurement of extinction coefficients in both the UV-visible and IR.

However:

1. only weakly absorbing solvents can be employed
2. neither OTE nor OTTLE is surface-sensitive
3. the OTTLE cell is prone to oxygen leakage, hence, reductive electrochemistry usually demands that the cell be encased in some form of environmental protection.

2.2.6   The electrochemical quartz crystal microbalance

The quartz crystal microbalance has a long history of application as a means of determining film thickness in vacuum deposition techniques and more recently as a means of detecting trace constituents in the gas phase. In essence, it is an extremely sensitive sensor capable of measuring mass changes in the nanogram range.

The principle of operation of a quartz balance can be easily described: if an AC potential is placed across a quartz piezoelectric crystal, the crystal will oscillate spatially and the amplitude of this oscillation is greatest at the resonance frequency of the crystal. This resonance frequency, f_0, is a function of several parameters, including the mass of the crystal, and the mass of foreign material placed on the crystal. In fact, the change in the resonance frequency, Δf_0, on placing some foreign material on it is given approximately by:

$$\Delta f_0 = - \{f_0^2/\rho N\}\{\Delta m[1 + \Delta m/m]\} \qquad (2.184)$$

where ρ is the density of the crystal, N is the frequency constant for the particular crystal type, m is the mass per unit area of the oscillating quartz and Δm is the change in the mass per unit area on attaching the foreign material. Inserting all of the relevant constants into equation (2.184), and assuming that $\Delta m \ll m$, gives the Sauerbrey equation:

$$\Delta f_0 = - 2.26 \times 10^{-6} f_0^2 \Delta m \qquad (2.185)$$

i.e. there is a linear relationship between the change in resonance frequency of the crystal and the increase in mass of anything deposited on it.

The first application of the quartz crystal microbalance in electrochemistry came with the work of Bruckenstein and Shay (1985) who proved that the Sauerbrey equation could still be applied to a quartz wafer one side of which was covered with electrolyte. Although they were able to establish that an electrolyte layer several hundred angstroms thick moved essentially with the quartz surface, they also showed that the thickness of this layer remained constant with potential so any change in frequency could be attributed to surface film formation. The authors showed that it was possible to take simultaneous measurements of the *in situ* frequency change accompanying electrolysis at a working electrode (comprising one of the electrical contacts to the crystal) as a function of the applied potential or current. They coined the acronym EQCM (electrochemical quartz crystal microbalance) for the technique.

A typical quartz crystal is shown in Figure 2.108. The crystal is coated on both sides, usually in the form of a central circular spot, typically $0.22\,\mathrm{cm^2}$ of evaporated gold, $c.\ 900\,\text{Å}$ thick. The electrical inputs to the oscillator

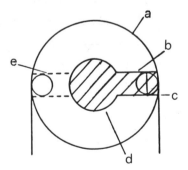

Figure 2.108   Piezoelectric quartz crystal with vacuum-deposited electrodes. Gold is deposited on both sides of the quartz wafer a in the pattern shown. Electrical connection is made to the electrode flag b by means of a spring clip with attached lead c. The circular gold electrode in the crystal centre d opposes an identical electrode on the other side of the crystal. Electrical connection to the opposing electrode is made to the electrode flag e. Reprinted from *Electrochimica Acta*, **30**, S. Bruckenstein and M. Shay, Experimental aspects of use of the quartz crystal microbalance in solution, pp. 1295–1300 (1985), with kind permission from Pergamon Press Ltd., Headington Hill Hall, Oxford OX3 0BW, UK.

circuit make contact via an additional gold patch, as shown in the figure. One of the gold spots acts also as the working electrode in the potentiostat circuit. As much as possible of the contact patch is covered with an inert insulating material, such as silicon rubber.

The mass sensitivity of the quartz crystal is at a maximum at the centre and decreases to zero at the edge. In general, for the purpose of mass calculation, the experimental mass-sensitive region is chosen to be the geometric working electrode area plus any of the exposed patch.

The Sauerbrey equation predicts a mass sensitivity per unit area of $0.226\,Hz\,cm^{-2}\,ng^{-1}$. For a typical crystal the exposed area is $c.\ 0.25\,cm^2$ and the absolute mass sensitivity is $0.904\,Hz\,ng^{-1}$. The resolution of modern frequency counters is easily $\pm 0.1\,Hz$ in $10\,MHz$, giving a theoretical mass resolution of $c.\ 9 \times 10^{-10}\,g$; in practice this is usually found to be closer to $\pm 2\,ng$.

Δf_0 is measured as the difference between the oscillation frequency of the working electrode crystal and a reference crystal since this mode of operation effectively removes any contributions due to experimental drift.

A schematic representation of the lower half of an EQCM cell is shown in Figure 2.109. The crystal is clipped or glued to the bottom of the electrochemical cell. Within the cell are the reference and counter electrodes, and a purging device to allow N_2-saturation of the electrolyte.

An example of the very high sensitivity of the EQCM to mass change at the surface of the crystal is provided in the work of Bruckenstein and Shay (1985) on the electro-oxidation of gold in perchloric acid.

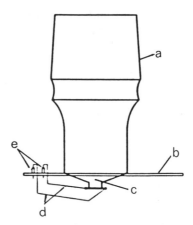

Figure 2.109   Lower half of electrochemical cell with attached crystal. a = ground glass joint for connection to Luggin capillary, auxillary chamber and solution bubbler, b = an epoxy-glass plug-board cemented to the glass, c = quartz crystal attached to a 0.40 inch O.D. glass tube, d = rigid leads connecting the crystal to the plug-board jacks, e = plug-jacks to connect the crystal to the oscillator circuit board. Reprinted from *Electrochimica Acta*, **30**, S. Bruckenstein and M. Shay, Experimental aspects of use of the quartz crystal microbalance in solution, pp. 1295–1300 (1985), with kind permission from Pergamon Press Ltd., Headington Hill Hall, Oxford OX3 0BW, UK.

The exact nature of the processes that occur when a gold electrode is electrochemically cycled into the oxide-formation region remain controversial. Figure 2.110(a) shows simultaneous voltammetric and mass-potential curves at a gold EQCM electrode in nitrogen-saturated perchloric acid and Figure 2.110(b) the corresponding charge- and mass-potential plots. The charge measurements were performed by analog integration of the current.

Figure 2.110(a) was obtained while the solution near the electrode was agitated via N_2-bubbling: in Figure 2.110(b), the solution was quiescent. The difference between the initial and final frequencies at 0.02 V in the two figures was interpreted by the authors in terms of a net mass change accompanying the gold dissolution/redeposition processes that occur during cyclic voltammetry at gold. Gold dissolves during the anodic and cathodic sweeps in the potential region where the oxide is present, with some redeposition occurring

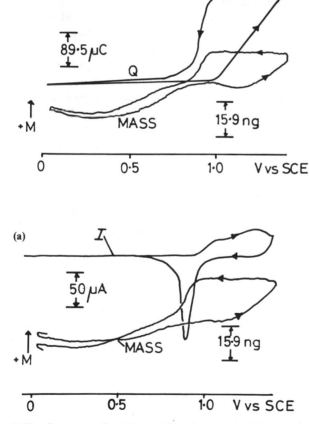

Figure 2.110   (a) Simultaneous cyclic voltammetric and mass-potential curves at gold. Potential scan rate = 50 mV/s, 0.2 M perchloric acid. (b) Simultaneous charge-potential and mass-potential curves at gold. Potential scan rate = 50 mV/s, 0.2 M perchloric acid. From Bruckenstein and Shay (1985).

after the reduction of the oxidised surface. The amount of gold loss into solution clearly depends on any solution turbulence adjacent to the electrode, hence there is more loss in Figure 2.110(a) than in Figure 2.110(b).

As can be seen from the figures, however, the change in mass between 0.02 V and 0.8 V in the anodic sweep is not accompanied by the passage of *any* faradaic charge. The authors conclude that this mass change can be attributed to the re-organisation of the double layer at the reduced gold surface, i.e. the potentials employed are positive of the pzc of gold under these experimental conditions, hence the adsorption of ClO_4^- is a plausible explanation of the observed mass increase even though perchlorate is not normally envisaged to absorb at electrode surfaces.

At potentials $\geqslant 1.10$ V, the mass of the electrode starts to increase until the scan is reversed at 1.39 V. The mass then remains constant until c. 0.98 V before decreasing and levelling off at 0.2 V. The gain in mass during the cycle 1.10 V $\rightarrow 1.39$ V $\rightarrow 1.10$ V, corresponding to oxide formation, is 20.1 ± 0.4 ng; this was 1.8 ng higher than that calculated from the charge under the oxide reduction peak. The difference was attributed to gold dissolution and the authors concluded that the measured mass change corresponds to the formation of one monolayer of adsorbed oxygen.

From Figures 2.110(a) and 2.110(b) it can be seen that the mass-potential curve lags behind the charge-potential and current-potential plots. During the anodic scan about 40% of the charge required to form the monolayer of oxide passes before significant mass change occurs. In addition, during the cathodic sweep, c. 20% of the charge required to reduce the monolayer passes before appreciable mass change. The authors proposed the following mechanism to explain their results:

1. $$Au(H_2O)_{ads} \longrightarrow AuOH + e^-$$

which will give rise to a negligible mass change, followed by:

2. $$Au–OH \longrightarrow Au{=}O + H^+ + e^-$$

During the second process, place exchange occurs resulting in the oxide atom moving below the plane of the surface:

3. $$Au{=}O \longrightarrow O{=}Au$$

4. $$O{=}Au + H_2O \longrightarrow O{=}Au(H_2O)_{ads}$$

The rate of the second electron transfer, and the subsequent place exchange, are not significant until c. a monolayer of adsorbed OH has been formed.

2.2.7   *FTIR and related techniques*

Having established whether an electrochemical reaction occurs, the potential at which it occurs and its rate, the next step must be to identify the products, and any intermediates, of the reaction. For this requirement the molecular information provided by vibrational spectroscopy would be ideal. As was seen above SERS and SERRS are purely surface techniques; while the normal Raman effect is too weak to provide information on solution species, resonance Raman is possible only if the species of interest happen to absorb at the same wavelength as that of the exciting laser. Consequently, the most suitable approach would be one employing *in situ* IR and, as was discussed above, there are several IR techniques that can be employed *in situ*. Of these the fast potential modulation techniques (EMIRS and dispersive ATR) are primarily surface techniques confined to fast, reversible reactions. As was seen above IRRAS is not confined to the investigation of surface species but it is essentially an expensive means of carrying out experiments more easily and cheaply performed using the straightforward FTIR approach. Thus FTIR provides the most effective method of studying the near-electrode region.

The *in-situ* FTIR approach can be divided into the transmittance, external reflectance and internal reflectance methods.

2.2.7.1   Transmittance.   This was discussed in section 2.2.5; the application of the transmittance IR approach to the study of electroactive solution species is effectively limited to non-aqueous solvents.

2.2.7.2   External reflectance.   As was seen above, in this approach the strong electrolyte absorption is minimised by employing a thin layer configuration and the possible severe restrictions imposed on the diffusion of species to and from the electrode in such an arrangement can lead to some difficulties.

Thus, in an electrochemical cell the electrolyte has a small but finite resistance, R_e, resulting in a potential drop, V_d, between the working and reference electrodes. From Ohm's law, $V_d = IR_e$, where I is the current flowing across the working electrode/electrolyte interface. As a result of this resistance, the measured potential V_m is related to the real potential, V_r, by:

$$V_m = V_r - V_d$$

or:

$$V_m = V_r - IR_e \qquad (2.186)$$

Under normal circumstances, i.e. in strong aqueous electrolyte, R_e is very small and $V_m \approx V_r$. However, in a thin layer cell severe restriction on free ionic diffusion may result in a large resistance R_e; thus, V_m may be several tenths of a volt lower (anodic current) or higher (cathodic current) than V_r at large currents. This effect can increase in severity on traversing the working electrode causing a potential gradient across the surface.

As an example of this phenomenon consider Figure 2.111(a) which shows spectra collected from a Pt electrode immersed in N_2-saturated 0.1 M NaH_2PO_4/1 M MeOH (pH 4.4) at potentials > 0.3 V vs. SCE normalised to the reference taken at -0.5 V. The figure shows the $C\equiv O_{ads}$ absorption as

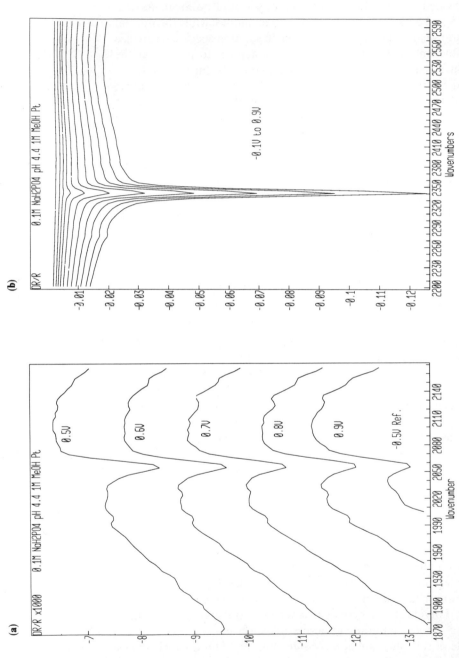

Figure 2.111 *In situ* FTIR spectra of a platinum electrode immersed in 0.1 M NaH₂PO₄/NaOH pH 4.4 electrolyte containing 1 M CH₃OH. Spectra were collected every 100 mV after the reference spectrum taken at −0.5 V vs. SCE. (a) The C≡O$_{ads}$ region for the potential range 0.5 V to 0.9 V, (b) the CO₂ region, showing spectra collected over the full spectral range. From P.A. Christensen and A. Hamnett, unpublished results.

discussed above in the section on EMIRS. The feature is monopolar indicating that methanol does not dissociatively chemisorb at the reference potential. At potentials > 0.3 V the $C\equiv O$ absorption changes frequency and intensity only very little, yet at these potentials, the electrode is covered with a layer of oxide which would completely replace any of the more weakly adsorbed $C\equiv O$. The possible answer to this problem lies in Figure 2.111(b) which shows the CO_2 gain feature due to the electro-oxidation of the methanol during the same experiment. From the intensity of the CO_2 feature it can be deduced that a high turnover of methanol is occurring at these potentials, suggesting that a relatively high current is being passed and hence that a large IR drop is present. Thus, it appears that this IR drop is such that only the $C\equiv O_{ads}$ at the edges is oxidised, resulting in the small coverage-dependent decreases in the band frequency and intensity, while the remainder of the surface is experiencing a potential much lower than that measured. Hence, spectra collected at higher potentials in regions where appreciable current flows require great care in interpretation.

In addition to IR problems, the restriction placed on diffusion by the thin layer configuration results in the rapid depletion of electroactive species within the layer, with effectively no concomitant replenishing diffusion from the bulk electrolyte outside of the thin layer. Consider, for example, the reduction of a species O^{n+} at an electrode:

$$O^{n+} + n\,e^- \longleftrightarrow O$$

with the kinetics of the process sufficiently fast to render diffusion the rds. For a thin layer $c.$ 5 μm thick and electrode 7 mm in diameter, diffusion from the bulk into the layer occurs over the radius of the electrode, 3 mm. From equation (2.112) this gives a diffusion time of $c.$ 3 h. In contrast, species diffusing from the extremes of the thin layer, i.e. near the cell window, to the electrode can do so in 25 ms. Hence, the thin layer will become depleted of O^{n+} in this time. This calculation is rather approximate though diffusion times into the thin layer of up to an hour are commonly observed in practice. Thus, on the timescale of a typical FTIR experiment, 30 s to 3 min, all of the electroactive species in the thin layer may be depleted. If a faradaic reaction involving the generation or consumption of protons is taking place, such as the formation of surface oxides or hydrides on a metal, then this can lead to large pH swings in the layer (see below).

Before in situ external reflectance FTIR can be employed quantitatively to the study of near-electrode processes, one final experimental problem must be overcome: the determination of the thickness of the thin layer between electrode and window. This is a fundamental aspect of the application of this increasingly important technique, marking an obstacle that must be overcome if it is to attain its true potential, due to the dearth of extinction coefficients in the IR available in the literature. In the study of adsorbed species this determination is unimportant, as the extinction coefficients of the absorption bands of the surface species can be determined via coulometry.

However, in the case of near-electrode phenomena diffusion complicates the simple determination of the quantitative aspects of the reaction and extinction coefficients are essential. A typical approach that may be envisaged is as follows.

The thin-layer configuration and its associated diffusion problems means that it is possible to oxidise (or reduce) all of the electroactive species in the thin layer before they can be replenished to any marked degree. Consider, for example, the O^{n+}/O couple, with a standard redox potential well within the 'electrochemical window' of the solvent, so that the current in the absence of the couple is small and can easily be accounted for. With the electrode pushed against the window the potential is stepped cathodic enough to ensure the rapid reduction of the O^{n+} and the current measured as a function of time, the concentration such that the time for the current to reach zero, or a steady residual value, is small. If the area under the I/t curve is A ampere seconds, then the charge passed $Q = A$ coulombs. Thus, the number of moles of O^{n+} reduced, N_O, is given by:

$$N_O = A/nF$$

If the known concentration of O is $[O^*]$, then:

$$[O^*] \times Ad = A/nF$$

where A is the area of the electrode and d is the thickness of the thin layer (in m^2 and m, respectively for $[O^*]$ in $mol \cdot m^{-3}$). Thus:

$$d = A/nFA[O^*] \tag{2.187}$$

The thin-layer thickness calculated in this manner may well have some serious errors associated with it due to sample purity, errors in the weighing out of the solution, the diffusion of species near the thin layer into it within the timescale of the experiment, etc.

One example of the application of *in situ* FTIR to the study of the near-electrode region concerns the study of the electro-oxidation of ethylene glycol (EG) at a platinum electrode in base. This work clearly illustrates the relative ease with which the products of an electrochemical reaction can be detected and identified, and a mechanism deduced.

Figures 2.112(a) and (b) show cyclic voltammograms of the reflective Pt working electrode immersed in 1 M NaOH in (a) the absence and (b) the presence of 1 M EG.

The oxidation reaction was investigated using EG concentrations of 0.06 M, 0.2 M, 0.5 M and 1.0 M. In each case, the reference spectrum was collected at -0.85 V after which the potential was stepped to successively higher values. At each step another spectrum was collected and all the spectra were normalised to the reference. A typical result is shown in Figure 2.113(a) which shows the spectra obtained using the 0.5 M EG. The spectra were dominated by a broad structured loss feature extending down from 3000 cm^{-1} to below 1000 cm^{-1}. This was attributed by the authors to the loss of OH^-

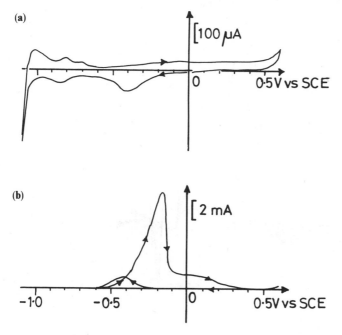

Figure 2.112 Cyclic voltammograms of a Pt electrode immersed in N_2-saturated 1 M NaOH solution, and (a) in the absence and (b) in the presence of 1 M ethylene glycol. Scan rate 100 mV s^{-1}, the voltammograms were taken using the IR spectroelectrochemical cell with the electrode pulled back from the cell window. From Christensen and Hamnett (1989).

from the optical path and the assignment was confirmed by admitting 1 M NaOH solution into the spectroelectrochemical cell, with the thin layer containing pure water. The Pt electrode was held at open circuit and the build-up of the OH$^-$ absorptions followed as a function of time (see Figure 2.113(b)). This technique is the most convenient and accurate method of assigning unknown absorptions. The other bands present were assigned in an analogous fashion. In this way, the products of the oxidation were identified as CO_3^{2-}, glycolate and oxalate (see Table 2.2). In addition, CO_2 was observed in the experiments employing 1 M EG indicating that a large pH swing had taken place in the thin layer.

Table 2.2 Species observed during the oxidation of ethylene glycol in aqueous NaOH ($\bar{\nu}$/cm^{-1})

CO_2	CO_3^{2-}	Glycolate	Oxalate
2340			
		1578	1600
	1410	1413	
		1360	
		1321	1310
		1075	

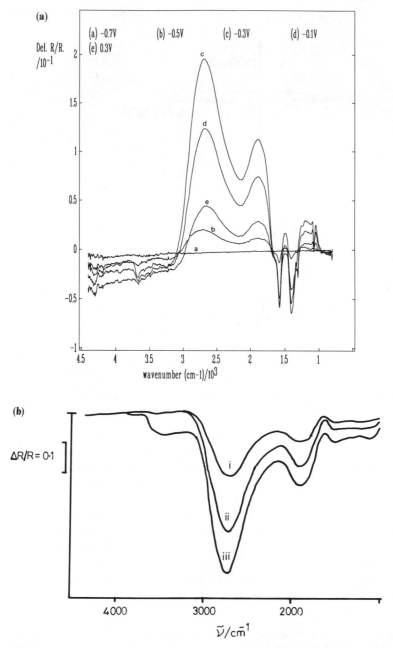

Figure 2.113(a) *In situ* FTIR spectra of a Pt electrode immersed in N_2-saturated 1 M NaOH/0.5 M ethylene glycol. The spectra were collected at potentials increasingly more positive of the reference potential, which was taken at -0.85 V vs. SCE. (a) -0.7 V, (b) -0.5 V, (c) -0.3 V, (d) -0.1 V and (e) 0.3 V. (b) *In situ* FTIR spectra of the diffusion of OH^- into the thin layer of the spectroelectrochemical cell. A thin layer of pure water was formed by pushing the Pt electrode against the CaF_2 window of the cell and a reference spectrum collected at open circuit voltage. 1 M NaOH was then admitted into the cell, spectra collected at regular intervals and normalised to the reference spectrum of the pure water. Time is increasing from i to iii. From Christensen and Hamnett (1989).

The variation in intensity of the various product features with potential are shown in Figure 2.114(a)–(d) for the initial concentrations of EG used.

As can be seen from the plots, the oxidation of EG takes place in two distinct potential regions separated by a region within which the reaction appears to be inhibited. Between $-0.6\,V$ and $-0.3\,V$ all three oxidation products are produced, with the exception of the lowest concentration of EG where the authors detected no glycolate. At higher potentials, further

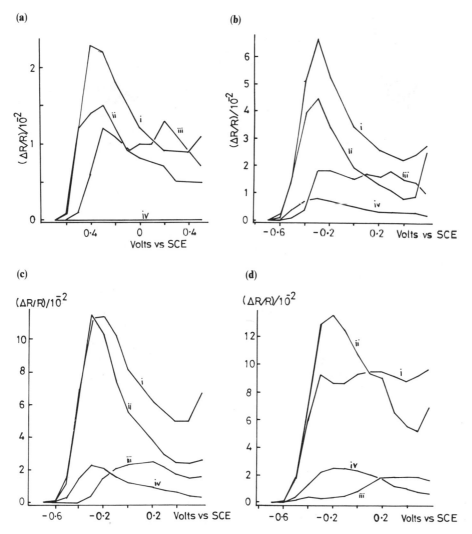

Figure 2.114 Plots of the intensities of (i) the $1410\,cm^{-1}$ band, (ii) the OH^- feature near $1910\,cm^{-1}$, (iii) the $1310\,cm^{-1}$ oxalate band and (iv) the glycolate band near $1075\,cm^{-1}$, vs. potential observed in the FTIR experiments on the oxidation of ethylene glycol. Initial ethylene glycol concentrations were (a) 0.06 M, (b) 0.2 M, (c) 0.5 M and (d) 1.0 M. From Christensen and Hamnett (1989).

conversion takes place to CO_3^{2-} and CO_2, and the OH^- feature follows the same behaviour.

The authors ascribed the large loss of solution OH^- primarily to its consumption in the various oxidation processes:

$$5OH^- + (CH_2OH)_2 \longrightarrow CH_2OH-COO^- + 4H_2O + 4e^-$$
$$5OH^- + CH_2OH-COO^- \longrightarrow (COO^-)_2 + 4H_2O + 4e^-$$
$$(COO^-)_2 + 4OH^- \longrightarrow 2CO_3^{2-} + 2e^- + 2H_2O$$

with some loss due to the operation of Donnan exclusion, whereby the thin layer is unable to support the larger build-up of nett negative charge and so ejects OH^- into the bulk of the solution.

On the basis of the results, the authors postulated the following mechanism:

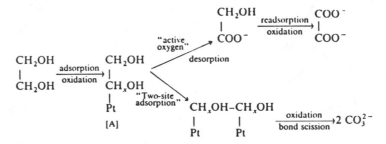

From Figure 2.114, it can be seen that the oxalate production was always delayed somewhat with respect to both CO_3^{2-} and glycolate. The authors postulated that this was as a result of the desorption and then subsequent re-adsorption, of glycolate.

At negative potentials EG can compete successfully with glycolate for adsorption sites, unless the EG concentration is too low, and this accounts for the lack of glycolate at the lowest EG concentration. The simultaneous appearance of glycolate and carbonate was taken by the authors as strong evidence for the existence of a common intermediate, identified as A in the scheme. It seems that A may desorb oxidatively *or* the second carbon may become bound to the surface if a suitable neighbouring adsorption site is available. Once this last process has taken place, carbon–carbon bond scission occurs with complete oxidation to CO_3^{2-}.

The inhibition of the oxidation process between $-0.3\,V$ and $0.4\,V$ must be due to the poisoning of active sites, since there are no obvious adsorption products, and so the poison must be present in very low amounts. Similar inhibition was observed by the authors in acid solution. In order to increase the sensitivity of their system, the authors carried out a PDIR experiment and observed adsorbed CO which, they concluded, was the main poison in acid and base. At potentials $> 0.4\,V$, the oxidation recommences as a result of the freeing up of active sites as the poison is oxidised.

The major progress that has been made in *in situ* IR spectroscopy to date has been achieved with commercial IR instruments. However, an important limitation of the technique is that it has proven impossible to obtain reliable data in the spectral range below $c.\ 800\,cm^{-1}$ in aqueous solution; a spectral region rich in information concerning the electrode-adsorbate bonds. This limitation is a result of the steep rise in the absorption coefficient of water in this region. Serious thought has been given to the best way forward on this aspect of probing the electrode/electrolyte interface and it has been concluded that the problem lies in the weakness of commercial IR sources. The search for new sources has already resulted in the use of the radiation from synchrotron facilities and, in the future, it may be possible to employ sources such as the free electron laser.

2.2.8 Mass spectrometry techniques – DEMS

Mass spectrometry (MS) is an extremely powerful method of chemical analysis and the possibility of measuring electrochemical reaction products via MS was first suggested by Grambow and Bruckenstein (1977). The technique of differential electrochemical mass spectroscopy (DEMS) was later perfected and pioneered by Wolter and Heitbaum (1984).

Figure 2.115 shows a schematic representation of the DEMS apparatus. In essence, the electrochemical cell is separated from a mass spectrometer by a porous, non-wetting PTFE membrane of very small pore size. The working electrode is then deposited as a porous metal layer on the thin

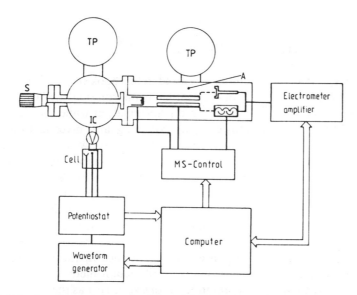

Figure 2.115 Experimental system for differential electrochemical mass spectroscopic (DEMS) measurements with automatic data acquisition. TP = turbo pump, IC = inlet chamber, A = analysis chamber, S = screw mechanism to control aperture between both chambers. After Iwasita and Vielstich (1990).

membrane and the electrode + membrane then fixed onto a frit at the entrance to the spectrometer with the coated side facing into solution. Volatile solution products, and gaseous products, can then pass through into the spectrometer. The use of turbo pumps minimises the delay between the generation of the product and its subsequent detection in the mass spectrometer. This means that cyclic voltammograms and mass intensity profiles can be obtained simultaneously for scan rates of $\leqslant 50\,\text{mV/s}$.

Wolter and Heitbaum proposed that the area under a particular MS peak, A, is directly proportional to the faradaic charge passed, Q, according to:

$$A = (K^*/n)\xi Q \qquad (2.188)$$

where ξ is the faradaic efficiency of the process resulting in the generation of the product under study, n is the number of electrons necessary to generate one mole of the product and K^* is a constant that includes all the spectrometer parameters.

The simplest case is one for which there is an adsorbed product. Equation (2.188) can then be used to establish the stoichiometric composition of the adsorbed species providing that they can be oxidised to volatile products and that the system can be calibrated, i.e. K^* determined. Thus, for adsorbed CO, $n = 2$. The CO can be oxidised off as CO_2 with the MS and current response being measured as a function of potential, allowing K^* to be calculated.

2.3
Ex situ emersion techniques

2.3.1 Structural and analytical methods

The in-situ spectroscopic methods discussed in this book owe their importance of two factors: first, because of the molecular information they can provide, and second, because they can obtain this information while the electrode is covered with electrolyte. However, by confining any investigation of the electrode surface, or double layer, to in situ techniques, we do deny ourselves access to a range of techniques that have immense application in the very areas that have only been sparsely covered by in situ methods such as electrochemical SEXAFS and X-ray diffraction. Thus, there is still a considerable void in the data obtainable with respect to the elemental analysis of adlayers or adsorbates on an electrode, or indeed of the electrode surface itself, or of the short- or long-range order in such systems. The most notable of these techniques that have revolutionised surface science in the years since the 1960s are the electron spectroscopic and diffraction methods. To a certain extent, they are not as convenient in terms of size or cost as the more straightforward in situ technique and they do require the study of the well-defined surfaces provided by single crystals; polycrystalline surfaces give results complicated by the presence of several exposed facets.

These *ex situ* techniques require ultra high vacuum (UHV) to operate since:

1. The data from those methods that involve electron bombardment or emission will suffer from the scattering of the electrons if any gas molecules are present.
2. More importantly, it is vital to be certain that the surface structure is not the result of a reconstruction due to adsorption of any contaminant from the air. Similarly, elemental analysis is only meaningful when the elements determined come from the sample and not exposure to the atmosphere.

Contamination from the gas phase will be proportional to the number of collisions a surface undergoes. Hence, from kinetic theory, the number of collisions per unit time and unit area, Z_c, is given by:

$$Z_c \approx P/(2\pi mkT)^{1/2} \qquad (2.189)$$

where P is the pressure of the gas and m is the mass of a gas molecule. At 298 K, this expression can be written as:

$$Z_c \approx 2.03 \times 10^{21} P/\sqrt{M} \, \text{cm}^{-2} \, \text{s}^{-1} \qquad (2.190)$$

where M is the molecular mass of the gas and the pressure is in mm Hg. For air, $M \approx 29$. At 760 mm Hg, $Z_c \approx 3 \times 10^{25} \, \text{cm}^{-2} \, \text{s}^{-1}$: each 1 cm^2 has approximately 10^{15} atoms and each atom is thus struck 10^{10} times in a second. At a pressure of 10^{-6} mm Hg, the pressure in a conventional high vacuum apparatus, each atom is struck every 0.1 s. However, at 10^{-9}–10^{-10} mm Hg, the pressure in an UHV system, each atom is struck only once every few hours.

Such a high vacuum imposes strict requirements on the equipment. The apparatus must be entirely metal and capable of being heated to 300°C to drive off any adsorbed water or gases. No grease is employed, and the sample is usually cleaned by Ar^+ bombardment.

UHV techniques are usually classified in terms of the electron/photon method, as is shown in Table 2.3 which lists the common electron bombardment and emission techniques that have been employed in electrochemical studies. A detailed description of UHV surface analysis techniques is beyond the scope of this book: there are many excellent reference texts that can be consulted for this purpose (see further reading list). It is sufficient to note that methods involving electron bombardment or emission are inherently surface-sensitive as a result of the low pathlength, or escape depth, of electrons in condensed media. In addition, Table 2.3 briefly describes the type of information each method provides.

This section considers only those techniques employed via the *emersion* approach as being the closest to *in situ* observation. However, as has been discussed briefly in other sections, the use of the emersion approach is still controversial and a source of much debate. Thus, in order to employ UHV-based techniques via emersion the electrode has to be transferred from the electrochemical cell to the vacuum chamber. Ideally, there should be no

Table 2.3 UHV techniques that are commonly employed to study the electrode surface via emersion

Technique	$e^-/h\nu$ in/out classification	Information obtained
Ultraviolet Photoelectron Spectroscopy/X-ray photo-electron spectroscopy	Photon in/electron out	Direct measurement of the absolute binding energy and widths of core (X-ray) and valence (UV) bands. The core levels do not participate in bonding, hence each element gives a characteristic XPS spectrum: electron spectroscopy for chemical analysis (ESCA). ESCA gives the elemental composition of the surface of a solid sample (except H), the relative amounts of each element present, its oxidation state and some information on the chemical environment around each element. In addition, it is capable of providing an estimate of the depth of a deposited overlayer
Auger electron spectroscopy	Electron in/electron out	Again, AES spectra arise from core levels, hence are characteristic for a given element. Surface elemental analysis (except H and He) with very high sensitivity, able to detect $<1\%$ of a monolayer. Usually employed first to check that the surface is free from contamination
Low energy electron diffraction Reflection high energy electron diffraction	Electron in/electron out	Surface crystal structure information on the lateral arrangement of any adatoms *in the unit cell* – not the exact position. RHEED has a lower surface sensitivity than LEED, giving surface and bulk structural information (due to the greater penetration depth of its electron beam) as well as some indication of surface roughness. In addition, the higher energy of the probing electrons in RHEED can result in sample damage

changes in the electrode as a result of the transfer: at worst, any such changes should be known and accounted for.

There are three commonly employed approaches to the 'transfer problem':

1. The electrochemical cell is built into the UHV chamber, the solution is removed by suction or blowing and direct transfer is then effected.
2. The electrode is removed from the cell and transferred via a glove box.
3. No direct transfer is involved, two electrodes are treated in identical ways and studied in parallel by the electrochemical and UHV methods.

Emersion has been shown to result in the retention of the double layer structure: i.e. the structure including the outer Helmholtz layer. Thus, the electric double layer is characterised by the electrode potential, the surface charge on the metal and the chemical composition of the double layer itself. Surface resistivity measurements have shown that the surface charge is retained on emersion. In addition, the potential of the emersed electrode, E, can be determined in the form of its work function, Φ, since E and Φ represent the same quantity: the electrochemical potential of the electrons in the metal. Figure 2.116 is from the work of Kotz *et al.* (1986) and shows the work function of a gold electrode emersed at various potentials from a perchloric acid solution: the work function was determined from UVPES measurements. The linear plot, and the unit slope, are clear evidence that the potential drop across the double layer is retained before and after emersion. The chemical composition of the double layer can also be determined, using AES, and is consistent with the expected solvent and electrolyte. In practice, the double layer collapses unless (i) potentiostatic control is maintained up to the instant of emersion and (ii) no faradaic processes, such as O_2 reduction, are allowed to occur after emersion.

Thus, emersion results in the retention of the double layer of some $10\,\text{Å}$. However, the substrates commonly employed fall into two categories, depending on whether any additional electrolyte is left behind on the surface or not.

Thus, all of the excess electrolyte rapidly drains off a *hydrophobic* electrode. This so called 'dry' emersion has proved to be the exception rather than the

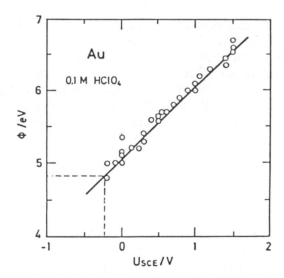

Figure 2.116 Work function Φ of an emersed gold electrode in UHV as a function of emersion potential (0.1 M $HClO_4$). Dashed lines indicate Φ-value for emersion at $0\,\text{V}$ vs. NHE. From Kolb, Lehmpfuhl and Zei in *Spectroscopic and Diffraction Techniques in Interfacial Electro-chemistry*, eds. C. Gutierrez and C. Melendres, Nato ASI Series, Series C: Mathematical and Physical Sciences, Vol. 320, Chapter 11, Kluwer Academic Publishers, Dordrecht, 1990.

rule. It appears that flatness and cleanliness are the essential requirements to obtain dry emersion: the latter is generally tested via AES.

Emersion resulting in substantial amounts of electrolyte remaining on the (*hydrophilic*) electrode is much more commonly observed. When the solvent evaporates from the surface, the electrolyte is left behind as small crystallites. These can distort LEED and RHEED patterns and, more importantly, can render quantitative evaluation of the surface cation and anion concentrations by ESCA impossible.

Several solutions to this problem have been employed:

1. If low concentrations of the electrolyte are used, this minimises the interference. However, as only the compact portion of the double layer is retained, the diffuse portion is lost. Hence, this restricts the electrolyte concentration to $\geqslant 0.01$ M.
2. Completely volatile electrolytes can be used, such as HF. However, this severely restricts the range of systems that can be studied.
3. The electrolyte is washed away with a volatile solvent.
4. The excess electrolyte is blown off the electrode with a stream of pure Ar.
5. The excess electrolyte is sucked off using a suitably-positioned capillary tube.

The last two methods have proved to be highly effective.

2.3.2 Mass spectrometric techniques

Electrochemical thermal desorption mass spectroscopy (ECTDMS) was first developed by Wilhelm *et al.* in 1987 (see Iwasita and Vielstich (1990) and references therein). The technique involves the emersion of a foil working electrode and subsequent direct transfer into a UHV chamber (see Figure 2.117). Any adsorbed species are then desorbed by heating the foil with an intense white light source, the temperature of the foil being monitored by a thermocouple, which gives a constant rate of heating. Any desorbed species are then detected with a quadrupole mass spectrometer. As many adsorbates are highly sensitive to the presence of O_2, it becomes even more essential to exclude adventitious air.

As was discussed above, it is essential to determine the effect, if any, that the emersion process has on the double layer. To do this, Wilhelm and colleagues have performed the definitive type of blank experiment. CO was adsorbed onto the Pt working electrode from sulphuric acid electrolyte. After adsorption, the CO-saturated solution was replaced with pure electrolyte. The potential of the electrode was then ramped in order to oxidise off the adsorbate, as CO_2, and the voltammogram so obtained is shown in Figure 2.118(a). The experiment was then repeated: CO was adsorbed as before, but the electrode was emersed and transferred into the UHV chamber, before being re-immersed and the potential ramp applied. The voltammogram so

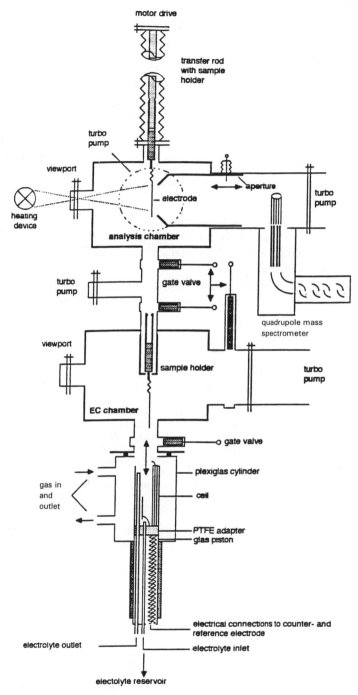

motor drive

transfer rod
with sample
holder

turbo
pump

viewport

aperture

turbo
pump

heating
device

electrode

analysis chamber

turbo
pump

gate valve

quadrupole mass
spectrometer

viewport

sample holder

turbo
pump

EC chamber

gate valve

gas in
and
outlet

plexiglas cylinder

cell

PTFE adapter
glas piston

electrical connections to counter- and
reference electrode

electrolyte outlet

electrolyte inlet

electrolyte reservoir

Figure 2.117 Experimental system for electrochemical thermal desorption mass spectroscopy
(ECTDMS).

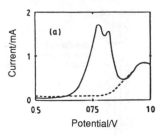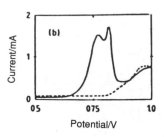

Figure 2.118 Test for the survival of adsorbed CO on Pt after transfer in the UHV. $I - E$ curves during the oxidation of CO_{ads} in 0.1 M H_2SO_4 (a) immediately after elimination of bulk CO, (b) after elimination of bulk CO and transfer in the UHV. After Iwasita and Vielstich (1990).

obtained is shown in Figure 2.118(b). A comparison of the two voltammograms shows that only minor changes occurred during the emersion process: the charges under the voltammograms in Figure 2.118(a) and (b) are the same to within $\pm 5\%$.

The predominant application of ECTDMS has been in the study of methanol oxidation and this will be discussed in the next chapter.

References

Albery, W.J. (1975) *Electrode Kinetics*, Clarendon Press, Oxford.
Albery, W.J. and Hitchman, M. (1971). *Ring-disc Electrodes*, Clarendon Press, Oxford.
Albrecht, M.G. and Creighton, J.A. (1977) *J. Am. Chem. Soc.*, **99**, 5215.
Allendoerfer, R.D., Martincheck, G.A. and Bruckenstein, S. (1973) *Anal. Chem.*, **47**, 890.
Beden, B., Bewick, A., Kunimatsu, K. and Lamy, C. (1981) *J. Electroanal. Chem.*, **121**, 343.
Beden, B., Collas, N., Lamy, C., Leger, J.M. and Solis, V. (1985) *Surface Sci.*, **162**, 795.
Bedzyk, M.J., Bommarito, G.M., Caffrey, M. and Penner, T.L. (1990) *Science*, **248**, 52.
Binning, G., Quate, C.F. and Gerber, Ch. (1986) *Phys. Rev. Lett.*, **12**, 930.
Birke, R.L. and Lombardi, J.R. (1988) in *Spectroelectrochemistry: Theory and Practice*, Gale, R.J. (ed.), Plenum Press, New York, Chapter 6.
Blum, L., Abruna, H.D., White, J., Gordon, II, G.J., Borges, G.L., Samant, M.G. and Melroy, O.R. (1986) *J. Chem. Phys.*, **85**, 6732.
Bruckenstein, S. and Shay, M. (1985) *J. Electroanal. Chem.*, **188**, 131.
Carroll, J.B. (1983) Ph.D. Thesis, State University of New York at Buffalo.
Chen, C.Y., Davoli I., Ritchie, G. and Burstein, E. (1980) *Surface Sci.*, **101**, 363.
Christensen, P.A. and Hamnett, A. (1989) *J. Electroanal. Chem.*, **260**, 347.
Compton, R.G. and Coles, B.A. (1983) *J. Electroanal. Chem.*, **144**, 87.
Compton, R.G., Coles, B.A. and Day, M.J. (1986) *J. Electroanal. Chem.*, **200**, 205.
Corrigan, D.A. and Weaver, M.J. (1986) *J. Phys. Chem.*, **90**, 5300.
Feenstra, R.M., Stroscio, J.A., Tersoff, J. and Fein, A.P. (1987) *Phys. Rev. Lett.*, **58**, 1192.
Feldberg, S.W. (1984) *J. Am. Chem. Soc.*, **106**, 4671.
Fleischmann, M., Hendra, P.J. and McQuillan, A.J. (1974) *Chem. Phys. Lett.*, **26**, 163.
Fleischmann, M., Oliver, A. and Robinson, J. (1986) *Electrochim. Acta*, **31**, 899.
Gokhshtein, A.Y. (1970) *Electrochim. Acta*, **15**, 219.
Golden, W.G., Dunn, D.S. and Overend, J. (1981) *J. Catal.*, **71**, 395.
Goodwin, R.D., Gilbert, J.C. and Bard, A.J. (1975) *J. Electroanal. Chem.*, **59**, 163.
Grahame, D.C. (1947) *Chem. Rev.*, **41**, 441.
Grambow, L. and Bruckenstein, S. (1977) *Electrochim. Acta*, **22**, 377.
Greenler, R.G. (1966) *J. Chem. Phys.*, **44**, 310.
Griffiths, P.R. and deHaseth, J.A. (1986) *Fourier Transform Infrared Spectrometry*, Wiley-Interscience, New York.

Hamers, R.J., Tromp, R.M. and Demuth, J.E. (1986) *Phys. Rev. Lett.*, **56**, 1972.

Hamnett, A. and Hillmann, A.R. (1988) *J. Electrochem. Soc.*, **135**, 2517.

Heinze, J., Storzbach, M. and Mortensen, J. (1987) *Ber. Bunsenges. Phys. Chem.*, **91**, 960.

Hupp, J.T., Larkin, D. and Weaver, M.J. (1983) *Surface Sci.*, **125**, 429.

Itaya, K., Sugawara, S., Sashikata, K. and Furuya, N. (1990) *J. Vac. Sci. Technol.*, **A8**. 515.

Iwasita-Vielstich, T. (1990) in *Advances in Electrochemical Science and Engineering, Gerischer, H.* and Tobias, C.W. (eds), VCH Publishers Inc., New York, Vol. 1, Chapter 3.

Jeanmarie, D.L. and Van Duyne, R.P. (1977) *J. Electroanal. Chem.*, **84**, 1.

Jiang, X.C., Seo, M. and Sato, N. (1991) *J. Electrochem. Soc.*, **138**, 137.

Kolb, D.M. (1988) in *Spectroelectrochemistry: Theory and Practice*, Gale R.J. (ed.), Plenum Press, New York, p. 150.

Kondrikov, N.B., Orlov, V.V., Ermakov, V.I. and Fioshin, M.Ya. (1972) *Elektrokhimiya*, **8**, 920.

Kotz, R. Neff, H. and Muller, K. (1986) *J. Electroanal. Chem.*, **215**, 331.

Kunimatsu, K., Seki, H., Golden, W.G., Gordon, II, J.G. and Philpott, M.R. (1986) *Langmuir*, **2**, 464.

Larkin, D., Guyer, K.L., Hupp, J.T. and Weaver, M.J. (1982) *J. Electroanal. Chem.*, **138**, 401.

Li, F.B., Hillman, A.R. and Lubetkin D. (1992) *Electrochim. Acta*, **37**, 2715.

Maki, A.H. and Geske, D.H. (1959) *J. Chem. Phys.*, **30**, 1356.

Manne, S., Hansma, P.K., Massie, J., Elings V.B. and Gewirth A.A. (1991) *Science*, **251**, 185.

Melendres, C.A., You, H., Maroni, V.A. and Nagy, Z. (1991) *J. Electroanal. Chem.*, **297**, 549.

Mengoli, G., Musiani, M., Fleischmann, M., Mao, B. and Tian, Z.Q. (1987) *Electrochim. Acta*, **32**, 1239.

Murray, C.A. and Allara, D.L. (1982) *J. Chen. Phys.*, **76**, 1290.

Paliteiro, C. (1985) D.Phil Thesis, Oxford.

Piette, L.H., Ludwig, R. and Adams, R.N. (1962) *Anal. Chem.*, **34**, 916.

Randles, J.E.B. (1948) *Trans. Far. Soc.*, **44**, 327.

Ranjith, S.K.A., Gamage, S.U. and McQuillan, A.J. (1990) *J. Electroanal. Chem.*, **284**, 229.

Russell, J.W., Overend, J., Scanlon, K., Severson M. and Bewick, A. (1982) *J. Phys. Chem.*, **86**, 3066.

Seo, M., Makino, T. and Sato, N. (1986) *J. Electrochem. Soc.*, **133**, 1138.

Seo, M., Jiang, X.C. and Sato, N. (1987) *J. Electrochem. Soc.*, **134**, 3094.

Sevcik, A. (1948) *Czech. Chem. Commun.*, **13**, 349.

Shirakawa, H., Louis, E.J., MacDiarmid, A.G., Chiang, C.K. and Heeger, A.J. (1977) *J. Chem. Soc. Chem. Comm.*, 578.

Uosaki, K. and Koinuma, M. (1992) *Far. Diss.* **94**.

Waller, A.M. and Compton, R.G. (1989) in *Comprehensive Chemical Kinetics*, Compton, R.G. and Hamnett, A. (eds), Elsevier, Amsterdam, Chapter 7, Vol. 29.

Wilhelm, S., Vielstich, W., Buschmann, H.W. and Iwasita, T. (1987) *J. Electroanal. Chem.*, **229**, 377.

Wolter, O. and Heitbaum, J. (1984) *Ber. Bunsenges. Phys. Chem.*, **88**, 2.

Further reading

Albery, W.J. (1975) *Electrode Kinetics*, Clarendon Press, Oxford.

Albery, W.J. and Hitchman, M. (1971) *Ring Disc Electrodes*, Clarendon Press, Oxford.

Arvia, A.J. (1990) in *Spectroscopic and Diffraction Techniques in Interfacial Electrochemistry*, Gutierrez, C. and Melendres, C. (eds), Nato ASI Series, Series C: Mathematical and Physical Sciences, Vol. 320, Kluwer Academic Publishers, Dordrecht, Chapter 16.

Ashley, K. and Pons, S. (1988) in *Chem. Rev.*, **88**, 673.

Banwell, C.N. (1983) *Fundamentals of Molecular Spectroscopy*, McGraw-Hill, Maidenhead.

Bard, A.J. and Faulkener, L.R. (1980) *Electrochemical Methods*, John Wiley and Sons, New York.

Beden, B. and Lamy, C. (1988) in *Spectroelectrochemistry: Theory and Practice*, Gale, R.J. (ed.), Plenum Press, New York, Chapter 5.

Behm, R.J., Garcia, R. and Rohrer, H. (1990) *Scanning Tunnelling Microscopy and Related Methods*, NATO ASI Series, Series E: Applied Sciences, Vol. 184, Kluwer Academic Publishers, Dordrecht.

Bewick, A. and Pons, S. (1985) in *Advances in Infrared and Raman Spectroscopy*, Clark, R.J.H. and Hester, R.E. (eds), Vol. 12, Wiley-Heyden, p. 1.

Birke, R.J. and Lombardi, J.R. (1988) in *Spectroelectrochemistry: Theory and Practice*, Gale, R.J. (ed.), Plenum Press, New York, Chapter 6.

Bockris, J.O'M. and Reddy, A.K.N. (1970) *Modern Electrochemistry*, Plenum, New York.

Bruckenstein, S. and Shay, M. (1985) *Electrochim. Acta*, **30**, 1295.

Cataldi, T.R.I., Blackham, I.G., Briggs, G.A.D., Pethica, J. and Hill, H.A.O. (1990) *J. Electroanal. Chem.*, **290**, 1.

Chang, R.K. (1990) in *Spectroscopic and Diffraction Techniques in Interfacial Electrochemistry*, Gutierrez, C. and Melendres, C. (eds), Nato ASI Series, Series C: Mathematical and Physical Sciences, Vol. 320, Kluwer Academic Publishers, Dordrecht, Chapter 5.

Christensen, P.A. (1992) *Chem. Soc. Rev.*, **21**, 207.

Christensen, P.A. and Hamnett, A. (1989) in *Comprehensive Chemical Kinetics*, Vol. 29, Compton R.G. and Hamnett, A. (eds), Elsevier, Amsterdam, Chapter 1.

Compton, R.G. and Hamnett, A. (1989) *Comprehensive Chemical Kinetics*, Vol. 29, Elsevier, Amsterdam.

Compton, R.G. and Waller, A.M. (1988) in *Spectroelectrochemistry: Theory and Practice*, Gale, R.J. (ed.), Plenum Press, New York, Chapter 5.

Fleischmann, M. and Hill, I.R. (1984) in *Comprehensive Treatise of Electrochemistry*, Vol. 8, White, R.E., Bockris, J.O'M., Conway, B.E. and Yeager, E. (eds), Plenum Press, New York, Chapter 6.

Foley, J.K. and Pons, S. (1985) *Anal. Chem.*, **57**, 945A.

Greef, R. (1984) in *Comprehensive Treatise of Electrochemistry*, White, R.E., Bockris, J.O'M., Conway, B.E. and Yeager, E. (eds), Vol. 8, Plenum Press, New York, Chapter 5.

Greef, R. (1989) in *Comprehensive Chemical Kinetics*, Compton, R.G. and Hamnett, A. (eds), Vol. 29, Elsevier, Amsterdam, Chapter 10.

Griffiths, P.R. and deHaseth, J.A. (1986) *Fourier Transform Infrared Spectrometry*, Wiley-Interscience, New York.

Gutierrez, C. and Melendres, C. (1990) *Spectroscopic and Diffraction Techniques in Interfacial Electrochemistry*, Nato ASI Series, Series C: Mathematical and Physical Sciences, Vol. 320, Kluwer Academic Publishers, Dordrecht.

Hamers, R.J. (1989) *Ann. Rev. Phys. Chem.*, **40**, 531.

Hammond, J.S. and Winograd, N. (1984) in *Comprehensive Treatise of Electrochemistry*, Vol. 8, White, R.E., Bockris, J.O'M., Conway, B.E. and Yeager, E. (eds), Plenum Press, New York, Chapter 8.

Hansen, W.N. (1973) in *Advances in Electrochemistry and Electrochemical Engineering*, Vol. 9, Delahay, P. and Tobias, C.W. (eds), John Wiley and Sons, New York, Chapter 1.

Heineman, W.R., Hawkridge, F.M. and Blount, H.N. (1984) in *Electroanalytical Chemistry, A Series of Advances*, Vol. 13, Bard, A.J. (ed.), Marcel Dekker Inc., New York, Chapter 1.

Hester, R.W. (1989) in *Comprehensive Chemical Kinetics*, Compton, R.G. and Hamnett, A. (eds) Vol. 29, Elsevier, Amsterdam, Chapter 2.

Hollas, J.M. (1988) *Modern Spectroscopy*, John Wiley and Sons, Chichester.

Howarth, O. (1973) *Theory of Spectroscopy*, Thomas Nelson and Sons Ltd, London.

Iwasita, T. and Vielstich, W. (1990) in *Advances in Electrochemical Science and Engineering*, Vol. 1, Gerischer, H. and Tobias, C.W. (eds), VCH Publishers Inc., New York, Chapter 3.

Kastening, B. (1984) in *Comprehensive Treatise of Electrochemistry*, Vol. 8, White, R.E., Bockris, J.O'M., Conway, B.E. and Yeager, E. (eds), Plenum Press, New York, Chapter 7.

Kruger, J. (1973) in *Advances in Electrochemistry and Electrochemical Engineering*, Delahay, P. and Tobias, C.W (eds.), Vol 9, John Wiley and Sons, New York, Chapter 4.

Macdonald, J.R. (1987) John Wiley and Sons, New York.

McLauchlan, K.A. (1972) *Magnetic Resonance*, Clarendon Press, Oxford.

Melendres, C.A. (1990) in *Spectroscopic and Diffraction Techniques in Interfacial Electrochemistry*, Gutierrez, C. and Melendres, C. (eds), Nato ASI Series, Series C: Mathematical and Physical Sciences, Vol. 320, Kluwer Academic Publishers, Dordrecht, Chapter 6.

Miles, R. (1983) *Surf. and Inter. Anal.*, **5**, 43.

Muller, R.H. (1973) in *Advances in Electrochemistry and Electrochemical Engineering*, Delahay, P. and Tobias, C.W. (eds), Vol. 9, John Wiley and Sons, New York, Chapter 3.

Nichols, R.J. (1992) in *Adsorption of Molecules at Metal Electrodes*, Lipkowski, J. and Ross, P.N. (eds), VCH Publishers Inc., New York, Chapter 7.

Parker, V.D. (1986) in *Electroanalytical Chemistry, A Series of Advances*, Vol. 14, Bard, A.J. (ed.), Marcel Dekker Inc., New York, Chapter 1.

Pettinger, B. (1992) in *Adsorption of Molecules at Metal Electrodes*, Lipkowski, J. and Ross, P.N. (eds), VCH Publishers Inc., New York, Chapter 6.

Plieth, W., Kozlowski, W. and Twomey, T. (1992) in *Adsorption of Molecules at Metal Electrodes*, Lipkowski, J. and Ross P.N. (eds), VCH Publishers Inc., New York, Chapter 5.

Robinson, J. (1988) in *Spectroelectrochemistry: Theory and Practice*, Gale, R.J. (ed.), Plenum Press, New York, Chapter 2.

Sauerbrey, G. (1955) *Z. Phys.*, **155**, 206.

Sawyer, D.T. and Roberts, J.L. (1974) *Experimental Electrochemistry for Chemists*, John Wiley and Sons, New York.

Sonnenfeld, R., Schneir, J. and Hansma P.K. (1990) in *Modern Aspects of Electrochemistry*, White R.E., Bockris, J.O'M. and Conway, B.E. (eds), Plenum Press, New York, p. 1.

Southampton Electrochemistry Group (1990) *Instrumental Methods in Electrochemistry*, Ellis Horwood, Chichester.

Waller, A.M. and Compton, R.G. (1989) in *Comprehensive Chemical Kinetics*, Compton, R.G. and Hamnett, A. (eds), Vol. 29, Elsevier, Amsterdam, Chapter 7.

West, A.R. (1985) *Solid State Chemistry and its Applications*, John Wiley and Sons, Chichester.

3 Examples of the application of electrochemical methods

This chapter is intended to provide an idea of the application of the various techniques described in the previous chapter to the problems commonly encountered by the electrochemist. Constraint on space precludes an exhaustive treatment of the various systems to be described below; for such a treatment, the reader is encouraged to consult the further reading list at the end of the chapter.

3.1 The cyclic voltammogram of platinum in acid solution

Platinum is the most catalytically active substrate employed by the electrochemist and it shows no dissolution between the potentials of hydrogen and oxygen evolution in aqueous solution. Hence, it finds common application as the working electrode in the study of a wide range of systems. The cyclic voltammogram of a platinum electrode in aqueous $1\,M\,H_2SO_4$ is shown in Figure 3.1. Provided that the solutions are free of organic and inorganic impurities, the hydride and oxide adsorption peaks are well defined and we consider them now separately.

3.1.1 The hydride region

Prior to the advent of cyclic voltammetry, the most common approach to the investigation of a new electrochemical system involved the measurement of the potential of the electrode of interest, as a function of time, during the passage of a fixed current. The results of such an experiment were usually plotted as a graph of charge passed (i.e. current × time) vs. the potential. Bowden (1928, 1929) studied the Hg, mercury amalgam and Pt/Hg interfaces with this technique, and postulated the existence of adsorbed hydrogen at potentials near the hydrogen potential (i.e. $0\,V$ vs. RHE), probably in an amount equivalent to a single layer. This conclusion was hotly contested by several workers including Frumkin and Slygin (1934) who disputed the existence of adsorbed hydrogen or oxygen on thermodynamic grounds. To investigate further the possibility of such adsorption Bowden embarked on a study of the potential-charge behaviour of a platinum-foil electrode, choosing a Pt substrate on the basis that such measurements would not be complicated by the dissolution of the substrate. The currents employed were

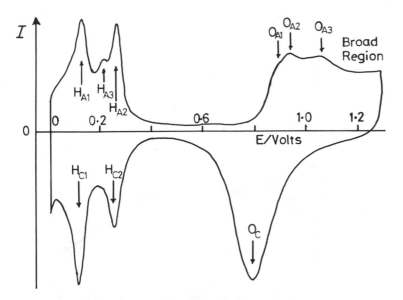

Figure 3.1 Cyclic voltammogram of a Pt electrode in 0.5 M H_2SO_4 at 25°C. The scan rate is 0.1 V s^{-1} (see text for details).

reasonably high, to minimise interference by cutting down on experimental running time, and a series of experiments were performed in H_2SO_4 using platinum electrodes pre-treated in various ways. Figure 3.2(a) shows an E vs. Q plot obtained from a Pt electrode in H_2-saturated electrolyte that had only been employed as a cathode, i.e. had never been exposed to oxygen once immersed in the electrolyte. Thus, after the electrode had been evolving H_2 for a sufficient length of time (with the potential of the Pt $= -0.6$ V vs. SCE), the polarity of the current was suddenly reversed, such that the potential increased as shown before reaching an equilibrium value of $+1.6$ V with the evolution of O_2. In order to measure the potential accurately, particularly in the early portion of the trace, Bowden employed a movie camera to photograph the scale of the galvanometer employed as the monitoring device. The reading on the scale was read from each of the stills, with a time resolution of a then impressive 1/300th of a second. He postulated that such a potential excursion must take the state of the surface from a monolayer of adsorbed hydrogen to a monolayer of adsorbed oxygen, as PtO. The total charge passed during this postulated process was 3 mC, the experiment lasting *c.* 0.1 s.

Figure 3.2(b) shows a second experiment, also performed in H_2-saturated electrolyte. Anodic current was passed such that the measured equilibrium potential was $+1.9$ V, with O_2 evolution. The polarity of the current was then reversed and the potential monitored with time, as before. The length of the plateau at 0.37 V was found to depend on the time for which the electrode had been held while passing the anodic current. Excluding this

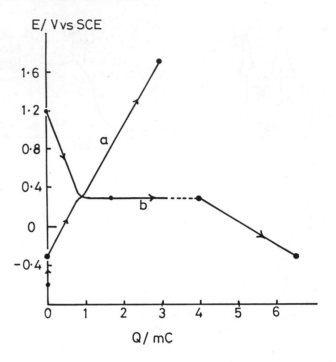

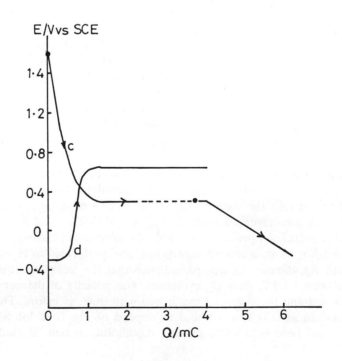

plateau, or arrest as the author termed it, the current passed, on traversing the potential range from O_2 evolution to H_2 evolution, was 3.2 mC.

The initial rapid changes in the potential observed in both the experiments employing cathodic currents and those using anodic currents are due to the recharging of the double layer that follows a switch in the polarity of the current.

Similarly, Figures 3.2(c) and (d) depict the same experiments as those in (a) and (b) except that they were carried out in O_2-saturated solution. In the complete (initial) absence of H_2, i.e. after switching from anodic to cathodic current, the potential falls linearly and rapidly to $+0.9$ V and then more slowly to 0.37 V where the potential is arrested for a period, before dropping to the equilibrium potential. Again, the length of the arrest was dictated by the time held while evolving oxygen. Excluding the plateau, 2.5 mC of current was passed during the experiment.

Bowden reasoned that passing from the H_2 potential to the O_2 potential, or vice-versa, clearly required c. 3 mC of charge and that it was this charge that was employed in the stripping of one adsorbed layer and the subsequent formation of the other. The 'arrests' at $+0.37$ V for passage from the H_2 potential to the O_2 potential (and vice-versa) in H_2-saturated solution, and at 0.62 V in O_2-saturated solution (see Figures 3.2(a)–(d)), were clearly dependent on the time spent at the anodic potential. He explained this in terms of the formation of a thicker oxide layer with an equilibrium potential of 0.62 V of the oxide in O_2-saturated solution and 0.37 V the equilibrium potential of the oxide in H_2-saturated solution.

As was discussed in section 2.1.1, electrocapillarity measurements at mercury electrodes, which have well-defined and measurable areas, allow the double-layer capacitance, C_{DL}, to be obtained as $F\,m^{-2}$. Bowden assumed that the overpotential change at the very beginning of the anodic run in H_2-saturated solution was a measure of the double-layer capacity. The slope of the E vs. Q plot in this region was taken as giving $1/C_{DL}$, and this gave 2×10^{-5} F. He then assumed that, under these same conditions, the double-layer capacity, in $F\,m^{-2}$, of the mercury electrode is the same. This gave the *real* surface area of the electrode as $3.3\,cm^{-2}$, as opposed to its *geometric* area of $1\,cm^2$.

Figure 3.2   The variation in the potential, as a function of charge passed, of a Pt electrode immersed in aqueous sulphuric acid. The current density was $2.5 \times 10^{-2}\,A\,cm^{-2}$ and the potential axis is vs. the RHE. In (a) the electrode was set first to generate H_2, such that the solution became saturated with H_2 and no O_2 was present. The polarity of the current was then switched from cathodic to anodic and the trace recorded. The trace shown in (b) was recorded after that in (a) and on reversing the polarity of the current again to cathodic. During the time taken for the polarity to be reversed the potential fell from $+1.9$ V to $+1.2$ V. The trace in (c) was obtained *via* a similar type of experiment except starting from oxygen-saturated solution. Thus, trace (c) was recorded after holding the current anodic to generate O_2 before reversing the polarity. The trace in (d) was obtained after that in (c) by passing cathodic current in the O_2-saturated solution for a few seconds before switching the electrode to open circuit. After Bowden (1928, 1929).

Thus, 9×10^{-4} C/real cm^2 are involved in stripping a monolayer of H_{ads} and forming a monolayer of O_{ads}.

The diameter of the Pt atom can be calculated from the density as 2.5×10^{-8} cm; postulating that the adsorption following the metal packing (rather than forming a compact layer on the surface, with each adsorbate atom touching another) gave 1.6×10^{15} H atoms/real cm^2. To remove this number of atoms from the surface according to:

$$H_{ads} \longrightarrow H^+ + e^- \qquad (3.1)$$

requires 2.7×10^{-4} C, and to deposit an equivalent number of O atoms according to:

$$H_2O \longrightarrow O_{ads} + 2H^+ + 2e^- \qquad (3.2)$$

requires $2 \times 2.7 \times 10^{-4}$ C. This gives a total charge of 8.1×10^{-4} C, in reasonable agreement with the observed value of 9×10^{-4} C.

Bowden concluded that at the hydrogen potential the surface of the Pt was covered with a monolayer of adsorbed hydrogen. In addition, his work suggested the presence of an oxide layer at the O_2 potential that could grow to a thickness greater than a monolayer. In support of his postulate that the adsorbed hydrogen followed the metal packing he cited evidence from the electron diffraction work of Davisson and Germer (1927) (the early origin of LEED) who found that gas atoms adsorbed on Ni followed the spacing of the Ni atoms.

Hence, Bowden's work firmly emplanted the concept of hydrogen and oxygen adsorption, with the maximum coverage of the former being a monolayer. However, it was assumed that the coverage increased smoothly from zero as the potential was decreased. Consequently, the next development was the realisation that there was more than one form of adsorbed hydrogen. The first studies to identify two distinct forms of adsorbed hydrogen were those of Eucken and Weblus (1951) and Breiter et al. (1956), both using AC impedance. They showed that the pseudo-capacitance, $C = -F(\partial n_H/\partial E)$, where n_H is the amount of adsorbed hydrogen per unit area, which when plotted against potential, E, revealed two distinct maxima, as shown in Figure 3.3. These maxima corresponded to weakly and strongly adsorbed forms of hydrogen.

Studies on the adsorption of hydrogen from the gas phase had provided strong evidence for the existence of two forms of adsorbed hydrogen and the AC impedance studies were supported by the results of the new LSV and CV techniques. The early measurements using the voltammetry methods were hampered by the use of impure electrolytes which resulted in ill-defined hydrogen adsorption and desorption peaks; but the realisation of the need for a clean electrochemical system soon resulted in the routine observation of the now familiar twin H_{ads} peaks.

Further confirmation of the existence of two well-defined forms of adsorbed hydrogen was provided by the work of Breiter (1964) who modelled the

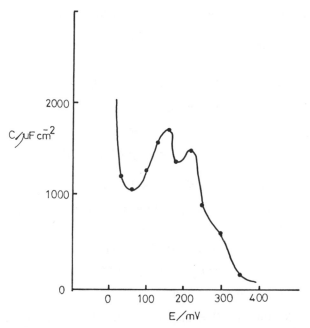

Figure 3.3 Capacity of a smooth Pt electrode at 40 Hz in 1 N K_2SO_4 + 0.1 N H_2SO_4. After A.N. Frumkin in *Advances in Electrochemistry and Electrochemical Engineering*, eds. P. Delahay and C.W. Tobias, Vol. 3, Chapter 5, John Wiley and Sons, New York, 1963. Copyright © 1963, reprinted by permission of John Wiley and Sons, Inc.

current/potential profile of platinum in the hydrogen adsorption region in terms of Langmuir and Frumkin isotherms.

Breiter's work was based on the following logic. The processes taking place during the cathodic sweep into the potential region of H_2 evolution are:

$$H^+ + e^- \longrightarrow H_{ads} \tag{3.3}$$

$$2H_{ads} \longleftrightarrow H_2 \tag{3.4}$$

Giving the overall process:

$$2H^+ + 2e^- \longleftrightarrow H_2 \tag{3.5}$$

This will obey the Nernst equation:

$$E = E^0 - (RT/2F)\log_e\{(p/p^0)/[H^+]^2\} \tag{3.6}$$

where p is the pressure of the hydrogen gas in equilibrium with the electrode and p^0 is the pressure corresponding to 1 atmosphere. It is assumed that equation (3.4) is always in equilibrium and the adsorbed hydrogen atoms obey the Langmuir Isotherm, i.e. (i) the maximum coverage is a single layer, (ii) all sites on the surface are identical, and (iii) the enthalpy of adsorption is independent of coverage. The author also assumed that each of the two forms of H_{ads} occupied two distinct regions up to a maximum coverage of a monolayer in each region, and hence individually obeyed the Langmuir isotherm, to give two overlapping isotherms.

The rates of the adsorption and desorption in equation (3.4) can be written:

$$V_{ads} = k_{ads}(p/p^0) \cdot (1 - \theta)^2$$
$$V_{des} = k_{des}\theta^2$$

as the processes are at equilibrium, $V_{des} = V_{ads}$, and:

$$k_{ads}/k_{des} = K = \theta^2/(p/p^0) \cdot (1 - \theta)^2$$

or:

$$p/p^0 = \theta^2/K(1 - \theta)^2 \tag{3.7}$$

In terms of the chemical potentials of the adsorbate, μ_a, and gas, μ_g, we can write:

$$\mu_a = \mu_a^0 - RT \log_e \{\theta/(1 - \theta)\}$$
$$\mu_g = \mu_g^0 - RT \log_e \{p/p^0\}$$

again, at equilibrium $\mu_g = 2\mu_a$ and:

$$RT \log_e \{p/p^0\} = (2\mu_a^0 - \mu_g^0) + 2RT \log_e \{\theta/(1 - \theta)\} \tag{3.8}$$

Rearranging:

$$\log_e \{p/p^0\} - 2 \log_e \{\theta/(1 - \theta)\} = (2\mu_a^0 - \mu_g^0)/RT$$

or:

$$\log_e \{p/p^0\} - \log_e \{\theta/(1 - \theta)\}^2 = (2\mu_a^0 - \mu_g^0)/RT$$
$$\log_e [\{p/p^0\}/\{\theta/(1 - \theta)\}^2] = (2\mu_a^0 - \mu_g^0)/RT$$
$$\{p/p^0\}/\{\theta/(1 - \theta)\}^2 = \exp\{(2\mu_a^0 - \mu_g^0)/RT\} \tag{3.9}$$

The right hand side of equation (3.9) is comprised entirely of constants, hence we can write:

$$p/p^0 = \theta^2/K(1 - \theta)^2 \tag{3.10}$$

with $K = \exp\{-(2\mu_a^0 - \mu_g^0)/RT\}$, which is the Langmuir isotherm.

The pressure at any potential E can be calculated from the Nernst equation and, at $\theta = 1/2$ for either of the two forms of adsorbed hydrogen, from equation (3.8):

$$RT \log_e \{p(\theta = 1/2)/p^0\} = (2\mu_a^0 - \mu_g^0)$$

where $p(\theta = 1/2)$ is the pressure at half coverage. This can be calculated for each of the two forms of hydrogen from the Nernst equation, by reading off the peak potentials for the adsorbates from the voltammogram (see section on the voltammetry of adsorbed species). Thus, the constant $(2\mu_a^0 - \mu_g^0)$ term can be calculated and, by applying equations (3.6) and (3.8) individually to the strongly and weakly adsorbed hydrides, the coverage of each of the two forms of adsorbed hydrogen can be calculated as a function of the potential.

Breiter expressed the net current I as:

$$I = -N_w^0 F(d\theta_w/dt) - N_s^0 F(d\theta_s/dt) \tag{3.11}$$

where N_i^0 is the number of hydride atoms of type i corresponding to monolayer coverage of its sites (i.e. $N_w^0 + N_s^0 =$ the total possible coverage of the Pt surface by weakly and strongly adsorbed hydrogen). $d\theta/dt$ will be positive on the cathodic scan and negative on the anodic scan.

From equation (3.11):

$$(d\theta/dt) = (d\theta/dE)\cdot(dE/dt)$$
$$(d\theta/dt) = (d\theta/dE)v \tag{3.12}$$

where v is the scan rate. We can obtain E in terms of θ by replacing (p/p^0) by equation (3.7) in the Nernst equation:

$$E = E^0 - (RT/2F)\log_e\{(1/K)\cdot[\theta/(1-\theta)]^2\}$$

and rearranging we obtain:

$$E = E^0 - (RT/F)\log_e[\theta/(1-\theta)] + (RT/2F)\log_e K$$

Differentiating:

$$dE/d\theta = -(RT/F)\cdot[(1/\theta) + (1/\{1-\theta\})]$$

Simplifying and inverting:

$$d\theta/dE = -(F/RT)\cdot\theta(1-\theta)$$

From equation (3.12), therefore:

$$d\theta/dt = -(F/RT)\cdot\theta(1-\theta)v \tag{3.13}$$

The net current is thus:

$$I = N_w^0 F(F/RT)\cdot\theta_w(1-\theta_w)v + N_s^0 F(F/RT)\cdot\theta_s(1-\theta_s)v \tag{3.14}$$

The scan rate is negative on the cathodic scan and positive on the anodic scan. Clearly, on first entering the hydrogen adsorption region, $\theta_w = 0$ across the potential range at which the strongly bound hydride is formed, hence the first term on the RHS is 0. Similarly, on continuing into the region of weakly adsorbed hydrogen, $\theta_s = 1$, and the second term becomes zero. For species i, therefore, the maximum (peak) current, I_{ip}, will occur when $\theta_i = 1/2$ and is given by:

$$I_{ip} = N_i^0 Fv(F/RT)/4 \tag{3.15}$$

Replacing equation (3.15) into (3.14) gives:

$$I = 4I_{wp}\theta_w(1-\theta_w) + 4I_{sp}\theta_s(1-\theta_s) \tag{3.16}$$

As was shown above, the coverage of adsorbed hydrogen can be calculated at each potential and this thus allows the calculation of the net current as a function of potential via equation (3.16).

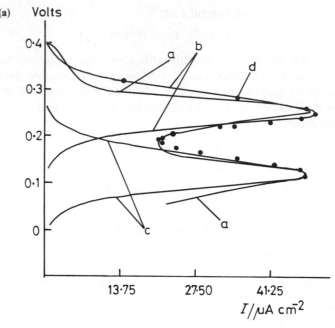

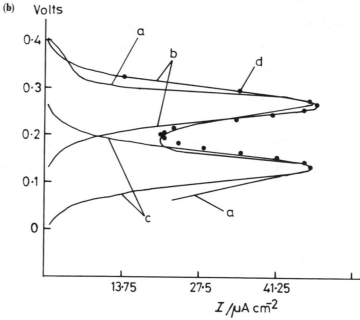

Figure 3.4 (a) Experimental $I - E$ curve obtained at 0°C and curves computed from the assumption of a Langmuir isotherm. a = experimental curve; b = partial curve for the strongly bound hydrogen, c = partial curve for the weakly bound hydrogen, d = computed net curve. (b) Experimental $I - E$ curve obtained at 70°C and curves computed from the assumption of a Frumkin isotherm. a = experimental curve; b = partial curve for the strongly bound hydrogen, c = partial curve for the weakly bound hydrogen, d = computed net curve. After Breiter (1964).

Hence, the author was able to model the I/E curves in the hydrogen adsorption region of Pt. The calculations were compared with experiment (using highly purified electrolytes), at various temperatures, and the result obtained at 0°C is shown in Figure 3.4(a).

As can be seen from the figure, the 'fit' of the model to the experimental data is reasonably good and this was found to be generally true at temperatures < 20°C. At temperatures > 50°C, the experimental data were found to fit a Frumkin isotherm, as shown in Figure 3.4(b). This isotherm is related to the Langmuir treatment save that it makes some allowance for the interaction between adsorbed species, i.e. it allows ΔH_{ads} to change with coverage through the parameter β. Thus, the chemical potential of the adsorbed species becomes:

$$\mu_a = \mu_a^0 - 2RT\beta - RT\log_e\{\theta/(1-\theta)\}$$

$\beta > 0$ is an attractive interaction, $\beta < 0$ repulsive and $\beta = 0$ results simply in the Langmuir isotherm. Under these conditions, the net current is given by:

$$I = I_{wp}(4 - 2\beta_w)\theta_w(1 - \theta_w)/[1 - 2\beta_w\theta_w(1 - \theta_w)]$$
$$+ I_{sp}(4 - 2\beta_s)\theta_s(1 - \theta_s)/[1 - 2\beta_s\theta_s(1 - \theta_s)] \tag{3.17}$$

The fit was again good, providing that the interaction parameter was positive for both types of adsorbed hydrogen, indicating an attractive interaction between adsorbed particles and thus a decreasing ΔH_{ads} with increasing θ.

The fact that the observed electrochemistry could be explained in terms of two overlapping but discrete isotherms was taken as strong evidence for the existence of two discrete forms of H_{ads}.

As the platinum CV became more familiar, and the concept of monolayer hydride adsorption accepted, discussion then arose as to the exact potential in the cathodic scan at which monolayer adsorption is complete; this has a major impact on the calculation of the coverage of Pt by organic adsorbates via the reduction in charge under the hydrogen adsorption/desorption peaks (see below). One suggested solution was postulated by Biegler *et al.* (1971) and is shown in Figure 3.5. The double-layer charging current was extrapolated back to 0 V, and the contribution to the current below *c.* 0.1 V from hydrogen evolution estimated as shown. The two forms of hydrogen were also classified in terms of 'strongly bound' and 'weakly bound'. The former has the higher enthalpy of adsorption and hence is the first to adsorb during the cathodic sweep and the last to be stripped off during the anodic sweep.

Having firmly established that there were two forms of hydrogen adsorbed on Pt, the question then arose as to their identity. Some of the strongest evidence with respect to this came from the *in situ* UV–visible reflectance studies of Bewick and Tuxford in the 1970s. The authors employed phase-sensitive detection to investigate the reflectivity changes observed on modulating the potential of the Pt electrode across the hydride region.

The full treatment of the reflectance results is beyond the scope of this

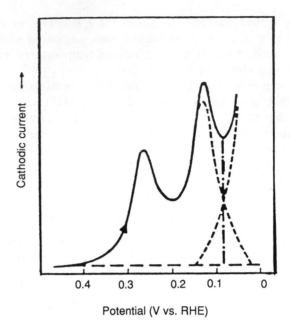

Figure 3.5 Schematic cathodic voltammograms in the hydrogen region. (—) total current, (---) double layer charging current, (- - -) extrapolated hydrogen adsorption and evolution currents, (—·—) limit for integration of the hydrogen adsorption charge. After Biegler *et al.* (1971).

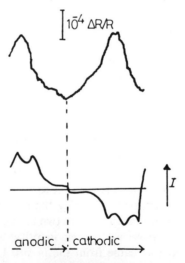

Figure 3.6 Reflectivity-potential curve (top) and corresponding current-potential cyclic voltammograms (bottom) for a platinum electrode in $1.0\,M\ H_2SO_4$. The reflectivity curve was taken at 546 nm using S-polarised light at a 70° angle of incidence. The potential limits for both the reflectivity and cyclic voltammetry experiments were $+0.535\,V$ and $-0.006\,V$ vs. NHE, and the scan rate was $26.46\,V\,s^{-1}$. From Bewick and Tuxford (1973).

book. Briefly, Figure 3.6 shows a reflectivity/potential curve obtained at constant wavelength whilst the potential was modulated between $-6\,mV$ and $+535\,mV$ vs. NHE. For comparison, the cyclic voltammogram obtained at the same frequency, 30 Hz ($26.46\,V\,s^{-1}$), is also shown. Taking only the cathodic sweep, the increase in reflectivity on passing through the double-layer region was explained by the authors in terms of an increase in the charge density at the electrode surface, giving a corresponding increase in the reflectivity, as would be expected.

The first form of hydride obtained on leaving the double-layer region is the strongly bound species. Its formation coincides with a relatively large increase in reflectivity. This increase in reflectivity with decreasing potential in the potential range covering the adsorption of the strongly bound form was found for all wavelengths $> c.\ 380\,nm$. Such an *increase* in reflectivity on forming a chemisorbed surface layer that would be expected to *absorb* light and so give a *decrease* in reflectivity was wholly unexpected. For a typical chemisorbed layer of a non-metal, the reflectivity equations would predict a decrease in reflectivity with increasing coverage and with increasing wavelength. This was found for the weakly absorbed form (see Figure 3.6). Using a three-layer model (metal/adsorbate/solution), the results obtained by the authors could only be explained if $n_{metal} \approx n_{layer}$ and $k_{metal} < k_{layer}$. The authors obtained values of n and k for the strongly adsorbed hydride layer via a best-fit to their results of reflectivity change vs. wavelength and these were compared to literature values of n and k for platinum. The results arc shown in Figure 3.7. n_{metal} and n_{layer} are indeed very close at all wavelengths,

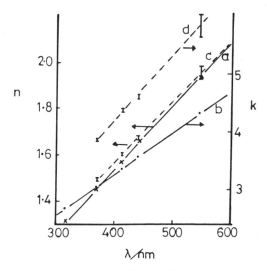

Figure 3.7 Curves (a) and (b): the variation in the optical constants of platinum with wavelength (from M.A. Barret and R. Parsons, *Symp. Far. Soc.*, **4** (1970) 72). Curves (c) and (d): the variation with wavelength of the calculated optical constants of the strongly-bound hydrogen layer (from Bewick and Tuxford (1973)).

whereas the values of k_{layer} are considerably greater than k_{metal} and the difference increases with wavelength. This was taken as evidence that the strongly bound hydrogen involves the H atom sitting within the electronic surface of the metal with its electron in the conduction band. The extra electron in the metal conduction band gives an increased value of k, and hence an increased reflectivity, whilst the presence of the H atom within the

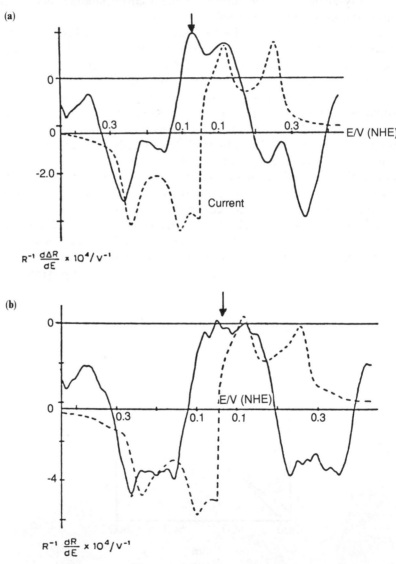

Figure 3.8 Current-potential linear sweep voltammogram and the differential reflectivity change in the hydrogen adsorption region at fixed wavelengths: (a) 2.34 μm and (b) 1.93 μm. The sweep rate was $15\,\text{mV}\,\text{s}^{-1}$, with a square wave modulation of $\pm 10\,\text{mV}$ at 8.5 Hz. From Bewick *et al.* (1981).

skin of the metal explains the fact that $n_{metal} = n_{layer}$. The data were also consistent with the weakly bound hydrogen simply forming a covalently bound layer on top of the platinum atoms.

The first papers reporting the EMIRS (see chapter 2) technique, by Bewick, Kunimatsu and colleagues (1980, 1981), were a more detailed UV−visible/near IR study of hydrogen adsorption on platinum. Two types of experiment were performed. In the first, the potential was modulated between two chosen values at c. 8.5 Hz, the wavelength slowly scanned at 6.3 μm s^{-1}, and the resulting variation in reflectivity recorded as a function of the wavelength (in microns, 10^{-6} m) via lock-in methods. An experiment employing this approach took up to c. 10 h of data collection time. In the second method, the electrode potential was linearly ramped between two limits and the reflected light intensity monitored at a single wavelength. The detector was 'locked-in' to a ± 10 mV modulation superimposed on the linear sweep.

Figures 3.8(a) and (b) show the linear sweep voltammograms and concomitant reflectivity changes monitored at 2.34 μm (4273 cm^{-1}) and 1.93 μm (5181 cm^{-1}). Under the conditions employed the reflectivity change was obtained as $(1/R) \cdot (\Delta R/dE)$. This gives spectra in which an *increase* in reflectivity results in a negative value of $(1/R) \cdot (\Delta R/dE)$ and a decrease in reflectivity (i.e. absorption) gives a positive response.

At all wavelengths, the formation of the strongly bound H_{ads} leads to a large increase in reflectivity, in agreement with the results of the earlier studies and their interpretation. However, the formation of the weakly bound hydrogen gives rather different results. At most wavelengths, typified by Figure 3.8(a), the value of the reflectivity change decays almost to zero in the region of the weakly-adsorbed H (arrowed in figure). However, at wavelengths near 1.95 μm (see Figure 3.8(b)) the reflectivity becomes positive, indicating optical absorption by the hydride. This was confirmed by measuring EMIRS spectra as a function of the potential modulation amplitude, as shown in Figures 3.9(a)–(f).

For modulation entirely in the weakly bound region, see Figures 3.9(a) and (b), the spectrum shows an absorption band centred at 1.95 μm. As the modulation is extended into the strongly bound region there is an increasing gain in reflectivity, with the absorption band superimposed. The authors assigned the 1.95 μm band to absorption by:

$$Pt-H-OH_2$$

on the basis of the following reasoning:

1. The band could not be due to Pt–H, as it occurs at much too high a frequency (see below).
2. It is not an overtone of Pt–H, since there was no trace of the fundamental on scanning to higher wavelengths.
3. There is a strong water combination band in this region so the band could be related to this, but it shows potential-dependent behaviour and hence must be a surface species.

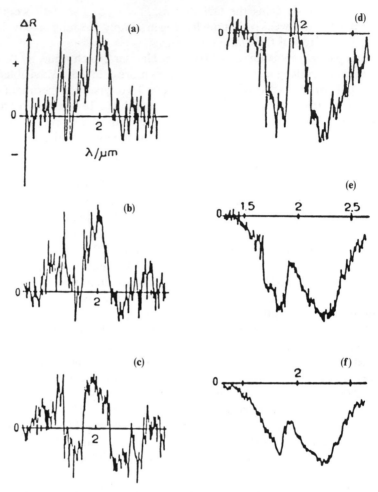

Figure 3.9   The effect of the modulation amplitude on the reflectivity change. Negative limit 0.05 V vs. NHE, positive limit: (a) 0.13 V; (b) 0.17 V; (c) 0.20 V; (d) 0.25 V; (e) 0.35 V; (f) 0.55 V. The relative scales of the figures are: (a) 1.0; (b)–(d) 1.25; (e) 2.5; (f) 5.0. From Bewick *et al.* (1981).

Thus, the spectroscopic work proved that the two forms of adsorbed hydrogen are very different. The weakly adsorbed form is definitely adsorbed above the Pt surface plane and so able to interact with the solution. In contrast, the strongly adsorbed form cannot interact with the solution, lying in the plane of the surface Pt atoms with its electron in the conduction band of the metal. This conclusion was supported by work that showed the strongly adsorbed form was not formed on close-packed surfaces.

Hence, it appeared that the story was complete and the adsorption of hydrogen on platinum understood. However, considerable controversy was arising even as the definitive experiments were being performed. In 1959, Schuldiner carried out kinetic measurements on the H_2-evolution reaction on platinum and concluded that the coverage of the metal surface by adsorbed hydrogen, the intermediate in the reaction, was extremely small at potentials immediately prior to the actual evolution. This, of course, is in apparently complete contrast to the conclusions reached on the basis of the experiments discussed above where a monolayer coverage of H_{ads} is predicted at $0\,V$ vs. RHE.

Schuldiner (1959) studied the effect of H_2 pressure on the hydrogen evolution reaction at bright (polished) Pt in sulphuric acid. The mechanism of the reaction was assumed to be as in equations (3.3) and (3.4). The step represented by equation (3.3) was assumed to be at equilibrium at all potentials and equation (3.4) represented the rate-determining step. The potentials were measured as overpotentials with respect to the hydrogen potential, i.e. the potential of the H^+/H_2 couple in the solution (0 V vs. RHE). The experiments were performed at 25°C, where the adsorption of both forms of hydride was assumed to follow the Langmuir isotherm. Thus, the current for the forward (cathodic) and reverse (anodic) reactions in equation (3.3) can be written as:

$$I_{ca} = -k_{ca}[H^+]_0(1-\theta)^2 \exp(-\beta\eta f) \tag{3.18}$$

$$I_{aa} = k_{aa}\theta^2 \exp([1-\beta]\eta f) \tag{3.19}$$

with $f = F/RT$. Equations (3.18) and (3.19) are simple modifications of the expressions used to derive the Butler–Volmer equation (see above). k_{ca} and k_{aa} are rate constants and $[H^+]_0$ is the surface concentration of H^+, which was assumed to remain constant over the current densities employed. As the two reactions are in equilibrium, $I_{ca} = -I_{aa}$, and hence:

$$k_{ca}[H^+]_0(1-\theta)^2 \exp(-\beta\eta f) = k_{aa}\theta^2 \exp([1-\beta]\eta f)$$

rearranging we obtain:

$$\theta^2/(1-\theta)^2 = K \exp(-\eta f) \tag{3.20}$$

where $K = k_{ca}[H^+]_0/k_{aa}$. At the equilibrium potential, where there is no net H_2 evolution, $\eta = 0$ and $\theta = \theta_0$, and:

$$\theta_0^2/(1-\theta_0)^2 = K \tag{3.21}$$

Otherwise, θ will change with the overpotential according to equation (3.20).

Turning now to the rate-determining step, the cathodic and anodic currents are:

$$I_{cev} = -k_{cev}\theta^2$$

$$I_{aev} = k_{aev}(p/p^0)(1-\theta)^2$$

where p and p^0 have the same meaning as that given above. The observed net H_2-evolution current at $\eta > 0$ is then:

$$I = I_{cev} - I_{aev} = -k_{cev}\theta^2 - k_{aev}(p/p^0)(1-\theta)^2 \tag{3.22}$$

At $\theta = \theta_0$, the net current is zero and the exchange current, I_0, is then:

$$I_0 = I_{aev} = -I_{cev}$$
$$I_0 = k_{cev}\theta_0^2 = k_{aev}(p/p^0)(1-\theta_0)^2$$

Thus:

$$k_{cev} = I_0/\theta_0^2 \tag{3.23}$$

$$k_{aev} = [I_0/(p/p^0)]\cdot(1-\theta_0)^2 \tag{3.24}$$

Replacing equations (3.23) and (3.24) in (3.22) gives:

$$I = I_0\{(\theta/\theta_0)^2 - [(1-\theta)^2/(1-\theta_0)^2]\} \tag{3.25}$$

which can be written in the form:

$$I = I_0[(1-\theta)^2/\theta_0^2]\cdot\{[\theta/(1-\theta)]^2 - [\theta_0^2/(1-\theta_0)^2]\} \tag{3.26}$$

Replacing $\theta/(1-\theta)$ and $\theta_0/(1-\theta_0)$ by equations (3.20) and (3.21), respectively, gives:

$$I = I_0[(1-\theta)^2(\{\exp[-2\eta f]\}-1)/(1-\theta_0)^2] \tag{3.27}$$

If the coverage of the adsorbed hydrogen intermediate is 1 at the hydrogen potential, i.e. $\theta_0 = 1$, then equation (3.27) would most certainly be inapplicable to the observed data. On the other hand, Schuldiner reasoned that if θ_0 was very low and the coverage only increased slowly as the potential was moved into the hydrogen evolution region, then equation (3.27) reduces to:

$$I = I_0(\{\exp[-2\eta f]\}-1)/(1-\theta_0)^2 \tag{3.28}$$

and hence a plot of the observed current against $(\{\exp[-2\eta f]\}-1)$ should be a straight line passing through the origin. Some of the data he obtained are plotted in Figure 3.10 and clearly equation (3.28) does apply: the coverage of the surface of the Pt by the hydride intermediate must be extremely low. Indeed, the author calculated that θ_0 was $c.\ 10^{-3}$ at a hydrogen pressure of 1 atm giving an exchange current of $1.6\ \mathrm{mA/cm^2}$. The coverage remained small even up to quite high current densities of $10\ \mathrm{mA/cm^{-2}}$. This was extremely convincing evidence that the coverage of the surface by the adsorbed hydrogen intermediate is nowhere near unity, in complete contrast to the accumulated wisdom of the time and supported by further evidence after.

These apparently mutually exclusive conclusions proved to be a rich source of controversy for over thirty years. Again, it was through the timely and elegant work of Nicholas and Bewick (1988) that the controversy was resolved by employing *in situ* FTIR to study the surface of a Pt electrode at potentials just into the region of hydrogen evolution. The spectra were obtained by accumulating interferograms at two potentials: the reference potential of 0.442 V vs. RHE in the double-layer region and a second potential E_s in the hydride region. The electrode was then switched alternately between the two values every 40 s and interferograms collected separately at each potential.

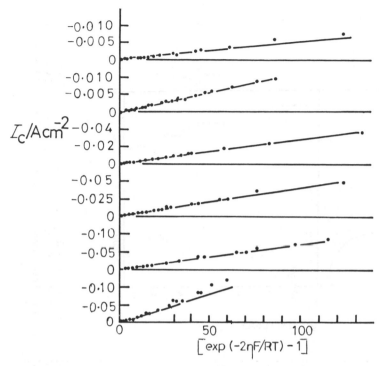

Figure 3.10 Plots of current density vs. $(\exp(-2\eta F/RT) - 1)$ for the evolution of hydrogen on bright platinum. From Schuldiner (1959).

The resultant co-added and averaged interferograms collected at E_s were transformed and normalised to those taken at the reference potential, and the experiment then repeated employing a lower value of E_s.

The 15 000 interferograms at each potential required to give the desired signal-to-noise needed several hours collection time. This posed two problems: instrumental drift leading to an unstable baseline and the maintenance of a clean surface. The former was overcome by a slow potential modulation, the potential being switched every 40 s; any drift in a particular 40 s period would be subtracted out as a result of the (hopefully) same drift in the sub-sequent period by the normalisation process. The problem of maintaining a clean surface was overcome by pulsing the potential of the Pt to $+1.0$ V for 1 s between every potential alteration, and so 'burning off' any adventitious organic impurity. Figure 3.11(a) shows IR spectra collected by the above method at various potentials from 0.027 V down to 0.242 V, and Figure 3.11(b) shows a linear sweep voltammogram taken during a cathodic sweep into the same potential region. At potentials $> c.$ 0.110 V the results were effectively as had been previously observed in the EMIRS experiments: no spectral bands were observed when E_s was in the double-layer region. On entering the region of strongly bound hydrogen a featureless gain of reflectivity was

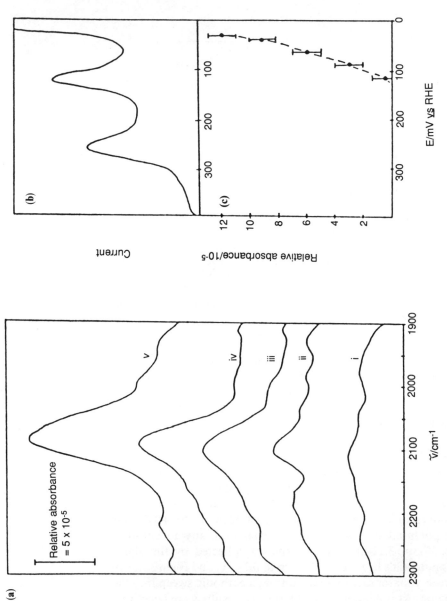

Figure 3.11 (a) Difference spectra from a polycrystalline Pt electrode in 1 M H_2SO_4, at potentials E_1 and E_2. E_1 was kept constant at $+442$ mV vs. RHE; the E_2 values used were: (i) $+242$ mV; (ii) $+67$ mV; (iii) $+82$ mV; (iv) $+42$ mV and (v) $+27$ mV. (b) The linear sweep voltammogram at 0.1 V s^{-1} on the negative scan over the hydrogen adsorption region for the system shown in (a). (c) The potential dependence of the intensity of the 2090 cm^{-1} band shown in (a). From Nichols and Bewick (1988).

observed, which had adsorption bands due to liquid water superimposed when E_s entered the weakly bound region. When E_s was made more negative than 0.110 V, a new (and extremely weak) absorption band appeared near 2090 cm^{-1}, the intensity of which increased as E_s was stepped to lower values (see Figure 3.11(c)). The data manipulation was such that the sign of the band indicated that it was due to species present at E_s but not at the reference potential of 0.442 V.

The authors were able to unequivocally assign the band to the vibration of a hydrogen atom singly coordinated on top of a single Pt atom, on the basis of IR measurements in the UHV. Confirmation of the assignment was achieved by repeating the experiment in D_2O/H_2SO_4. As expected, the band was observed to shift down to c. 1500 cm^{-1}, in agreement with the 1.39-fold decrease in the Pt–H stretch expected on deuteration.

The potential at which the Pt–H stretch appeared, and the correlation between its subsequent increase in intensity and the rise in the cathodic hydrogen evolution current, is extremely strong evidence that this form of H_{ads} is the intermediate in the H_2-evolution reaction as studied by Schuldiner (1959). This resolved the paradox between the kinetic results and the electrochemical measurements since Bowden. Clearly, the 'on-top' hydrogen is only present at extremely low coverage, presumably on active sites. The strongly and weakly bound hydrogen play no part in the reaction.

However, Bewick's results require the reassessment of the nature of the weakly bound form of adsorbed hydrogen. It has been since suggested that this form of hydrogen is multiply bonded, sitting between the surface atoms in a multi-coordination site.

3.1.2   The oxide region

As was discussed below, from the work of Bowden it was apparent that some form of adsorbed oxide resulted when the potential of a polycrystalline platinum electrode was rendered sufficiently anodic in aqueous solution. In contrast to galvanostatic charging measurements, such as those discussed in the previous section, cyclic voltammetry shows that there are at least three well-defined current peaks in the anodic sweep in the oxide region (O_{A1}, O_{A2} and O_{A3}, see Figure 3.1) followed by a broad, featureless almost capacitive region. However, the oxide layer so formed is reduced during the cathodic sweep in one single broad peak at somewhat more cathodic potentials than those at which the adsorbate is formed (O_C). This hysteresis strongly suggests some form of irreversible structural change that occurs during the anodic sweep and results in a more stable surface.

The first steps to the understanding of the nature of the surface oxide on platinum came with the advent of cyclic voltammetry and, as was noted above, the first major hurdle to overcome in trying to interpret the form of the Pt voltammogram was the purity of the electrochemical system. Early electrochemical measurements on platinum had been reported as showing that the charge passed during the formation of the oxide layer exceeded that passed during its subsequent removal by between 10–100%. However, Angerstein-Kozlowska *et al.* (1973) showed that this apparent inequality was almost certainly due to the presence of impurities and if the cyclic voltammetry was performed under extremely pure conditions, the charges were identical.

The early debate concerning the nature of the oxide film was intensified by the work of Reddy *et al.* (1968) who first employed ellipsometry in the study of the oxidation process.

The authors measured Δ and Ψ at a potential in the double-layer region where the electrode is film-free. Two methods were then employed; either (a) the potential was stepped to the desired value and Δ and Ψ measured, or (b) the potentiostatic control was rapidly switched to galvanostatic control, under which a constant anodic current was passed, with the potential being measured, as well as Δ and Ψ. Typical results are shown in Figure 3.12(a). By employing a literature value of n of 2.625, the authors were able to obtain a best fit to the observed values of Δ and Ψ using a value of k of 1.5. This allowed a calculation of the thickness of the oxide film as a function of potential, as shown in Figure 3.12(b). The plot does not show any structure that can be correlated with O_{A1} etc. and no change was observed in Δ or Ψ until potentials > 0.9 V, presumably as a result of the fact that the n and k values of the PtOH layer are not dissimilar enough to those of the solvent and metal to allow detection. At these higher potentials their results appeared to be in agreement with the current–voltage data, which showed that the charge passed under the voltammogram increases linearly with potential at potentials > 0.9 V. However, from Figure 3.1, it can be seen that oxide formation commences at *c.* 0.72 V. The authors concluded from their results that the coulometry reveals a partial monolayer at potentials < 0.9 V that is not seen by the ellipsometer, a conclusion borne out by later experiments. In addition, they postulated two different forms of oxygen: that present up to a monolayer and a second form present as a monolayer and above, with a 'drastic change' in the nature of the electrode surface at about 0.95 V. In support of this observation they cited work concerning the kinetics of the electrochemical oxidation of various materials on platinum. Thus, the rate of the oxidation of hydrogen, ethylene and other hydrocarbons shows a large decrease at potentials > *c.* 0.95 V. Indeed, as will be seen below, the rate of methanol electro-oxidation starts to decrease almost immediately the oxide layer starts to form and declines to zero at monolayer coverage.

The authors were also the first to postulate the existence of a place-exchange mechanism at Pt, suggesting that the path of oxide formation is:

$$Pt + H_2O \longleftrightarrow PtOH + H^+ + e^- \qquad (3.29)$$

$$PtOH \longrightarrow OHPt \text{ (rds)} \qquad (3.30)$$

$$HOPt \longrightarrow PtO + H^+ + e^- \qquad (3.31)$$

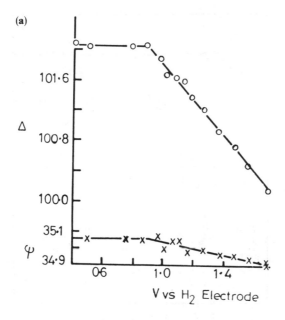

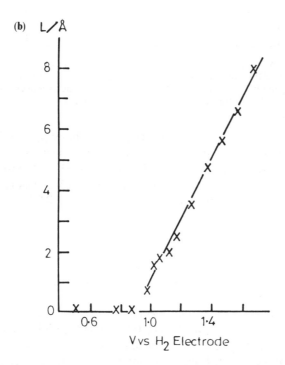

Figure 3.12 (a) Steady state values of Δ and Ψ as a function of potential. Platinum in aqueous sulphuric acid. (b) Dependence of oxide film thickness on potential. Platinum in aqueous sulphuric acid. From Reddy *et al.* (1968).

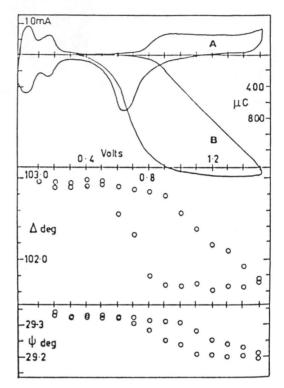

Figure 3.13 Curve A: Cyclic voltammogram for a Pt electrode in 0.5 M H_2SO_4. Electrodes are 1.13 cm^2, scan rate 0.1 cycles per second. Curve B: corresponding variation in the charge passed (in μC) vs. potential. Each point on the Δ and Ψ plots is the average of points obtained from four successive sweeps. From Greef (1969).

Notice the reversible step (3.29) and that the place-exchange is rate-determining. The inference was that at potentials > 0.95 V the oxygen-containing film takes on the characteristics of a true place-exchanged, or 'phase', oxide. As is discussed below, the conclusions of Reddy and colleagues (1968) on the properties of the oxide film at potentials > 1.0 V vs. RHE are essentially correct.

Later work by several research workers, and most particularly Greef (1969), showed that the very early stages of oxide film formation could be detected using more sensitive instrumentation and the changes in Δ, Ψ and the light intensity correlated very well with the charge passed both in the anodic and cathodic sweeps (see Figure 3.13).

Angerstein-Kozlowska and colleagues (1973) carried out cyclic voltammetry measurements on Pt in 0.05 M H_2SO_4, 0.5 M H_2SO_4 and 1 M $HClO_4$, and found that the voltammogram was unchanged, clearly showing that it was not a function of the anion. Figure 3.14 shows the charge passed in the oxide region, Q_{ox}, calculated from the area under the anodic sweep, as the ratio Q_{ox}/Q_H (where Q_H is the charge passed under the hydrogen adsorption) peaks

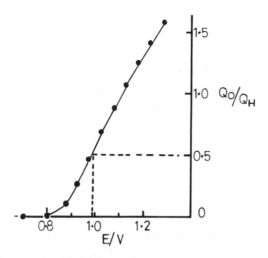

Figure 3.14 The integrated oxide formation charge, relative to the charge under the hydride formation region, Q_O/Q_H, as a function of anodic potential, for the cyclic voltammogram in Figure 3.1. From Angerstein-Kozlowska *et al.* (1973).

Table 3.1 Peak potential, charge passed and corresponding oxide coverage for the various oxide regions in the anodic sweep of the Pt cyclic voltammogram

Region	Charge to peak ($\mu C\,cm^{-2}$)	Peak potential (V vs. RHE)	Coverage to peak (based on e^-/Pt atom)	
O_{A1}	26–35	0.89	0.12–0.16	
O_{A2}	81–88	0.94–0.95	0.37–0.40	
O_{A3}	172	1.04–1.05	0.78	
Broad			1.10 V	1.00*
region			1.20 V	1.35
			1.30 V	1.65
			1.37 V	1.90

*Total coverage, in terms of e^-/Pt atom, up to indicated potential.

in the anodic or cathodic sweeps, corresponding to a monolayer (see above). For comparison, the platinum CV is also shown. Apart from a small change in slope above the critical ratio $Q_{ox}/Q_H = 1$, the line shows no structure that could correlate with O_{A1} etc., in agreement with galvanostatic charging measurements. Table 3.1 gives the characteristics of the various oxide peaks.

On the basis of the above observations, Angerstein-Kozlowska and colleagues (1973) reasoned thus: even at peak O_{A3}, less than 1 electron per Pt atom has passed, therefore, it is highly unlikely that the oxidation of the surface proceeds via a single two-electron process, i.e.:

$$Pt + H_2O \longrightarrow PtO + 2H^+ + 2e^- \qquad (3.32)$$

In addition, it is also unlikely that the oxidation proceeds via a sequential two-step mechanism, i.e.:

$$Pt + H_2O \longrightarrow Pt\text{--}OH + H^+ + e^- \tag{3.33}$$

$$Pt\text{--}OH \longrightarrow PtO + H^+ + e^- \tag{3.34}$$

since this would give rise to two current peaks in the voltammogram, at $\theta = 0.5$ and '$\theta = 1.5$'. Similarly, other such well-defined stoichiometric processes could be rejected. Thus the authors postulated that the resolution of the anodic sweep of the voltammogram into three distinct peaks at oxide coverages less than a monolayer could be attributed to the formation of distinct sublattices on the way to the production of a monolayer of OH_{ads}, by analogy with the results found for gas phase adsorption of O on Ni. The authors were able to resolve the voltammogram in the oxide region of the anodic sweep empirically in terms of three overlapping peaks corresponding to O_{A1} etc., as shown in Figure 3.15(a). The data are plotted as E vs. the pseudocapacitance C, calculated as i/v, ($C = dQ/dE = dQ/dt \times dt/dE$). The best 'fit' was obtained for a peak separation between O_{A1} and O_{A2} of 0.075 V,

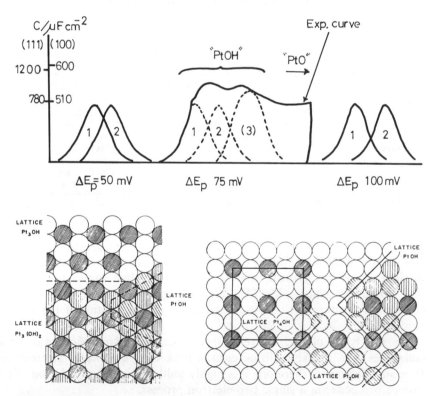

Figure 3.15 (a) Empirical resolution of the anodic I/V profile in Figure 3.1 into three pseudo-capacitance components plus the broad region beyond 1.2 V. (b) Significant sub-lattice structures of OH electrosorbed on the (100) or (111) crystal planes at Pt up to monolayer 'PtOH'. From Angerstein-Kozlowska *et al.* (1973).

while the shape and magnitude of O_{A3} was taken as evidence that interactions between the adsorbed species in this potential region are important.

The authors postulated that O_{A1}, O_{A2} and O_{A3} correspond to the following processes:

Charge under
peak ($\mu C\,cm^{-2}$)

$$4Pt + H_2O \longrightarrow Pt_4OH + H^+ + e^- \qquad\qquad 55 \qquad\qquad (3.35)$$

$$Pt_4OH + H_2O \longrightarrow 2Pt_2OH + 2H^+ + 2e^- \qquad 55 \qquad\qquad (3.36)$$

$$Pt_2OH + H_2O \longrightarrow 2PtOH + 2H^+ + 2e^- \qquad 110 \qquad\qquad (3.37)$$

'Pt_2OH' etc. do not represent stoichiometric species, but simply the surface site occupancy, as depicted in Figure 3.15(b). Some overlap of the surface arrangements would be expected, and should give rise to adsorbed OH species on neighbouring sites. This will result in unfavourable adsorbate–adsorbate interactions and hence place-exchange can result. As the surface occupancy tends towards monolayer PtOH, place-exchange also commences as above.

In the broad region beyond O_{A3}, the predominant faradaic process is the oxidation of the PtOH to PtO according to:

$$Pt-OH \longrightarrow PtO + H^+ + e^- \qquad\qquad (3.38)$$

The peak potentials for O_{A1}, O_{A2} and O_{A3} depend very little on sweep rate, indicating that they have the characteristics of a reversible process (see section on the voltammetry of adsorbed species). This is extremely interesting in the light of the pronounced hysteresis evident in the cyclic voltammogram and suggests that the purely faradaic processes taking place in O_{A1} to O_{A3} are reversible but there is an additional parallel process taking place that renders the overall change increasingly irreversible, in agreement with the place-exchange mechanism.

The authors then proceeded to try to determine at what potentials this irreversibility manifests itself. Figure 3.16 shows the work of Angerstein-Kozlowska and colleagues (1973) in which the anodic limit was progressively lowered. The actual shape of the voltammogram in the region of the anodic limit reversal is very revealing. On reversal, curves 1 to 3 in the figure, i.e. up to 0.85 V, show an immediate (vertical) drop in the current and a cathodic peak at almost the same potential as the anodic feature (the latter observation being exactly analogous to the case of hydrogen adsorption discussed above). This indicates considerable reversible character in the electrochemical process. From curves 3 to 8, the cathodic peak starts to move to less positive potentials, indicating increasing irreversibility. The cathodic peak potential then remains constant up to c. 1.2 V before moving again to more cathodic potentials. Thus, the irreversibility begins in the region of peak O_{A2}, i.e. at c. 0.95 V, well below monolayer coverage. This suggests that the process responsible for the irreversibility is in addition to the faradaic process and in parallel to it.

The next question to arise concerned the rate of this parallel process. In order to investigate this, the authors ramped the potential of the Pt electrode

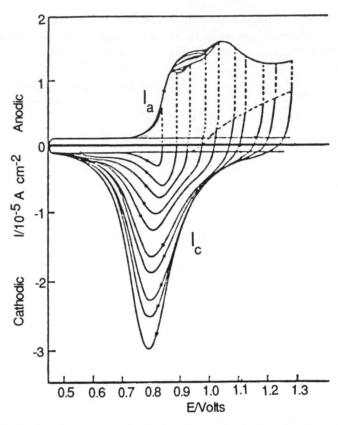

Figure 3.16 Cyclic voltammograms for the formation and reduction of surface oxide species on Pt in 0.5 M H_2SO_4 from 0.06 V vs. RHE to various reversal potentials in the anodic sweep. Scan rate was $0.10\,\text{V s}^{-1}$. From Angerstein-Kozlowska *et al.* (1973).

from 0.4 V (i.e. in the double-layer region) to a particular potential in the oxide region where it was held for various periods of time before being ramped back down. One such experiment is shown in Figures 3.17(a) and (b).

From Figure 3.17(a), where the anodic limit is 0.94 V, it can be seen that the multisweep cyclic voltammogram is almost reversible. However, with increasing holding time the cathodic peak potential shifts to progressively lower values and the charge under the cathodic wave increases appreciably. Similar behaviour was observed at all potentials $\geqslant$ the peak potential of O_{A2} but at higher holding potentials (see Figure 3.17(b)), while a larger amount of oxide was produced, the shift of the cathodic peak potential was less than that observed at the lower holding potentials. From the results, the authors concluded that:

1. the process leading to irreversibility appears when the anodic limit reaches O_{A2};
2. it occurs at all potentials in the 'PtOH' regions;

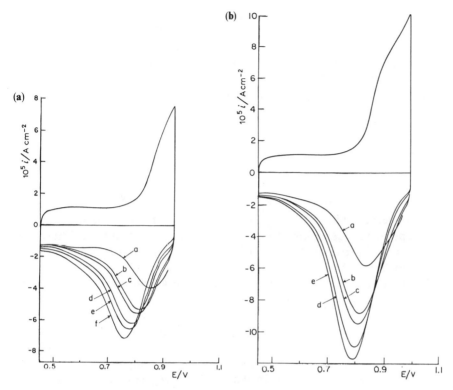

Figure 3.17 (a) Time effects in cathodic sweeps for reduction of surface oxide at Pt after various periods of holding at the anodic limit of 0.94 V. (a) Multisweep; (b) 0.5; (c) 1; (d) 5; (e) 10; (f) 30 minutes holding time. The scan rate was $0.10\,V\,s^{-1}$. (b) As in (a), except the anodic limit was 1.0 V. (a) Multisweep; (b) 0.5; (c) 1; (d) 5; (e) 10 minutes holding time. The scan rate was $0.10\,V\,s^{-1}$. From Angerstein-Kozlowska *et al.* (1973).

3. the film continues to grow on being held at anodic potentials positive of this point;
4. a different surface oxide is responsible for the observed electrochemistry at $\theta > 1$ from that at $\theta < 1$, i.e. place-exchange predominates at $\theta < 1$, while the oxidation of the (place-exchanged) PtOH to PtO is the most important process at $\theta > 1$.

It had long been realised that the shape of the platinum cyclic voltammogram in acidic solution is independent of scan rate, the peak currents merely increasing with increasing v, as expected. This suggested that the surface process responsible for the irreversibility in the oxide region, which was presumably some form of rearrangement of the surface atoms, could not be a generally slow process such as island formation. Island formation would involve the OH_{ads} species moving together via surface diffusion, and hence a cyclic voltammogram taken at 1 mV/s (where the process would have time to occur) would be different from that taken at 100 mV/s. An important observation in support of the formation of the phase oxide came with the realisation that the irreversibility commenced at the potential at which the Pt_2OH sublattice was formed. The authors showed that this arrangement of OH species, unlike the Pt_4OH sublattice, required some of the OH_{ads} to be on adjacent Pt atoms and this strongly suggested that the process leading to irreversibility was place-exchange. Place exchange would take place in order to minimise the unfavourable OH–OH interaction energy by rearrangement:

$$\begin{array}{ccc} OH & & Pt \\ | & \longrightarrow & | \\ Pt & & OH \end{array}$$

This would lower the free energy of the system, and hence would give rise to the observed shift to lower potentials of the cathodic peak, O_c.

Compelling evidence for the existence in support of this mechanism on platinum was provided by the STM work of Itaya and colleagues and is discussed in section 2.1.3.

The holding time experiments of Angerstein-Kozlowska and colleagues (1973), represented by Figures 3.17(a) and (b), not only demonstrated that the oxide layer continues to grow at a given potential but also that there are two processes involved in the growth. Thus, Figure 3.18 shows the variation in the cathodic peak potential with the extent of surface oxidation as calculated from the charge under the oxide stripping peak, O_c.

From the figure it can be seen that up to $Q_{ox}/Q_H = 1$ the peak potential, $E_{p,c}$, decreases. Between $Q_{ox}/Q_H = 1$–2, it remains constant, after which it

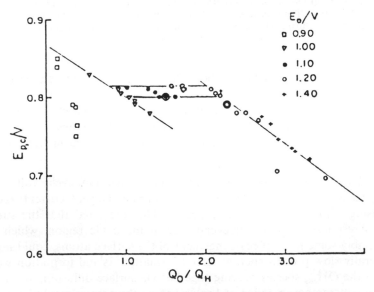

Figure 3.18 Dependence of cathodic peak potential $E_{p,c}$ on extent of surface oxidation Q_O/Q_H and on the corresponding potential. From Angerstein-Kozlowska *et al.* (1973).

decreases again. The authors postulated that the first sloping section of Figure 3.18, i.e. $Q_{ox} < 1$, corresponds to increasing coverage of PtOH and OHPt. At $2 > Q_{ox}/Q_H > 1$, a thickened layer of this same species is formed and, at $Q_{ox}/Q_H >$, the OH species are oxidised to O.

Angerstein-Kozlowska and colleagues (1973) also employed ellipsometry in their studies. In agreement with the work of Greef (1969) discussed above, they detected changes in Δ as soon as the oxide layer starts to form. Figure 3.19 shows their results, in which the change in Δ is plotted as a function of the oxide coverage. Δ was obtained during potential cycles to higher anodic potentials and compared to its value at the film-free surface.

As can be seen from the figure there are two inflections in the slope of the plot recorded during the anodic sweep, at $\theta_{ox} = 0.5$ and 1, suggesting changes in the nature of the oxide. Up to potentials near the first inflection, i.e. c. 0.98–1.02 V, the anodic and subsequent cathodic plots are almost superimposed, in broad agreement with the observed reversibility in the cyclic voltammetry experiments discussed above. At $\theta_{ox} \geqslant 0.6$ (1.02 V) the anodic and cathodic plots show increasing hysteresis. The authors interpreted the small degree of hysteresis in the anodic sweep up to 1.02 V as due to the formation of phase oxide OHPt, in parallel with the formation of the 'reversible' oxide PtOH. The increasing hysteresis at higher potentials was attributed to a corresponding increase in the contribution from the formation of the platinum (II) oxide in the phase oxide. On reversing the potential, the OHPt species are reduced first, giving a slope similar to that in the initial part of the anodic sweep corresponding to PtOH formation. The steeper final section arises as a result of the reduction of the OPt to Pt.

Conway and Gottesfeld (1973) performed electrically modulated reflectance

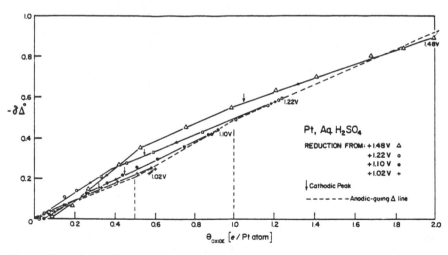

Figure 3.19 Relation between Δ and the degree of surface oxidation θ_{oxide} $(= Q_O/Q_H)$ in anodic-going and cathodic-going potential sweeps at $50\,\text{mV}\,\text{s}^{-1}$. From Angerstein-Kozlowska et al. (1973).

measurements at a single wavelength ($\lambda = 546\,nm$) and using P-polarised light. The potential modulation amplitude, ΔV, was 50 mV or 100 mV at a frequency of 32.5 Hz, superimposed on a slow potential ramp between potential limits straddling the oxide region. The reflectivity was measured as $\rho(V)$, where $\rho(V) = (\Delta R/R)/\Delta V$, and the lock-in detection system was only sensitive to *reversible* changes, i.e. as a result of reversible processes taking place at a fast enough rate to respond to the imposed potential modulation.

As can be seen from Figure 3.20, two distinct processes could be observed, a reversible process at lower anodic potential limits becoming increasingly irreversible as the anodic limit is advanced, in agreement with the voltammetry and ellipsometry experiments discussed above. For an anodic limit of $\leqslant 0.85\,V$, the optical response recorded during the cathodic sweep exactly traces that taken during the anodic sweep. For anodic limits $> 0.85\,V$, increasing hysteresis occurs. As the formation of the phase oxide becomes more complete, and the parallel oxidation of the OHPt to OPt increases, $\rho(V)$ decreases due to the increasing irreversibility in the optical response of the surface.

If the potential is held at anodic limits $> c.\ 0.8\,V$, a considerable time

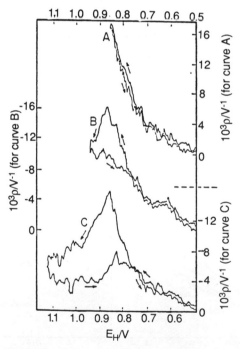

Figure 3.20 Profiles of $\rho(V)$ (in-phase component) against electrode potential E_H for anodic and cathodic-going potential sweeps taken to 3 different anodic end potentials: A = 0.86 V; B = 0.95 V; and C = 1.12 V vs. RHE, showing progressive diminution of reversibility. Modulation frequency = 32.5 Hz, amplitude = 100 mV peak-to-peak. The scan rate was $5\,mV\,s^{-1}$. From Conway and Gottesfeld (1973).

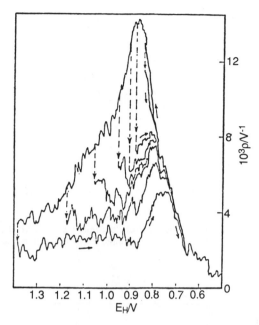

Figure 3.21 Time effects in $\rho(V)$ for 60 seconds holding at various potentials in the anodic-going sweep, with the corresponding $\rho(V)$ against potential profile in a following cathodic-going sweep at the same sweep rate. Modulation conditions and scan rate as in Figure 3.20. From Conway and Gottesfeld (1973).

dependence of $\rho(V)$ is observed (see Figure 3.21). Thus, if the potential is held at 0.8 V, $\rho(V)$ remains unchanged. However, holding the potential at values > 0.85 V results in a large drop in $\rho(V)$ until anodic limits $\geqslant 1.1$ V, at which point the change in $\rho(V)$ on holding is again reduced. These results were interpreted in terms of the surface rearrangement becoming more complete as θ_{OH} approaches 1, or as θ_O becomes significant beyond 1.1 V.

As a result of the accumulated evidence, including that discussed above, Conway and colleagues (1973) postulated the mechanism represented by equations (3.35) to (3.37), (3.30) and (3.31) above, with place-exchange occurring after the step represented by equation (3.35).

Under potential sweep conditions, up to c. 0.8 V, only the step represented by equation (3.35) occurs. However, the holding experiments show that the rearrangement represented by equation (3.30), to give the 'phase oxide', can occur slowly even at potentials as low as 0.8 V. At typical sweep rates, i.e. > 10 mV/s, the steps represented by equations (3.35) to (3.37) occur in the range 0.72 V $\pm$ 0.05 V to 0.9 $\pm$ 0.05 V, with the formation of the phase oxide, step (3.30), becoming significant beyond 0.9 V. The increasing irreversibility that becomes apparent at potentials > 0.9 V, i.e. beyond O_{A1}, is due to the increasing driving force for the place-exchange rearrangement caused by the increasing number of neighbouring OH species in the Pt_2OH and PtOH sublattices. The region at potentials > 1.1 V is dominated by the formation

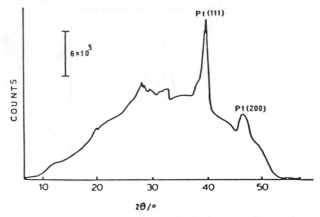

Figure 3.22 Reflection X-ray diffractogram from platinised Pt in 1 M H_2SO_4 in the double layer region. Data collection time 3×10^4 seconds. From Fleischmann and Mao (1987).

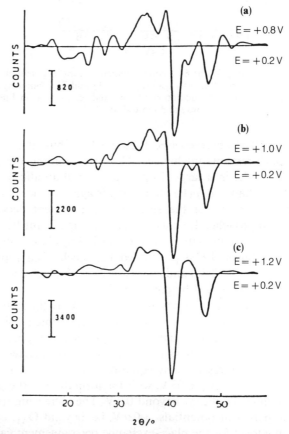

Figure 3.23 Reflection difference X-ray diffractograms from platinised Pt in 1 M H_2SO_4 in oxygen adsorption region. Potential modulated between 0.2 V vs. SCE and (a) 0.8 V, (b) 1.0 V, (c) 1.2 V. Data collection time 3×10^4 seconds. From Fleischmann and Mao (1987).

of the Pt(II) oxide via step (3.31). This occurs as a parallel process at lower potentials in the holding experiments (i.e. > 0.85 V).

In situ structural characterisation of the PtO layer was reported by Fleischmann and Mao (1987) using reflection X-ray diffraction. A platinised Pt electrode was employed in order to increase the surface area available for measurement. The diffractograms were recorded via the potential modulation approach as differences between the 'reference' diffractogram taken in the double-layer region (0.2 V vs. SCE) and a second diffractogram recorded at various potentials in the oxide region. The results are shown in Figures 3.22 and 3.23. Figure 3.22 shows the reflection X-ray diffractogram obtained at 0.2 V; the main features observed are the (111) and (200) diffractions of Pt superimposed on a broad background due to scattering by the electrolyte film. Figures 3.23(a) to (c) show the difference reflection X-ray diffractograms collected via modulation between 0.2 V and (a) 0.8 V, (b) 1.0 V and (c) 1.2 V. Peaks pointing down in the figure correspond to diffractions present at 0.2 V, any pointing up are due to diffractions present at the higher potential. Clearly, it can be seen that the only effect of oxide formation is the loss of Pt structure, this loss increasing with increase in the oxide formation potential. No discrete diffractions due to any platinum oxide were observed, which was interpreted by the authors in terms of the low crystallinity of the oxide adlayer.

With respect to the UHV-based techniques capable of providing chemical analysis, such as ESCA, AUGER, etc., several such studies have been performed. However, these studies were, by and large, performed on very thick oxide layers, formed after anodic oxidation of the Pt for many hours. Results from these studies thus have little bearing on the nature of the oxides formed on potential cycling. Part of the reason why these studies used such thick films lies in the considerable difficulty of detecting the thin oxide films formed during a potential sweep, even with relatively sensitive techniques.

Those experiments that did involve anodisation at more reasonable potentials, i.e. below oxygen evolution, suffered from an inability to characterise the initial PtOH species formed at coverages below a monolayer. Dickinson *et al.* (1975) systematically investigated the surface composition of a large number of Pt electrodes, polarised at various potentials in sulphuric acid, using XPS via the emersion approach. Figure 3.24 shows XPS spectra obtained from a Pt electrode after polarisation in sulphuric acid at 1.00 V and 1.5 V vs. SCE.

The two predominant features in Figure 3.24 are attributable to the 4f orbitals of the Pt electrode. The two peaks were deconvoluted as shown into a main peak and a smaller satellite peak. At potentials > 0.7 V vs. SCE, a peak at 77.1 eV was observed which was attributed to PtO. On the basis of these results, those of Kim *et al.* (1971), and the coulometric and ellipsometric data discussed above, Augustynski and Balsenc (1979) proposed that the signal attributed to the Pt 4f orbitals shifted via formation of PtO was only observed after the formation of the phase oxide, since it is only after this place exchange that the chemical environment of the Pt atoms is modified

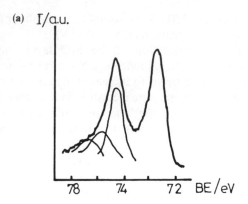

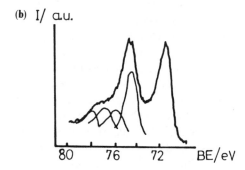

Figure 3.24 The platinum 4f electron peaks for a platinum electrode after polarisation for 15 minutes at (top) +1.00 V and (bottom) +1.50 V, in 0.5 M H_2SO_4. The deconvolutions are shown only for the $4f_{5/2}$ electron peaks for the sake of clarity. After Augustynski and Balsenc (1979).

sufficiently to cause the observed chemical shift. They also concluded that the PtOH species present below 1.1 V vs. NHE could not be distinguished.

3.1.3   *The voltammetry of single-crystal platinum electrodes*

The study of hydride and oxide adsorption on platinum has been substantially furthered in recent years by the investigation of carefully prepared single crystals. The surfaces of such crystals were reviewed in chapter 1 and it is evident that the Pt–Pt separation, as well as the morphology of the faces, shows considerable variation. It was not, therefore, immensely surprising to discover that the cyclic voltammograms of the various Pt surfaces in the hydride adsorption region also showed differences but the extent of these differences, especially on well-prepared surfaces with minimal surface defects, was a considerable shock. The cyclic voltammograms for the (100), (110) and (111) surfaces are shown in Figure 3.25 and the hydride region clearly extends

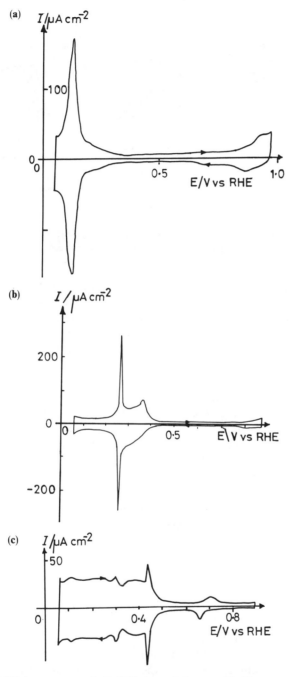

Figure 3.25 (a) Voltammogram of a Pt(100) electrode immersed in N_2-saturated 0.5 M H_2SO_4. (b) Voltammogram of a Pt(111) electrode immersed in N_2-saturated 0.5 M H_2SO_4. (c) Voltammogram of a Pt(110) electrode immersed in N_2-saturated 0.5 M H_2SO_4. From Sun and Clavilier (1987).

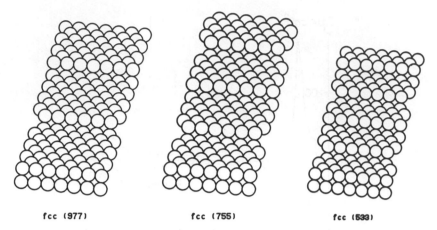

fcc (977)   fcc (755)   fcc (533)

Figure 3.26   A schematic representation of the face-centred cubic (977), (755) and (533) surfaces. From G.A. Somorjai, *Chemistry in Two Dimensions*, Cornell University Press, London, 1981.

over a substantial potential range with a structure radically different in all cases from that found on normal polycrystalline platinum.

Considerable controversy has surrounded this area, particularly in view of the very sharp features found in some of the cyclic voltammograms, and an immense effort has gone into the fabrication of single crystal surfaces intermediate in type between the three main low-index surfaces above. The importance of these intermediate surfaces lies in the fact that by selecting the right projection it is possible to prepare surfaces that contain *terraces* and *steps* but where these steps contain, at least in principle, no *kinks*. The basic type of arrangement is shown in Figure 3.26, which illustrates the surfaces (977), (755) and (533) intermediate between (100) and (111). By systematically traversing such intermediate surfaces, the identification of specific peaks associated with specific types of Pt site of the surface should be possible.

Work of this type has been carried out by Armand and Clavilier (1987), and by many others, and is extensively reviewed (see, for example, Parsons and Ritzoulis, 1991). An example is shown in Figure 3.27 in which voltammograms for the (110) and (111) surfaces are compared with intermediate surfaces of general structure $n(111) \times (111)$, where n is the number of rows of atoms of the (111) face in each terrace (see chapter 1). Starting with (111), three major changes are seen in the cyclic voltammograms: there is the growth of a peak at 0.11 V, the rapid disappearance of a sharp peak at 0.46 V and a steady decrease in hydride charge passed above 0.15 V. The appearance of the peak at 0.11 V is associated with the increasing density of step sites. These sites are the dominant structure on the (110) surface and the shift at (110) probably reflects the well-known hydride-induced reconstruction to the (2×1) surface for this projection. The structure above 0.15 V must be due to hydride atoms in *terrace* sites and the very sharp structure at 0.46 V,

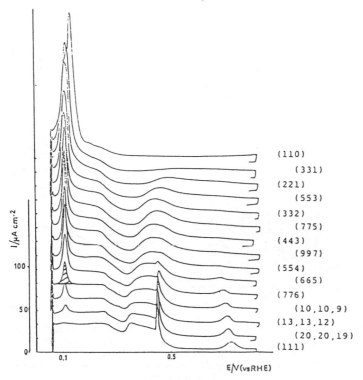

Figure 3.27   Set of positive sweeps of voltammograms of Pt $n(111) \times (111)$ in 0.5 M H_2SO_4, recorded at $50\,mV\,s^{-1}$. After Parsons and Ritzoulis (1991).

which is only found for $n \geqslant 20$, appears to be some type of long-range two-dimensional phase transition, possibly reflecting a change in long-range order as terrace hydride sites become depleted.

Very similar effects can be seen for the $[01\bar{1}]$ zone where the step structure is $n(111) \times (100)$. The result, given in Figure 3.28, shows the growth of a 0.25 V peak as we move from (111) to (100) and this peak is clearly due to adsorption of hydride on the (100) step sites. Finally, it is noteworthy that the two main peaks, at 0.1 V for the (110) steps and 0.25 V for the (111) steps, correspond closely to those seen on the voltammogram of polycrystalline Pt (see Figure 3.1). This suggests the cyclic voltammogram of polycrystalline Pt is dominated by structure derived from surface steps. The hydride on more extended terraces will still be present but as a relatively structureless continuum underlying the main peaks. The absence of any sharp structure in polycrystalline Pt at 0.46 V undoubtedly reflects the fact that very extended terraces are not expected on the polycrystalline surface.

These studies are of paramount importance if we are to gain a complete understanding of adsorption processes at electrodes such as platinum since, as we have seen above, we previously were able to determine the *number* of

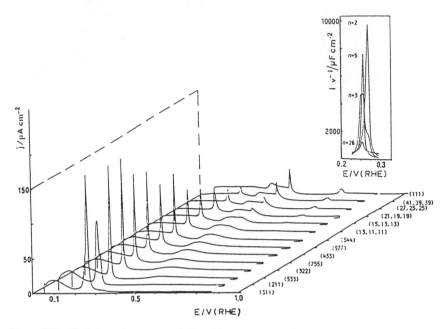

Figure 3.28 Positive-going sweep of the voltammograms of Pt(111) and Pt $n(111) \times (100)$ surfaces investigated with $n = 2 - 9$, 12, 14, 20, 26 and 40 in 0.5 M H_2SO_4. The sweep rate for all the experiments was $20 \, \text{mV s}^{-1}$. Displacements of the curves are arbitrary. The inset shows the evolution of the position of the step adsorption state for low values of n. After Parsons and Ritzoulis (1991).

atoms that an adsorbed species such as hydride is bound to, but not their *configuration*.

3.2
The electro-oxidation of methanol at platinum in acid solution

A fundamental piece of information concerning any chemical or electrochemical reaction is the identity of the products as this is the first step to determining a reaction scheme. The products of the electro-oxidation of methanol at Pt in acid solution, both at porous carbon-supported Pt and smooth Pt, have been long since identified, with CO_2 usually being the most significant product. However, the classical methods of identification, chiefly gas chromatography and standard chemical analysis, require long electrolysis times before sufficient products can be obtained for detection. Hence, it would seem appropriate to discuss the first real-time quantitative detection of the products of the electro-oxidation reaction measured during a cyclic voltammogram; a feat hitherto impossible.

Providing that the products of an electrochemical reaction are volatile, they can be detected in real-time by the on-line DEMS technique. Thus, Iwasita and Vielstich (1986) employed DEMS to monitor the products formed while running a cyclic voltammogram of the porous Pt DEMS electrode immersed in aqueous acidic methanol solution. As was discussed in chapter 2, the DEMS technique allows the volatile products of an electrochemical reaction to be analysed a fraction of a second after their production at scan rates $\leqslant 50\,\mathrm{mV/s}$. The current/potential response can thus be correlated to the mass/intensity signals obtained simultaneously.

Figure 3.29 shows the potential dependence of the m/e signals for H_2, $HCOOCH_3$ and CO_2 obtained during a cyclic voltammetry scan and the corresponding current response. From the figure it can be seen that the CO_2 mass signal exactly tracks the current, clearly indicating that it is a product of the *electrochemical* reaction at all potentials (rather than the result of a homogeneous chemical reaction of an intermediate in solution). The intensity of the CO_2 signal clearly shows that it is the major product. Both formic acid and formaldehyde have been suggested as products of the electrolysis of methanol under these conditions. However, both of these species react with methanol according to:

$$H_2CO + CH_3OH \longleftrightarrow CH_2(OCH_3)_2 + H_2O \qquad (3.39)$$

$$HCOOH + CH_3OH \longleftrightarrow HCOOCH_3 + H_2O \qquad (3.40)$$

hence they are detected as the acetal and ester. No signal attributable to H_2CO was observed by the authors but methyl formate was detected, as

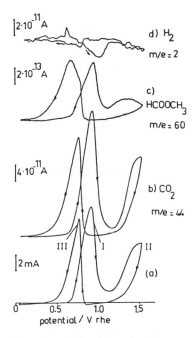

Figure 3.29 Cyclic current (a) and mass signal (b, c, d) voltammograms for 0.1 M CH_3OH in 1 M $HClO_4$ at porous platinum, and room temperature. The sweep rate was $20\,\mathrm{mV\,s^{-1}}$ and the surface roughness *c*. 50. After Iwasita and Vielstich (1990).

shown in Figure 3.29, indicating that HCOOH is a product under these conditions but H_2CO is not. Moreover, the authors proceeded to comment on the mechanism by which the formic acid was formed. The methyl formate mass signal tracks the current response over peak I indicating that it is the product of the electrochemical reaction at these potentials. However, whereas the current increases again in the oxide region, the methyl formate signal merely passes through a weak maximum. The authors interpreted this in terms of the removal of HCOOH itself at these potentials. In the region of the auto-oxidation current peak, III, the methyl formate mass signal shows a broader peak than the current and the maximum occurs at slightly lower potentials. In addition, the intensity of the mass signal is much higher than that expected for the corresponding current (assuming HCOOH and CO_2 are the only products and correlating the amounts of these produced with the charge under the cyclic voltammogram). This can be explained in terms of a *chemical* reaction at the freshly exposed Pt surface, i.e. after the reduction of the oxide layer in the cathodic sweep, according to:

$$CH_3OH + H_2O \longrightarrow HCOOH + 2H_2 \qquad (3.41)$$

Unfortunately, the noise on the $m/e = 2$ signal was too high to allow accurate detection of molecular hydrogen in order to prove the existence of reaction (3.41). However, the existence of this process was supported by DEMS results on the same system at open circuit where HCOOH was one of the two major products along with H_2CO.

The authors thus concluded that HCOOH is an intermediate in the electro-oxidation of methanol and a product of the chemical reaction of methanol with water. CO_2 is the major product of the electrochemical reaction.

3.2.1   Methanol oxidation at a smooth polished polycrystalline platinum electrode

The electrodes in the direct methanol fuel cell (DMFC) (i.e. the anode for oxidising the fuel and the cathode for the reduction of oxygen) are based on finely divided Pt dispersed onto a porous carbon support, and the electro-oxidation of methanol at a polycrystalline Pt electrode as a model for the DMFC has been the subject of numerous electrochemical studies dating back to the early years ot the 20th century. In this particular section, the discussion is restricted to the identity of the species that result from the chemisorption of methanol at Pt in acid electrolyte. This is principally because: (i) the identity of the catalytic poison formed during the chemisorption of methanol has been a source of controversy for many years, and (ii) the advent of *in situ* IR culminated in this controversy being resolved.

The DMFC is a potentially attractive alternative to the high temperature fuel cells currently available which are primarily based on H_2/O_2, since:

1. The process taking place at the anode:

$$CH_3OH + H_2O \longrightarrow CO_2 + 6H^+ + 6e^- \qquad (3.42)$$

has an E^0 of 0.029 V vs. NHE, allowing the DMFC a theoretical EMF of 1.201 V, comparable to that of the acid-based H_2/O_2 fuel cell (1.23 V).

2. The energy content per mole of methanol, in terms of ΔG^0 for the overall process:

$$CH_3OH + 3/2O_2 \longrightarrow CO_2 + 2H_2O \qquad (3.43)$$

is $c.$ 700 kJ mol^{-1}.

3. Methanol is a liquid at room temperature, rendering transport and storage relatively easy and safe.

4. The low temperature of operation of the DMFC, i.e. 60–120°C, ensures that no emission of NO_x or partially combusted hydrocarbons takes place.

However, the DMFC suffers from poor efficiency, this as a result of several factors:

1. CO_2 is produced as the primary product and this precludes the use of alkaline electrolytes due to the precipitation of CO_3^{2-} in the pores of the anode and consequent electrode fouling. Acid electrolytes lead to problems of corrosion and slow kinetics for the reduction of O_2 at the air cathode.

2. As was stated above, the most common catalyst employed at both the anode and the cathode is platinum. This leads to the problem of mixed potentials at both electrodes and a marked reduction in efficiency.

3. The electro-oxidation of methanol is slow at Pt at potentials near the E^0 of 0.029 V.

4. Pt is easily poisoned by adventitious impurities.

5. More seriously, Pt is poisoned by the products of the initial steps in the anodic reaction, i.e. the chemisorption of the methanol.

It is this last problem that will be discussed initially in this section.

Figure 3.30 shows the cyclic voltammogram of a polycrystalline platinum electrode in 1 M H_2SO_4 (a) in the absence of methanol and (b) in the presence of added methanol. The cyclic voltammogram has several characteristic features: the inhibition of the hydrogen adsorption/desorption peaks due to the chemisorption of methanol, a relatively low methanol oxidation current at potentials in the double layer region, a peak in the oxidation current near 0.85 V vs. NHE on the anodic scan, followed by a decline as the surface is blocked by the formation of the oxide. Increasing the potential still further drives the oxidation of the methanol on the oxide surface. On the return (cathodic) scan, stripping the oxide layer gives rise to the auto-oxidation current as methanol is oxidised on the freshly revealed bare Pt surface. Thus the cyclic voltammogram in Figure 3.30(b) clearly divides the methanol electro-oxidation into two potential regions: <0.85 V, where inhibition occurs via the formation of strongly adsorbed methanol fragments, and >0.85 V,

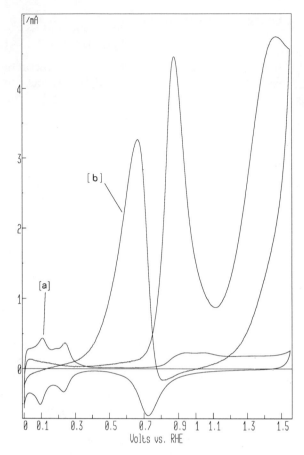

Figure 3.30 Cyclic voltammogram of a platinum electrode in N_2-saturated $1\,M\ H_2SO_4$ in (a) the absence and (b) the presence of $1\,M\ CH_3OH$. With thanks to J. Munk.

where the inhibition is due to the surface oxide. For many years, the identity of the species responsible for the inhibition at low potentials, i.e. before turnover of the methanol occurs, was a source of major controversy and it is this problem we will consider.

Methanol chemisorbs at platinum at potentials in the double-layer region to give species that block the activity of the surface with respect to the methanol oxidation reaction and it is these fragments that are responsible for the low currents observed in this potential region. If a platinum electrode is held at a potential in the double layer region in acidified aqueous methanol, and the solution replaced with fresh electrolyte prior to running a cyclic voltammogram, then the chemisorbed methanol fragments are oxidatively desorbed at potentials $> 0.5\,V$ (see below). Below this potential the coverage can be estimated from the reduction in charge under the hydrogen adsorption/ desorption peaks.

The identity of the strongly adsorbed species responsible for poisoning the platinum anode was eventually narrowed down to two possibilities:

$$CH_3OH \longrightarrow CO_{ads} + 4H^+ + 4e^- \tag{3.44}$$

$$CH_2OH \longrightarrow COH_{ads} + 3H^+ + 3e^- \tag{3.45}$$

via work such as that of Biegler and Koch (1987) who performed chronocoulometric studies upon the methanol oxidation reaction. These authors held the potential of the Pt electrode at 1.5 V vs. NHE in 1 M H_2SO_4 + 0.5 M CH_3OH to clean the surface of any adsorbed organic species. The potential was then stepped to a value in the double layer, and the current/time response recorded, as shown in Figure 3.31. After the initial rapidly decaying current due to the reduction of the oxide, an anodic current flows due to the methanol oxidation reaction that decreases due to the poisoning of the surface until reaching a steady-state value. The authors assumed that all of the charge passed during the potential step to a low-enough potential in the double layer region is used to generate the adsorbed species as the actual turnover of methanol to CO_2 at these potentials is comparatively slow. By comparing the charge required to oxidise the adsorbate off the surface, Q_{ox}, to that passed during the adsorption process, Q_{ads}, it is possible to postulate the identity of the adsorbate. Thus, the relevant oxidation processes are:

$$CO_{ads} + H_2O \longrightarrow CO_2 + 2H^+ + 2e^- \tag{3.46}$$

for which $Q_{ads}/Q_{ox} = 2$ (see equation (3.44)) and:

$$COH_{ads} + H_2O \longrightarrow CO_2 + 3H^+ + 3e^- \tag{3.47}$$

for which $Q_{ads}/Q_{ox} = 1$ (see equation (3.45)).

The authors found the charge ratio to be 2.3 and took this as c. 2, concluding that the adsorbate was CO.

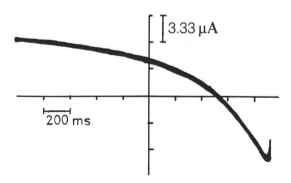

Figure 3.31 Current/time curve for methanol adsorption. The methanol concentration was 0.5 M. Adsorption occurred after holding the potential at +1.55 V for 20 ms, and then stepping the potential to 0.37 V. The step from 1.55 V to 0.37 V takes place 40 ms in from the right-hand edge. The x-axis scale is 200 ms cm^{-1}; the y-axis scale is 3.33 µA cm^{-1}. After Beigler and Koch (1967).

Table 3.2 The results of Bagotzky and Vassilyev (1967) on the adsorption of methanol at bulk platinum in acid solution at 0.4 V vs. RHE.

Time of adsorption (s)	Q_{ads} $(mC\,cm^{-2})$	Q_H^* $mC\,cm^{-2})$	Q_{ox} $(mC\,cm^{-2})$
0.7	0.027	0.023	0.030
1.0	0.097	0.082	0.090
15.0	0.120	0.111	0.119
120.0	0.150	0.149	0.154

* Reduction in charge under hydrogen adsorption peaks.

On the other hand, Bagotzky and Vassilyev (1967) also measured the charge passed during methanol adsorption at a potential in the double-layer region and the adsorbate oxidation. The results they obtained are given in Table 3.2 and strongly suggested that the adsorbate was COH (see equations (3.45) and (3.47)).

On the basis of such charge measurements other workers postulated CHO as the adsorbate since this would also give $Q_{ads}/Q_{ox} = 1$:

$$CH_3OH \longrightarrow CHO_{ads} + 3H^+ + 3e^- \tag{3.48}$$

$$CHO_{ads} + H_2O \longrightarrow CO_2 + 3H^+ + 3e^- \tag{3.49}$$

The absence of any *direct*, i.e. molecular, means of identifying the adsorbed species *in situ* rendered the controversy unresolvable and it remained undecided over the ensuing fifteen years. However, in 1981 Beden *et al.* published EMIRS spectra that were destined to have a major impact on this dispute, as discussed in section 2.1.6. This early paper concluded that $C{\equiv}O_{ads}$ is the dominant strongly adsorbed species (poison) and it is present at high coverage. Some Pt_2CO is also present but there is no evidence of COH_{ads} under the experimental conditions employed (non steady-state, potential perturbation at 8.5 Hz and with the dissociative chemisorption of methanol slow). The principal assignments of the paper were very quickly verified by Russell *et al.* (1982) using the IRRAS technique.

Although the detection of CO_{ads} by *in situ* IR was accepted as not ruling out the existence of other adsorbed species (particularly since the experiments were not quantitative in terms of coverage and the potential-modulation aspect of the technique could render it 'blind' to adsorbed species that do not exhibit a potential-dependent absorption frequency), it was generally accepted that the EMIRS data had ended the long controversy over the nature of the poison derived from methanol.

All that remained was to unambiguously correlate the loss of the electro-activity of a platinum electrode with the build-up of the $C{\equiv}O$ adsorbate and this was achieved in the extremely elegant work of Kunimatsu and Kita (1987).

The authors' first experiments were intended to give some quantitative appreciation of the coverage of the CO adsorbate arising from the chemisorption of methanol and were based on a comparison of the CO and CH_3OH

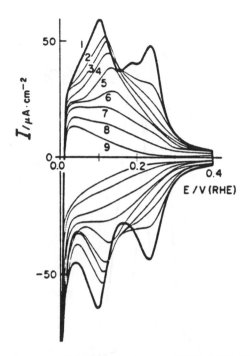

Figure 3.32   Cyclic voltammograms for Pt in solutions of 10^{-4}–$1.2\,M$ $CH_3OH/0.5\,M$ H_2SO_4. The sweep rate was $50\,mV\,s^{-1}$. Methanol was adsorbed at 0.4 V vs. RHE for controlled times to give the desired coverage θ. θ = (1) 0; (2) 0.17; (3) 0.22; (4) 0.32; (5) 0.49; (6) 0.57; (7) 0.64; (8) 0.76; (9) 0.86. From Kumimatsu (1986).

adsorption at potentials in the double-layer region. Thus, a Pt electrode was immersed in sulphuric acid, either saturated with CO gas or containing 1 M CH_3OH, at 0.4 V vs. RHE. The coverage of the methanol adsorbate was controlled by adjusting the methanol concentration and/or the adsorption time, while the coverage in the CO experiment was controlled by partial oxidation of the adsorbed layer. In both cases, the coverage was calculated from the reduction in charge under the hydrogen adsorption peaks (see Figure 3.32). An IRRAS spectrum was when collected and ratioed to a reference spectrum taken of the adsorbate-free surface at 0.8 V (see Figure 3.33(a)). The spectrum shows two normalised IRRAS spectra collected at 0.4 V vs. RHE, one of the $C{\equiv}O_{ads}$ arising from the adsorption of CO gas and one of the $C{\equiv}O_{ads}$ arising from the chemisorption of methanol. In both cases, the total coverage of the surface by adsorbed species was 0.86. The authors made two simple assumptions:

1. that the surface species resulting from the adsorption of CO will contain only C and O;
2. that the extinction coefficients of the $C{\equiv}O_{ads}$ arising from the methanol and CO adsorption processes would be the same.

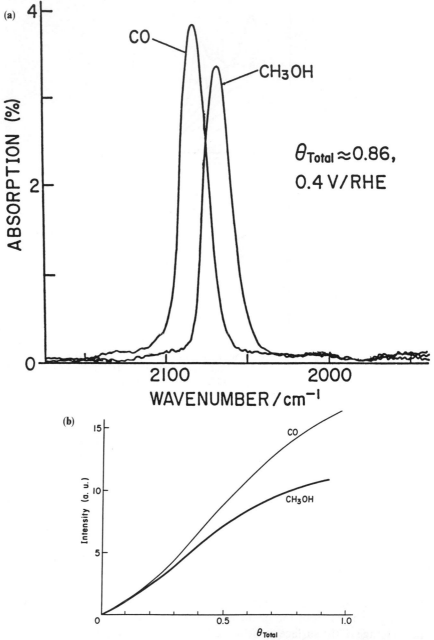

Figure 3.33 (a) Comparison of the polarisation IR spectra of the linear CO_{ads} derived from CO and CH_3OH on Pt in 0.5 M H_2SO_4 saturated with CO gas or containing 1 M CH_3OH. Adsorption was conducted at 0.4 V vs. RHE for 15 minutes. The spectra are referred to 0.8 V. The total adsorbate coverage of 0.86 is common to the two spectra. (b) Integrated infrared absorption intensities of the linear CO_{ads} derived from CO and CH_3OH plotted as a function of the total adsorbate coverage. From K. Kunimatsu, *Berichte der Bunsen-Gesellschaft für Physikalische Chemie*, 1990, **94**, 1025–1030.

Taking into account these assumptions it is clear from the figure that $C\equiv O_{ads}$ is not the only adsorbate present after methanol chemisorption. This observation is supported by Figure 3.33(b) which shows the integrated band intensities of the linear $C\equiv O_{ads}$ derived from methanol and CO as a function of the total coverage of adsorbed species. As can be seen from the plot, the intensity of the $C\equiv O_{ads}$ derived from methanol coincides with that from CO up to c. $\theta = 0.3$, after which the former band intensity becomes progressively lower than that due to CO, indicating that other methanolic fragments contribute at higher coverage.

From Figure 3.33(b), the coverage of $C\equiv O_{ads}$ at $\theta = 0.9$ is 0.6. Thus, the other adsorbed fragments constitute a coverage of 0.3, of which some will be due to the bridge-bonded form of CO_{ads}. The authors concluded that adsorbed CO is the predominant fragment on the surface resulting from methanol chemisorption. However, if methanol oxidation is occurring at a small number of 'active sites' on the surface, then the residual, i.e. non 'CO', adsorbates may still be the poisoning species. In order to disqualify this possibility, Kunimatsu and Kita (1987) investigated the potential dependences of the methanol oxidation current and the $C\equiv O_{ads}$ coverage.

A Pt electrode in acidic aqueous methanol was cleaned of adsorbed methanol fragments by pulsing the potential to 1.4 V vs. RHE for a few seconds prior to stepping the potential to various values between 0.4 V and 0.05 V for 15 min, after which IRRAS spectra were collected. The spectra are shown in Figure 3.34. As can be seen from the figure, the frequency shift is

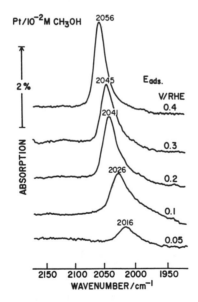

Figure 3.34 Dependence of the polarisation IR spectra of the linear CO_{ads} derived from methanol on Pt in 10 mM $CH_3OH/0.5$ M H_2SO_4 on the initial adsorption potential. From K. Kunimatsu, *Berichte der Bunsen-Gesellschaft für Physikalische Chemie*, 1990, **94**, 1025–1030.

much larger than $30\,cm^{-1}/V$, indicating coverage, rather than potential, dependence. The coverage of linear CO_{ads} decreases strongly on entering the hydrogen adsorption region, indicating that the chemisorption of methanol becomes increasingly difficult as the coverage of the more strongly adsorbed hydride increases. However, if the methanol is chemisorbed at 0.4 V, and the potential then stepped down, the $C\equiv O_{ads}$ is not replaced by H_{ads}, i.e. there is no recognisable intensity change, indicating the irreversible nature of the chemisorption process.

Figure 3.35 shows the potential dependence of the integrated band intensity of the linear CO observed in the experiment described above and the corresponding variation in the methanol oxidation current. The latter was monitored as a function of potential after the chemisorption of methanol under identical conditions to those employed in the IRRAS experiments. As can be seen from the figure the oxidation of the $C\equiv O_{ads}$ layer starts at c. 0.5 V and the platinum surface is free from the CO by c. 0.65 V. The methanol oxidation current shows a corresponding variation with potential, increasingly sharply as soon as the CO is removed; strong evidence in support of the hypothesis that the adsorbed CO layer established at 0.4 V acts as a catalytic poison for the electro-oxidation of methanol.

Definitive evidence of the correlation between the decrease in catalytic activity of Pt with respect to methanol oxidation and the appearance of adsorbed CO is to be found in further work by Kunimatsu and Kita (1987). Methanol was chemisorbed at 0.4 V vs. RHE for various times, after which cyclic voltammograms were taken, as shown in Figure 3.36(a). Clearly, as

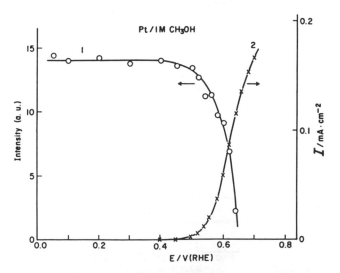

Figure 3.35 Potential dependence of the (1) integrated band intensity of the linear CO_{ads} derived from methanol at 0.4 V vs. RHE in 1 M $CH_3OH/0.5$ M H_2SO_4 and (2) the methanol electro-oxidation current observed after the adsorption of methanol at 0.4 V. From K. Kunimatsu, *Berichte der Bunsen-Gesellschaft für Physikalische Chemie*, 1990, **94**, 1025–1030.

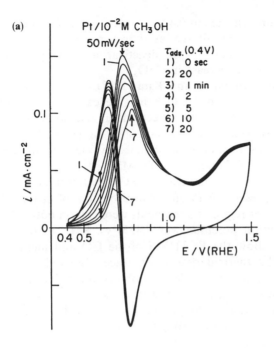

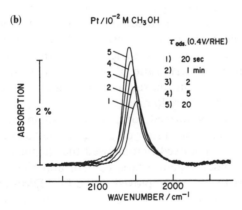

Figure 3.36 (a) Effect of the holding time τ_{ads} at 0.4 V on the cyclic voltammograms of methanol oxidation in 10 mM CH_3OH/0.5 M H_2SO_4. (b) Development of the linear CO_{ads} IR band while the potential was held at 0.4 V in 10 mM CH_3OH/0.5 M H_2SO_4. From K. Kunimatsu, *Berichte der Bunsen-Gesellschaft für Physikalische Chemie*, 1990, **94**, 1025.

the adsorption time is increased, the electrocatalytic activity with respect to methanol oxidation is decreased. In addition, a new current peak grows near 0.8 V in the anodic sweep due to the oxidation of the accumulated poison. In a separate experiment the Pt electrode was held, as before, at 0.4 V in acidic aqueous methanolic electrolyte with the electrode pulled back from the cell window. The electrode was then pushed against the window and an

IRRAS spectrum collected at 0.4 V. In this way the build-up of $C{\equiv}O_{ads}$ with adsorption time was monitored, as shown in Figure 3.36(b). From Figures 3.36(a) and (b) it is clear that the deactivation of the platinum electrode is closely related to the increase in $C{\equiv}O_{ads}$.

The work of Kunimatsu and Kita (1987) is very powerful evidence in favour of linearly adsorbed CO being the catalytic poison for methanol oxidation at a smooth platinum electrode in acid solution and has resulted in this hypothesis being generally accepted. However, there is some conflict between the IR results and those obtained by Vielstich and colleagues using chronocoulometry, ECTDMS and DEMS.

Wilhelm et al. (1987a) employed the ECTDMS technique to try to identify the adsorbates present on Pt after chemisorption of methanol at potentials in the double-layer region. Experiments were run on both CO and CH_3OH, and the adsorption potential was 0.45 V vs. RHE in both cases. Figures 3.37(a) and (b) show the ECTDMS of the CO and methanol adsorbates, respectively. The thermal desorption trace observed for CO adsorption (Figure 3.37(a)) closely resembles that observed for CO adsorption from the gas phase. Two peaks are observed in the $m/e = 28$ (CO) desorption trace: one near 415 K, corresponding to weakly adsorbed CO, and the other near 530 K, corresponding to the strongly adsorbed form. No significant amounts of desorbed hydrogen were observed.

The methanol ECTDMS (Figure 3.37(b)) shows a similar desorption trace for CO to that in Figure 3.37(a). In addition, a strong desorption peak for H_2 is observed with one desorption state near 455 K in the temperature region where CO desorbs. From the areas under the CO and H_2 desorption traces the authors obtained a $CO:H_2$ ratio of 1:0.42 and concluded that this was the results of two species on the surface, CO_{ads} and a C, O, H-containing adsorbate, the latter being predominant.

Iwasita et al. (1987) performed some chronocoulometric experiments to see if they supported the conclusion obtained on the basis of the ECTDMS work. A flow cell was employed that allowed the change of electrolyte with the electrode still under potentiostatic control and the experiments were performed both on Pt foil, as employed in the ECTDMS experiments, and on the porous platinum electrodes used in DEMS work. In the latter case, the chemisorption process was monitored to see if any volatile products were produced. The potential of the working electrode was cycled between the hydride- and oxide-formation regions until a stable voltammogram was obtained indicating a clean surface. The potential was then stepped to 0.33 V vs. the Pd/H_2 electrode (c. 0.4 V vs. RHE), where the amount of H_{ads} is negligible, and the current allowed to decline to zero. The electrolyte was then replaced by fresh electrolyte containing methanol and the current/time response recorded (see Figure 3.38(a)). The charge passed, Q_{ads}, was evaluated by integrating the I/t curve. The mass spectrometer showed no volatile products were produced during the chemisorption at the porous Pt electrode and the authors concluded that the only process taking place was the

(a)

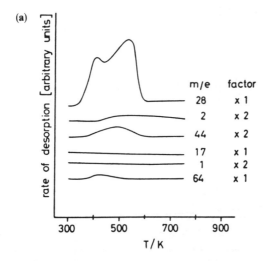

(b)

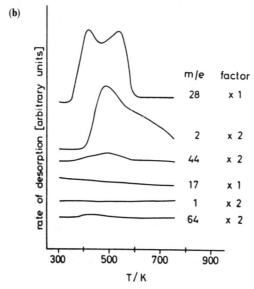

Figure 3.37 (a) ECTDMS of CO adsorbate on platinum from contact with (99% N_2 + 1% CO) saturated 0.05 M H_2SO_4. Desorption diagram for $m/e = 44$ (CO_2), $m/e = 28$ (CO), $m/e = 17$, $m/e = 2$ (H_2), $m/e = 1$ (H_2O) and $m/e = 64$ (SO_2). The temperature scan rate was $5 K s^{-1}$, and the adsorption was carried out at 0.455 V for 300 s. (b) ECTDMS of methanol adsorbate from contact with 5×10^{-2} M CH_3OH + 0.05 M H_2SO_4. Desorption diagram and conditions as in (a), except adsorption at 0.455 V was for 120 s. From Wilhelm *et al.* (1987a).

formation of the adsorbate, i.e. there was no turnover of methanol to CO_2 to contribute to the observed charge passed.

After the chemisorption the solution in the cell was again replaced, this time by the pure electrolyte, the potential stepped back to 0.02 V vs. Pd/H_2, and two cyclic voltammograms taken, as shown in Figure 3.38(b). The

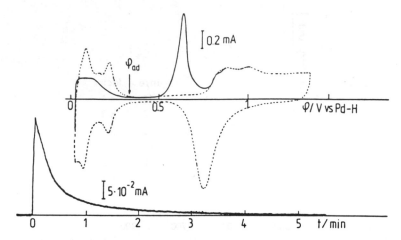

Figure 3.38 (a) Current-time transient during the adsorption of methanol from solution at 0.330 V vs. Pd/H in 0.01 M $CH_3OH + 0.5$ M H_2SO_4. (b) Cyclic voltammogram showing the oxidation of the adsorbate after replacing the solution by pure supporting electrolyte. $10\,mV\,s^{-1}$, same solution as in (a). From Iwasita *et al.* (1987).

adsorbate was oxidised off during the first sweep, giving an anodic peak as shown and the area under this peak gives Q_{ox}. As well as giving Q_{ox}, the first voltammogram showed that all of the methanol had been removed by the electrolyte exchange and the second voltammogram was used to check that the surface was free from methanol fragments. The area under the hydrogen adsorption peaks was used to calculate the coverage of the surface by the chemisorption products.

The authors performed the above experiment at various methanol concentrations and charge ratios, Q_{ads}/Q_{ox}, between 1 (COH or CHO) and 2 (CO) were obtained depending on the methanol concentration and coverage. By assuming that only $C{\equiv}C_{ads}$ and 'COH_{ads}' (i.e. either CHO or COH) are present on the surface, the authors were able to calculate their mole fractions as a function of total coverage and methanol concentration. Thus, if the mole fraction of CO is x and of COH is y, then:

$$x + y = 1,$$

and

$$2x + y = Q_{ads}/Q_{ox}$$

which can be solved for x and y. The mole fractions of the COH and CO adsorbates obtained in this way are plotted in Figures 3.39(a) and (b) for the smooth and porous Pt electrodes. It is clear from the figures that the mole fractions of the two adsorbates depend on the methanol concentration, with the $C{\equiv}O_{ads}$ predominating at high methanol concentrations, in agreement with the IR results. The mole fractions also depend on the coverage, itself a function of concentration and adsorption time.

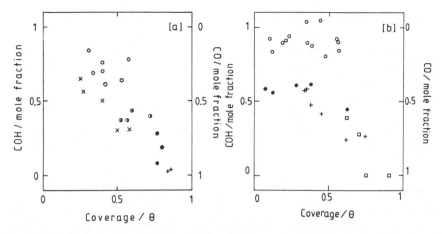

Figure 3.39   Mole fraction of COH and CO on Pt after methanol adsorption as a function of coverage, calculated from charge measurements on polished anealed and porous Pt electrodes. 0.5 M H_2SO_4 and CH_3OH concentrations (a) ($\bigcirc$) 5×10^{-3} M, ($\circleddash$) 5×10^{-2} M, ($\bullet$) 0.5 M (polished Pt); ($\times$) 5×10^{-3} M, ($+$) 0.5 M (annealed Pt). (b) ($\bigcirc$) 10^{-2} M, ($\bullet$) 2×10^{-2} M, ($+$) 5×10^{-2} M, ($\square$) 0.5 M (porous Pt). From Wilhelm *et al.* (1987b).

Vielstich and colleagues next set out to discriminate between COH and CHO as the hydrogen-containing adsorbate. To accomplish this the authors used DEMS to study the chemisorption of CD_3OD in protic aqueous acid. Because of the ease with which the alcoholic deuterium is exchanged, CD_3OD exists almost entirely as CD_3OH in protic aqueous solution. The CD_3OH can thus undergo chemisorption via two mechanisms to give a D- or H-containing adsorbate:

$$CD_3OH \longrightarrow CDO_{ads} + 2D^+ + H^+ + 3e^- \qquad (3.50)$$

$$CD_3OH \longrightarrow COH_{ads} + 3D^+ + 3e^- \qquad (3.51)$$

Oxidation of these adsorbates then gives:

$$(CDO)_{ads} + H_2O \longrightarrow CO_2 + D^+ + 2H^+ + 3e^- \qquad (3.52)$$

$$COH_{ads} + H_2O \longrightarrow CO_2 + 3H^+ + 3e^- \qquad (3.53)$$

Thus, if the adsorbate is the formyl species, oxidising it gives D^+ which can be detected as HD by stepping the potential of the working electrode sufficiently negative to reduce the D^+. For the experiment to be successful, the electrolyte must be replaced by pure protic solution after the chemisorption to prevent interference from the D^+ produced via the reactions in equations (3.50) and (3.51).

Iwasita *et al.* (1987) first optimised the DEMS system with respect to the detection of HD. CD_3OH was then chemisorbed at 0.356 V vs. Pd/H_2 for 400 s and the potential stepped to 0.975 V for 0.5 s to oxidise off the adsorbate. The CO_2 signal at $m/e = 44$ was then monitored and taken as proof that the potential and length of the oxidation step were sufficient to produce detectable quantities of products. The potential was then stepped to $-0.44\,V$ to reduce any D^+ ions to the volatile HD and the $m/e = 3$ signal carefully monitored. In this manner HD was observed but only in quantities that corresponded to the natural deuterium content in the system, i.e. the natural abundance of D in H_2O. The authors thus concluded that the hydrogen-containing adsorbate was COH.

In agreement with the DEMS results the ECTDMS work of Wilhelm et al. (1987b) also showed that the nature of the predominant adsorbate is a function of both coverage and methanol concentration. The authors performed ECDTMS measurements at two different methanol concentrations. The areas under the H_2 and CO desorption peaks were calibrated and allowed the number of desorbed H and CO particles, n_H and n_{CO}, to be calculated. n_H represents the number of COH species present on the surface and n_{CO} the total number of CO and COH adsorbates. From these values the mole fraction of $C{\equiv}O_{ads}$ and COH_{ads} can be calculated, as shown in Figure 3.40.

At high methanol concentration, such as those employed in an actual fuel cell, the chronocoulometric and ECTDMS results agree with the IR data in that the predominant adsorbed species is $C{\equiv}O$. However, the IR data were obtained at methanol concentrations ranging from $10^{-2}\,M$ to $1\,M$. At methanol concentrations of $10^{-2}\,M$ the predominant surface species detected by the chronocoulometry and ECTDMS experiments is COH and it is evident that the energy difference between COH_{ads} and $C{\equiv}O_{ads}$ must be slight.

It is evident from the above that the predominant adsorbed species can be affected by relatively small concentration changes. This strongly suggests that the energy *differences* between the various possibilities are likely to be very small. This is consistent with studies by Campbell and co-workers of

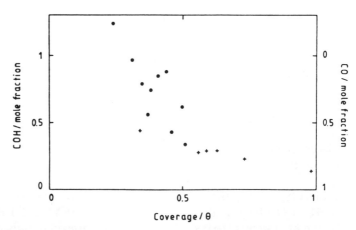

Figure 3.40   Mole fraction of methanol adsorbate from ECTDMS measurements as a function of coverage. $0.05\,M\ H_2SO_4$ and: ($\bullet$) $5 \times 10^{-3}\,M\ CH_3OH$, ($+$) $0.5\,M\ CH_3OH$. From Wilhelm *et al.* (1987b).

adsorbed CO at the metal/gas interface, which have shown that CO is highly mobile on platinum. This mobility is apparently through either a weakly adsorbed form of CO accessible by thermal excitation from the bonded form or is due to CO adsorbed on 'sterically unfavourable Pt sites'. Either of these is consistent with the observation by McCabe and Schmidt (1977) that strongly repulsive interactions exist between adsorbed carbonyls on adjacent sites with the desorption energy, E_d, decreasing by a factor of 2 as θ increases from zero to one. The decrease in E_d is especially marked on all crystalline surfaces at higher coverages, leading to adsorbed CO molecules that are substantially less stable and more mobile.

To explore the consequences of this mobility isotopic measurements were carried out (see Iwasita and Vielstich, 1990) to show that both the ^{13}CO and ^{13}COH formed on the surface by immersion of Pt in $^{13}CH_3OH$ do not exchange on re-immersion in $^{12}CH_3OH$ solution, and even the oxidation of $^{12}CH_3OH$ at lower potentials does not lead to replacement of the bulk of the surface adsorbed ^{13}CO (see Leung and Weaver, 1990). Indeed, Roth and Weaver (1991) showed that facile oxidation of solution CO could take place on Pt even at high coverages of adsorbed CO and the continuous supply of CO via the solution phase actually inhibits oxidation of the surface bound CO, a result interpreted by Lu and Bewick (1989) as a consequence of the presence of adsorbed CO primarily in extended patches. Oxidation of CO only takes place on the edges of these islands and if CO is present in solution the vacancies created are rapidly filled. Further evidence for these adsorbed CO patches has now been provided by STM: Vitus and co-workers have identified ordered adlayers of CO single-crystal Pt surfaces and it is evident from their work that such adlayers form ordered islands.

Considerable effort has gone into reconciling all the observations above with those data derived from purely kinetic studies. The framework for these studies was provided in a now classic paper by Bagotzky and Vassilyev (1967) who used pulse methods to determine the coverage by chemisorbed intermediates and attempted to derive functional expressions connecting current and coverage by intermediate and potential. Unfortunately the time dependence of the nature of the adsorbed intermediates makes this approach extremely difficult at lower potentials, though the authors report that in the region of coverage explored the current $I \sim \exp(\beta\xi\theta)\cdot\exp(\beta'FE/RT)\cdot[H^+]^{-1}$, where $\beta \sim 0.4$–0.5, $\beta' \sim 0.78$–0.86 and ξ is a number between 10 and 11, suggesting, in agreement with the gas-phase data, that adsorption of the intermediate is governed by a Temkin-type isotherm. Such behaviour is consistent with a rate-limiting step involving the oxidation of the intermediate. More recent kinetic data by Inada et al. (1990) are not consistent with the earlier data with a pronounced time dependency evident in the data below 0.55 V. In fact, after waiting for long periods at the test potential before taking readings, these authors find that $I \sim \exp(-k\theta)\cdot\exp(0.56EF/RT)$ for $E < 0.55$ V, which is consistent with the first adsorption process being rate limiting. Unfortunately, Inada et al., like Bagotzky and Vassilyev, did not

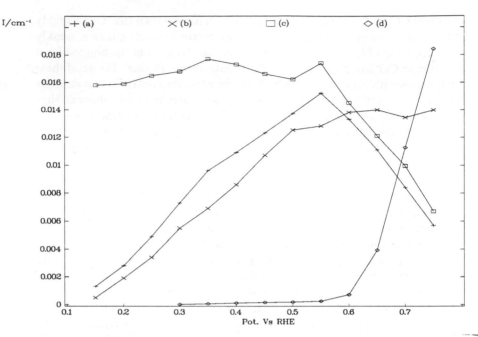

Figure 3.41 Integrated peak intensities for (a) CO gain (b) CO loss features and (c) total CO intensity compared to (d) CO_2 gain intensity ($\times 10^{-3}$) for type B experiment in 2.0M methanol. After Christensen *et al.* (1993).

independently assess the nature and coverage of adsorbed species by any spectroscopic method, relying primarily on electrochemical pulse methods to determine θ.

Unless the coverage of adsorbate is monitored simultaneously using spectroscopic methods with the electrochemical kinetics, the results will always be subject to uncertainties of interpretation. A second difficulty is that oxidation of methanol generates not just CO_2 but small quantities of other products. The measured current will show contributions from all these reactions but they are likely to go by different pathways and the primary interest is that pathway that leads *only* to CO_2. These difficulties were addressed in a recent paper by Christensen and co-workers (1993) who used *in situ* FTIR both to monitor CO coverage and simultaneously to measure the rate of CO_2 formation. Within the reflection mode of the IR technique used in this paper this is not a straightforward undertaking and the effects of diffusion had to be taken into account in order to help quantify the data obtained.

Two types of potential step experiments were carried out in this work: those in which the working electrode was electrochemically cycled in acid electrolyte alone (type A) before methanol was admitted with the electrode held at a very low potential, and those in which the electrode was cycled in acid electrolyte in the presence of methanol (type B) before the spectroscopic measurements were taken. The results from experiment A showed that CO coverage increased to a constant level above which CO_2 evolution was observed. In the initial CO_2 evolution region a Tafel slope of 90 mV was observed, with CO coverage almost constant, and this was ascribed to oxidative attack by water on CO being the limiting:

$$H_2O + CO_{ads} \longrightarrow (COOH)_{ads} + H^+ + e^-$$

followed by the rapid

$$(COOH)_{ads} \longrightarrow CO_2 + H^+ + e^-$$

If the coverage of CO is constant, then a Tafel slope of 120 mV would be expected but there will also be a contribution to the Tafel slope from the fact that at higher potentials CO_{ads} itself will become more susceptible to nucleophilic attack.

Type B experiments were more revealing in that CO_2 evolution began at a lower potential (see Figure 3.41) than for type A experiments and over a range of about 100 mV the current was almost independent of potential (i.e. the Tafel slope became very large). Throughout this region the current due to CO_2 evolution was always larger than for type A measurements and it was concluded that the roughening of Pt known from STM data to take place when Pt is cycled into the phase oxide region could be stabilised in the presence of methanol, presumably by adsorption of CO. These rough surfaces obviously possess a larger number of active sites, sufficient in fact that the rate limiting process for methanol oxidation in this region is the diffusion of adsorbed CO from the islands identified by Lu and Bewick (1989) to the active sites where they are instantly oxidised. By contrast, attack on the CO on electrodes of type A must be occurring at the *edges* of the islands; it certainly cannot be taking place in the middle of the islands since this would not account for the isotopic substitution data.

The observation that the state of the platinum surface has a major influence on the electrochemical kinetics for methanol oxidation is in good agreement with the observations of Pletcher and Solis (1982), and the picture of the platinum surface that emerges from this study is a subtle and interesting one. Clearly, the structure of the surface can itself be altered during the electrochemistry. Any notion that the surface somehow maintains itself as a mathematical entity must be put to one side; the platinum surface is an active player in the methanol oxidation process and its structure can be optimised and stabilised by the adsorbed CO. The CO itself is also dynamic, though many of the CO molecules adsorbed in the centre of the islands will not participate in the reaction, the edges of these islands will offer a highly dynamic environment. The dangers of relying entirely on current/voltage measurements, however sophisticated, is the final message from these studies. Particularly when the surface is as dynamic as the one envisaged here such studies need back up from spectroscopic and structural techniques if they are to be convincing.

3.3
The electrochemical reduction of CO_2

Carbon dioxide and water are the major waste products from most natural and industrial processes and hence are found in large quantities in the environment. If an efficient and cheap means could be found, the reduction of CO_2 could provide a potentially rich source of carbon for utilisation in the production of, for example, 'synthetic' hydrocarbon fuels to replace petroleum, formic and oxalic acids for the chemical industries and foodstuffs such as glucose.

It has long been accepted that the most difficult step in the reduction of CO_2 to CH_3OH, etc. is the initial activation of the molecule itself. Electrochemical reduction provides a relatively simple means of activating the molecule and hence has been investigated in depth almost ever since the birth of electrochemistry as a branch of applied science. However, a cheap and efficient electrochemical method of reducing CO_2 to useful products continues to elude scientists, primarily as a result of several major problems:

1. CO_2 is an extremely stable molecule, with a free energy of formation of $-396 \, kJ \, mol^{-1}$. Furthermore, the lowest unoccupied molecular orbital is of very high energy so a high overpotential is required for the first reduction step. In addition to being costly in terms of energy input, such large overpotentials result in major interference in aqueous solution from the competing hydrogen evolution process.
2. CO_2 has a low solubility in most solvent media. In addition, the slow conversion of bicarbonate to carbon dioxide in aqueous electrolytes gives rise to low current densities in aqueous alkaline media as well.
3. The carboxylic acids that are common products (formic and oxalic) are difficult to extract from the reaction solution.
4. *Unequivocal* mechanisms to describe the various observed reactions have not been established.

As a result, the study of CO_2 reduction is very much an ongoing process and, consequently, it is impossible even to approach a comprehensive treatment here. Nevertheless, there are several aspects of the topic that highlight the role of the various techniques that have been applied and these are discussed briefly below. For a full treatment of the subject the interested reader is directed to any one of several excellent reviews referred to at the end of this section.

3.3.1   The direct reduction of CO_2 in non-aqueous solvents

The competition by hydrogen evolution in the CO_2 reduction reaction has been minimised in two main ways: (i) in aqueous solutions by employing metals with large hydrogen overpotentials (e.g. Pb, Hg, etc.) as cathodes, and (ii) by employing aprotic solvents. In this section, we will consider the latter approach, with particular respect to the reduction of CO_2 at Pt and Au.

An early study on CO_2 reduction in non-aqueous solvents was carried out by Haynes and Sawyer (1967) who employed chronopotentiometry, controlled potential coulometry and galvanostatic methods to study the reduction of CO_2 at Au and Hg in dimethylsulphoxide (DMSO).

Figure 3.42 shows chronopotentiograms taken of a Hg electrode in DMSO saturated with N_2/CO_2 mixtures of different composition. The mercury working electrode was switched in at the currents indicated in the plot and the potential monitored as a function of time. The appearance of the double wave at intermediate CO_2 concentrations was interpreted as being due to the presence of two different solvated CO_2 species: CO_2 solvated by DMSO and by water. This was based on the observation that behaviour similar to that represented by the figure was seen if the CO_2 concentration was held constant and the water concentration varied. The reduction of CO_2 at gold, on the other hand, was found to give well-defined single-step chronopotentiograms at all CO_2 concentrations. The redox potentials for the various coupled involving CO_2 all have values more positive than -1.0 V vs. NHE, hence from Figure 3.42 it can clearly be seen that the reduction of CO_2 requires a large overpotential. Under such conditions, the net (cathodic) current, I_c, is given by (see section on Butler–Volmer):

$$I_c = nFAk_c[CO_2]_0 \qquad (3.54)$$

which can be written:

$$I_c = nFA[CO_2]_0 k^0 \exp(-\beta n_c f[E - E^0]) \qquad (3.55)$$

Where E^0 in a complex reaction mechanism must be defined with great care. n, as always, is the number of electrons taking part in the overall reaction

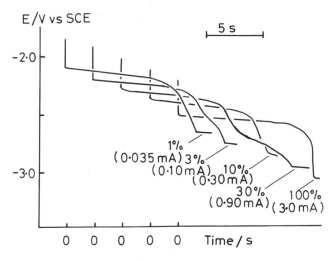

Figure 3.42 Chronopotentiograms for reduction of carbon dioxide in dimethyl sulphoxide at a mercury electrode. Percentage by volume of CO_2 in N_2 to saturate solution is noted on the curves. From Haynes and Sawyer (1967). Copyright 1967, American Chemical Society. Reprinted with permission.

and n_c is the number of electrons taking part in the rate-determining step. The problem of determining the value of the CO_2 concentration at the surface was circumvented by the authors by employing the concept of the transition time, τ. This is the time taken in a chronopotentiometric experiment for the concentration of the reactant at the electrode to fall to zero (see Figure 3.43). This time is then marked by a rapid change in the potential of the working electrode as the nature of the reaction at its surface is forced to change when the system is under galvanostatic control. The variation of potential in the immediate vicinity of the transition time is often quite large and hence τ can generally be measured with a high degree of precision. It can be shown that by incorporating the experimentally observed transition time equation (3.55) is modified as follows:

$$I_c = nFA[CO_2^*]k^0(1 - [t/\tau]^{1/2})\exp(-\beta n_c f[E - E^0]) \qquad (3.56)$$

where $[CO_2^*]$ is now the concentration of CO_2 in the bulk of the solution and t is the time from the start of the chronopotentiometric experiment, having values running from 0 to τ. The authors determined $[CO_2^*]$ via standard quantitative analysis techniques, and t and τ were measured from curves such as those shown in Figure 3.42.

From chronopotentiometric curves such as those shown, the authors obtained plots of:

1. the potential, E vs. $\log_e[1 - (t/\tau)^{1/2}]$, at constant $[CO_2^*]$ and I_c;
2. the potential at the start of the experiment (i.e. at time $t = 0$) vs. $\log_e I_c$ at various bulk CO_2 concentrations;
3. the potential at $t = 0$ vs. $\log_e[CO_2^*]$ at various current densities.

All the plots give βn_c from the slope and $\log_e k^0$ from the intercept. A typical plot obtained via method (1) is shown in Figure 3.44.

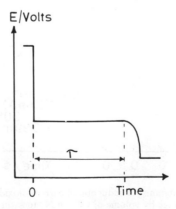

Figure 3.43 Schematic illustration of the transition time τ in a constant-current (chronopotentiometric) electrolysis experiment (see text for details).

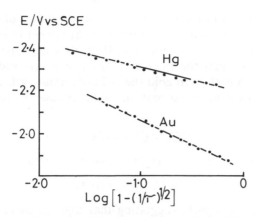

Figure 3.44 Analysis of chronopotentiograms for the reduction of CO_2 in dimethyl sulphoxide at mercury and gold electrodes. The solutions were saturated with 1% CO_2 (by volume in N_2), the current was 17 μA (Au) or 35 μA (Hg), the electrode areas were 0.225 cm^2 (Au) and 0.222 cm^2 (Hg). From Haynes and Sawyer (1967). Copyright 1967 by the American Chemical Society. Reprinted with permission.

Table 3.3 The values of βn_c and $\log_e k^0$ calculated from the chronopotentiometric experiments of Haynes and Sawyer (1967)

Cathode	βn_c	$\log_e k^0$
Hg	0.64	−3.18
Au	0.30	−1.40

The values of βn_c and $\log_e k^0$ obtained from the various experiments were then averaged and the results are shown in Table 3.3. Both the βn_c values are close enough to 0.5 to support the basic argument that the first step, the formation of the anion radical, is rate-limiting.

On both Hg and Au, CO was a product, as shown by GC measurements, with formate also a product, as detected by quantitative analysis. The formation of CO from a process giving 1 e$^-$ per molecule of CO_2 is, at first sight, somewhat contradictory since the formation of CO from CO_2 is nominally a two-electron process. This led the authors to postulate that the mechanism involved the coupling of two of the initially produced $CO_2^{\cdot-}$ radicals leading to disproportionation:

$$CO_2 + e^- \longrightarrow CO_2^{\cdot-} \tag{3.57}$$

$$2CO_2^{\cdot-} \longrightarrow O{=}C^-{-}O{-}CO_2^- \tag{3.58}$$

$$O{=}C^-{-}O{-}CO_2^- \longrightarrow CO + CO_3^{2-} \tag{3.59}$$

This consumes 2 e$^-$ per 2 CO_2 molecules, i.e. 1 e$^-$ per CO_2, in agreement with the results. This mechanism requires that CO_3^{2-} is not electroactive at the potentials employed and this indeed was found to be the case by the authors. The presence of carbonate as a product could only be inferred indirectly from pH titration experiments; a direct means of detecting all the major products only became available with the advent of *in situ* FTIR.

On the basis of the coulometric and pH titration experiments, the authors were able to verify that the rate-determining step involved the transfer of the first electron and that on gold the yield of formate was closely linked to the amount of water in the solvent. In the latter case, the reaction sequence was postulated to be:

$$CO_2 + e^- \longrightarrow CO_2^{\cdot -} \tag{3.60}$$

$$CO_2^{\cdot -} + H_2O \longrightarrow HCO_2^- + HO^{\cdot} \tag{3.61}$$

$$HO^{\cdot} + CO_2^{\cdot -} \longrightarrow HCO_3^- \tag{3.62}$$

There have been reports suggesting that high yields of oxalate can be obtained on both Hg and stainless steel, particularly at lower potentials. Such reports created a great deal of interest as they provided a potentially efficient synthetic route to C_2 chemistry via an abundant and cheap starting material.

Further work supported these observations but seemed to draw a clear distinction between metals such as Au, Pt and Ag other cathodes in that the predominant products at the former substrates appeared to be only those observed by Haynes and Sawyer (1967). In contrast, oxalate formation was favoured by most other metals.

The natural assumption made by a large number of researchers in the field of electrochemical CO_2 reduction was that the intermediate was $CO_2^{\cdot -}$, as postulated by Haynes and Sawyer (1967). The observation of oxalate as a major product in addition to, or in competition with, the formation of CO, CO_3^{2-}, HCO_3^- and $HCOO^-$, increased the attention focused on the reactive intermediate and the mechanisms by which it reacted. However, controversy has arisen over whether the subsequent reaction of the $CO_2^{\cdot -}$ was via dimerisation (the EC mechanism) or via attack on another CO_2 molecule (the ECE mechanism). In addition, the existence of such species as $CO_2^{\cdot -}$ (ads) and $HCOO^{\cdot}$ (ads) have also been suggested but, as we shall see, these are not now thought to play a major role on simple metals.

In order to study the identity and nature of the intermediate, Aylmer-Kelly *et al.* (1973) employed modulated specular reflectance spectroscopy. They studied the reduction reaction at a lead cathode in both aqueous and non-aqueous electrolytes. A phase-sensitive detection system was employed by the authors, locked-in to the frequency of the potential modulation. The potential was modulated at 30 Hz between the reference potential of $-1.0\,V$ vs. Ag/AgCl and a more cathodic limit.

In non-aqueous solvents the spectrum obtained consisted of two strong bands, as shown in Figures 3.45(a) and (b), using propylene carbonate and acetonitrile as the solvent. The two peaks could either both be ascribed to $CO_2^{\cdot -}$, or to $CO_2^{\cdot -}$ and a second radical species. The higher energy band, i.e. 285 nm in propylene carbonate or 270 nm in acetonitrile, was readily

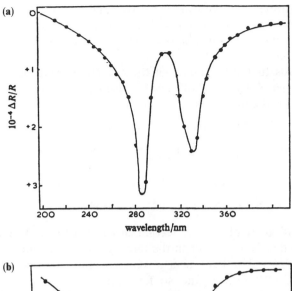

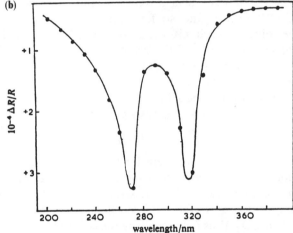

Figure 3.45 (a) Spectral dependence of the optical response on pulsing a lead cathode in 0.4 M tetramethylammonium perchlorate in propylene carbonate saturated with CO_2 from $+1.0\,V$ to $+0.2\,V$ at 30 Hz. The reference electrode was $Li/0.5\,M$ Li^+. (b) As in (a) except solvent changed to acetonitrile. From Aylmer-Kelly *et al.* (1973).

attributed to the anion radical in solution on the basis of pulse radiolysis experiments on the irradiation of formate. This band shifts to 250 nm in water, as would be expected on changing to a hydrogen bonding solvent. Thus, the transition responsible for the $CO_2^{\cdot-}$ absorption is the $\sigma^* \leftarrow n$, involving a non-bonding electron on the oxygen and an antibonding σ orbital on the carbon. Hydrogen bonding will involve the non-bonding electron, lowering its energy and hence causing a large shift. The decreased dipole moment of the excited state will cause a blue shift of the absorption on moving to more polar solvents. In addition, adding a small amount of Li^+

causes this feature to move to lower wavelengths, as expected on the basis of strong ion-pairing interactions. Thus, this part of the study established unequivocally the importance of solution intermediates but did *not* rule out adsorbed species.

On the basis that the product of the reduction in aprotic solvents was oxalate, Aylmer-Kelly and colleagues reasoned that the two most likely mechanisms for the fate of the anion radical were:

$$CO_2 + e^- \longrightarrow CO_{2\,ads}^{\cdot -} \tag{3.63}$$

$$2CO_{2\,ads}^{\cdot -} \longrightarrow C_2O_4^{2-} \tag{3.64}$$

and:

$$CO_2^{\cdot -} + CO_2 \longrightarrow {}^-O_2C\text{--}CO_2^{\cdot} \tag{3.65}$$

$${}^-O_2C\text{--}CO_2^{\cdot} + e^- \longrightarrow C_2O_4^{2-} \tag{3.66}$$

The dimerisation is, of course, second order in $CO_2^{\cdot -}$ while the solution coupling reaction is first order in the radical anion. The authors postulated that the absorption band at longer wavelengths in Figure 3.45(a) and (b) was due to the $C_2O_4^{\cdot -}$ radical and so favoured the solution mechanism. In addition, if the solution mechanism is correct, and given that the constants for the subsequent reactions of the radical anion are all very fast, then providing that the concentration of CO_2 in the solution remains unchanged, a steady state must be established at any given potential in which the rate of loss of the radical anion is equal to the rate of its formation. Thus, at low overpotentials where $[CO_2^*]$ is effectively constant:

$$I_c/nFA = k_2[CO_2][CO_2^{\cdot -}]. \tag{3.67}$$

By employing an extinction coefficient for the anion radical obtained from the pulse radiolysis experiments, the concentration of the radical could be calculated, and plotted against I_c. The straight line plot so obtained was taken as strong evidence for the ECE mechanism, i.e. the solution phase attack of $CO_2^{\cdot -}$ on CO_2, thus fully resolving the controversy over the identity and state of the intermediate. From the slope of the plot the authors obtained the rate constant k_2 as $7.5 \times 10^3 \, \text{dm}^3 \, \text{mol}^{-1} \, \text{s}^{-1}$.

The authors also reported that a Tafel plot of the reduction current gave a slope of 107 mV, indicating $n_c = 1$, in agreement with their proposed mechanism.

An important parameter required to further understand the reduction process was the $E^{0\cdot}$ value of the $CO_2/CO_2^{\cdot -}$ couple. However, up until 1977, attempts to detect the re-oxidation of the radical failed. This was almost certainly due to the very low concentrations of the radical actually present, as shown by the work of Aylmer-Kelly *et al.* (1973). Extrapolating E vs. t curves, such as those depicted in Figure 3.42, back to $t = 0$ gives some indication of E^0 for the reduction process and by such an extrapolation Haynes and Sawyer (1967) reported standard potentials of -2.11 V vs. SCE at Hg and -1.91 V vs. SCE at Au for the $CO_2/CO_2^{\cdot -}$ couple, both values being dependent on the residual water concentration. However, it is the work of Lamy *et al.* (1977) that is taken as the definitive study in this regard.

The technique employed by Lamy and colleagues was rapid-scan cyclic voltammetry in extremely dry DMF. In order to try and increase the lifetime of the $CO_2^{\cdot-}$ species the experiments were performed in the presence of active alumina suspensions. Aylmer-Kelly *et al.*, 1973 had calculated the rate constant for reaction of the radical with water as a fast $5.5\,dm^3\,mol^{-1}\,s^{-1}$ and it was also hoped that reducing the solvation of the radical by water would increase the coulombic repulsion between radicals and so reduce dimerisation).

Typical results of these authors are shown in Figure 3.46. A small faradaic re-oxidation wave was observed at very high sweep rates, i.e. up to $4400\,Vs^{-1}$. The standard potential of the $CO_2/CO_2^{\cdot-}$ couple was evaluated as the middle of the interval between the cathodic and anodic peak potentials at $4400\,Vs^{-1}$, giving -2.21 V vs. SCE.

From the section on voltammetry, the potential at which the (cathodic) current peak is observed, E_p, is related to the scan rate v by:

$$E_{p,c} = \{E^0 - (RT/\beta nF)[0.78 + \log_e(D^{1/2}/k^0) + \log_e(\beta nFv/RT)^{1/2}]\}$$

Thus, a plot of $E_{p,c}$ vs. $\log_e v$ gives β from the slope and k^0 from the intercept. Using $E^0 = -2.21$ V, $n = 1$ and $D = 10^{-5}\,cm^2\,s^{-1}$, the authors calculated a transfer coefficient of 0.4 and a standard electron transfer rate constant of $6 \times 10^{-3}\,cm\,s^{-1}$.

For most metals (Hg, Au, Pt, Pb) and solvents (DMSO, CH_3CN, propylene carbonate) investigated it is again found that the first electron transfer process is rate limiting and, in solution, the nature of the products from the reaction is a strong function of the conditions under which the reaction is performed.

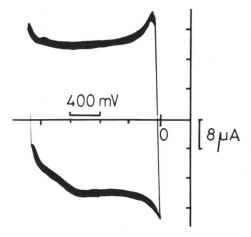

Figure 3.46 Cyclic voltammetry of CO_2 in DMF-active alumina suspension. Re-oxidation pattern at $4400\,Vs^{-1}$ and $4 \times 10^{-3}\,M\,CO_2$. After Lamy *et al.* (1977).

The problem, from the point of view of the investigation of the mechanism, is that since the first electron transfer step is rate limiting, conventional kinetic techniques for studying the reaction are useless as a means of elucidating the overall mechanism and conventional spectroscopic techniques have also proved to be of little help. Gressin and colleagues (1979) were therefore driven to the careful study of the nature of the products as a function of reaction conditions. This work is seminal in the study of the direct reduction of CO_2 since the authors were able to produce a general scheme for the reaction of $CO_2^{\bullet-}$ that is now widely accepted.

The authors studied the reduction of CO_2 at lead and mercury cathodes in DMF and monitored the distribution of products as a function of current, CO_2 concentration, water concentration and changing the solvent to DMSO. The product analysis was performed using standard quantitative analysis on the electrolyte and the gas phase above.

The authors proposed the scheme shown in Figure 3.47 as a working model based on the main products observed in aqueous and non-aqueous solvents.

As was suggested by Aylmer-Kelly et $al.$ (1973) Gressin and co-workers (1979) also postulated that the formation of oxalate could occur via two possible routes: the dimerisation of $CO_2^{\bullet-}$ (route 2 in Figure 3.47) or the attack of the $CO_2^{\bullet-}$ on a second molecule of CO_2 followed by reduction (route 3). Both of these routes involve coupling via C–C bond formation. Coupling can also occur to give the C–O–C linkage (route 3'), the route responsible for the formation of CO and CO_3^{2-} (steps 5, 5' and 6).

Increasing the water content in the DMF electrolyte was found to increase the formation of formate via steps 7, 8, and 8'. It also leads to increased

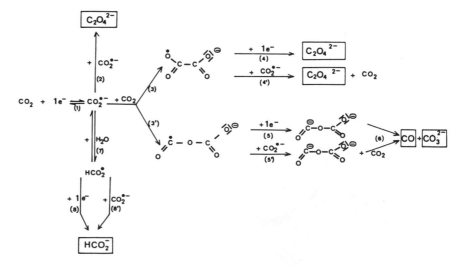

Figure 3.47 A summary of the processes taking place during the electrochemical reduction of CO_2. From Gressin et $al.$ (1979).

glycolate (CH_2OHCOO^-). This was interpreted by the authors in terms of the following mechanism:

$$H_2O + (COO^-)_2 \longrightarrow (COOH)(COO^-) + OH^- \qquad (3.68)$$

The monoprotonated dianion, unlike the dianion itself, can then be reduced:

$$(COOH)(COO^-) + 4e^- + 3H_2O \longrightarrow (CH_2OHCOO^-) + 4OH^- \qquad (3.69)$$

Increasing the current density should favour the dimerisation of the radical anion (route 2) while increasing the CO_2 concentration should favour route 3. If oxalate formation occurs via route 3, then changing the CO_2 concentration and the current density should affect both the CO and $C_2O_4^{2-}$ yields equally whereas if oxalate formation is via route 2, the effect of changing these same two conditions with respect to the CO yield will be opposite to that observed in the yield of oxalate.

Similarly, if formate formation occurs via route 7, then there should be competition between formate formation and oxalate/CO and CO_3^{2-} formation, i.e. between routes 2, 3 and 3'. Increasing the water concentration will *decrease* oxalate and *increase* formate. Further, the proportion of formate should *decrease* with increasing current.

The authors found that:

1. The yield of CO at Pb was consistently low.
2. Increasing the water concentration decreased the yield of formate, with the formation of CO remaining constant (Hg) or increasing (Pb).
3. At mercury, increasing the current density increased the yield of oxalate and decreased the yield of formate. At lead, changing the current density seemed to have little effect, with the major product (i.e. $\geqslant 85\%$) remaining oxalate.
4. At Hg, increasing the CO_2 concentration decreased the yield of oxalate, marginally increased the yield of formate and drastically increased the yield of CO. Again, the major product at Pb was unchanged, $\geqslant 88\%$ oxalate.

From these results it was concluded that:

1. $CO_2^{\cdot-}$ is specifically adsorbed on Pb, leading to the almost complete suppression of routes 3 and 3' and hence only a small production of CO.
2. Oxalate is found predominantly via the dimerisation of the radical (cf. the results of Aylmer-Kelly *et al.* (1973) with CO formation being via route 3' followed by step 5'.

Thus, the work of Gressin and co-workers (1979) provided a reasonable scheme to describe the various routes by which $CO_2^{\cdot-}$ can react.

As was discussed above, it became generally accepted that CO_2 reduction at Pt in aprotic solvents did not result in oxalate. In view of this fact, and the potential industrial and commercial importance of the production of oxalate from CO_2, a paper by Desilvestro and Pons in 1989 generated a

Table 3.4 Features observed by Desilvestro and Pons (1989) during the reduction of CO_2 at Pt in 0.1 M tetrabutylammonium tetra fluoroborate (TBAF)/CH_3CN, using *in situ* FTIR spectroscopy

Gain features (cm^{-1})	Loss features (cm^{-1})
2970	2341
2931	2293
2882	2257
1687	1096
1647	1065
1479	1029
1387	918
1361	754
1330	659
1304	
1275	
978	
888	
840	
828	
683	

great deal of interest by claiming the electrochemical reduction of CO_2 to oxalate at Pt in acetonitrile at potentials as high as -1.2 V vs. SCE. The technique they employed was *in situ* FTIR and the bands they observed during the reduction are detailed in Table 3.4. The authors based their assignment on an experiment in which a solution of tetrabutylammonium oxalate (TBAO), was *oxidised* at the Pt electrode and the *loss* features compared to the product features observed in the actual experiment. The two sets of data did not match, so leading the authors to explain the differences in terms of the poor solvation of the oxalate.

This system was subsequently investigated by Christensen *et al.* (1990) also using *in situ* FTIR, who observed identical product features (see Figure 3.48). In order first to compare directly the IR spectrum of oxalate generated *in situ*, the authors took advantage of the surface reactivity of Pt and the poor diffusion of species to and from the thin layer. Thus, a solution of oxalic acid in the electrolyte was placed in the spectroelectrochemical cell, the potential of the platinum working electrode stepped to successively lower values and spectra taken at each step. The spectra were all normalised to the reference spectrum collected at the base potential of 0 V vs. SCE. As a result of the deprotonation of adventitious water:

$$H_2O + Pt + e^- \longleftrightarrow Pt{-}H + OH^- \tag{3.70}$$

a large increase in pH occurs in the thin layer, maintained by the poor diffusion rate of the OH^- out into the bulk of the electrolyte. This then results in the sequential deprotonation of the oxalic acid:

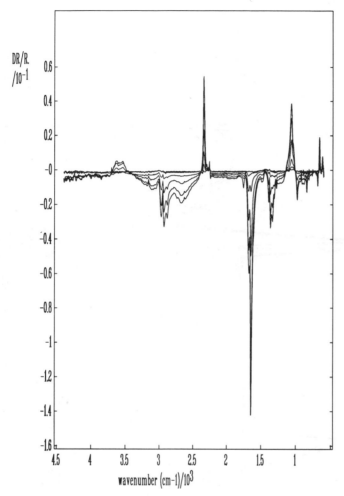

Figure 3.48 Reflectance spectra collected off a Pt electrode immersed in CO_2-saturated CH_3CN/0.1 M tetrabutylammonium tetrafluoroborate. The reference spectrum was taken at the base potential of -0.8 V vs. SCE. The potential was then stepped down to successively lower values, further spectra collected and normalised to the reference spectrum. The spectra were collected at -1.0 V, -1.2 V, -1.4 V, -1.6 V, -1.8 V and -1.9 V. The spectrum at -1.0 V showed little or no features, bands then grew in intensity as the potential was stepped down. From Christensen *et al.* (1990).

at $E < 0$ V:

$$(COOH)_2 + OH^- \longrightarrow (COO^-)(COOH) + H_2O \qquad (3.71)$$

at $E < -1.0$ V:

$$(COO^-)(COOH) + OH^- \longrightarrow (COO^-)_2 + H_2O \qquad (3.72)$$

These processes are shown in Figures 3.49(a) and (b). The loss features near 1750 cm^{-1} and 1180 cm^{-1} can be attributed to the acid and the bands

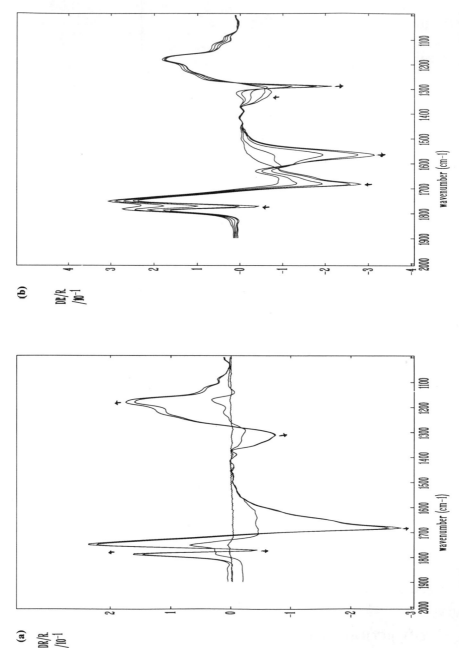

Figure 3.49 Reflectance spectra collected off a Pt electrode immersed in N_2-saturated $CH_3CN/0.1\,M$ tetrabutylammonium tetrafluoroborate containing $0.1\,M$ oxalic acid. The reference spectrum was taken at a base potential of $0\,V$. The potential was then stepped down to successively lower values, further spectra taken and normalised to the reference. (a) $-0.2\,V$, $-0.4\,V$, $-0.8\,V$ and $-1.0\,V$. (b) $-1.0\,V$, $-1.2\,V$, $-1.4\,V$, $-1.6\,V$ and $-1.8\,V$. The arrows show the direction of change of the various bands as the potential was stepped down. From Christensen *et al.* (1990).

near $1760 \, cm^{-1}$, $1690 \, cm^{-1}$ and $1320 \, cm^{-1}$ were attributed to the (COO^-) (COOH) species. At potentials $< -1.0 \, V$ (see Figure 3.49(b)) the second deprotonation occurs, giving gain features near $1290 \, cm^{-1}$ and $1570 \, cm^{-1}$ which are characteristic of the oxalate anion.

This, more direct, method of comparing the oxalate spectra to that of the reduction product clearly indicated that the latter was not oxalate.

Figure 3.50 shows IR spectra of tetrabutylammonium carbonate (TBAC) in CH_3CN/TBAF. The spectra were collected at various times after the

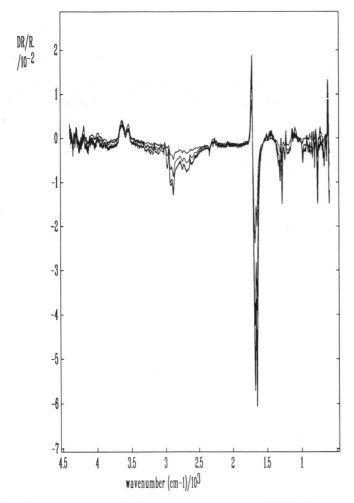

Figure 3.50 Reflectance spectra of a Pt electrode immersed in *c.* 0.05 M tetrabutylammonium carbonate/CH_3CN + 0.1 M tetrabutylammonium tetrafluoroborate at open circuit, showing the development of the IR bands of various solvated carbonate species with time. Spectra were taken at various times after the admission of the carbonate solution and normalised to the reference spectrum taken in the absence of the carbonate. The spectra cover the time period 0–13 minutes. From Christensen *et al.* (1990).

solution was run into the cell with the Pt electrode pressed against the window, the cell and thin layer having previously contained just the pure electrolyte. The spectra were all normalised to the reference spectrum taken immediately prior to the admission of the carbonate solution. The figure shows results obtained with a reasonably dry sample of the carbonate but all of the bands showed some variation with respect to the water content present in the system, indicating the important role of water solvation in the IR spectrum of CO_3^{2-}.

The authors rationalised the large variation in the absorption frequencies of the carbonate bands thus: CO_3^{2-} had D_{3h} symmetry, with the delocalisation of the negative charges giving rise to a bond order of $c.$ $1\frac{1}{3}$ for each of the three C–O bonds in solvents where the delocalised species can be solvated. Thus, in water, the degenerate v_3 vibration has a frequently of $c.$ 1400 cm^{-1}. However, if the delocalisation of the charge is lost by, e.g. forming two covalent C–O bonds to give an organic carbonate, then the degeneracy of v_3 is lifted as the bond orders of the three covalent bonds becomes 2:1:1. In dimethyl carbonate, where the CH_3–O bonds are strongly covalent, the C=O stretch rises to 1870 cm^{-1} and the C–O stretches move down to 1260 cm^{-1}. In carbonate complexes, the C=O stretch occurs around 1450 cm^{-1} (unidentate, bond orders $c.$ 1:1.5:1.5) or 1610 cm^{-1} (bidentate, bond orders $c.$ 1:1:2). Acetonitrile has a poor solvation ability towards anions with localised charge due to its inability to act as a hydrogen-bond donor. Hence, for anions with a large (charge)^2/radius ratio, i.e. carbonate, the only way for the anion to be stabilised is via ion-pair formation. The various forms of solvated carbonate that may be envisaged are shown in Figure 3.51. In the absence of water the TBA$^+$ cations will effectively 'lock in' the asymmetry via the electrostatic interaction between the negative charges locked onto the O atoms and the positive charge on the TBA$^+$.

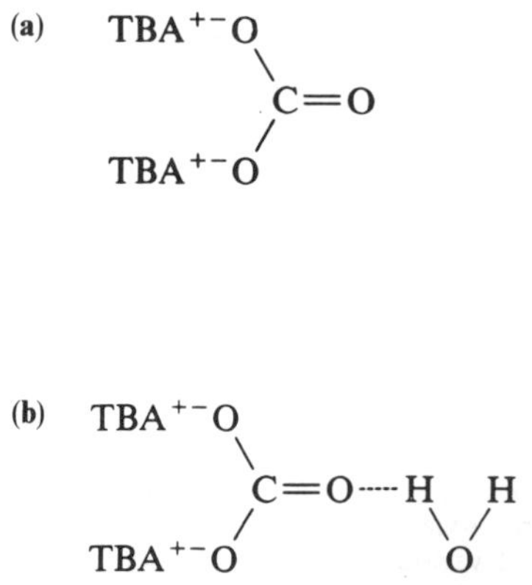

(a)   TBA^{+-}O
$$\begin{array}{c} \text{TBA}^{+-}\text{O} \\ \diagdown \\ \qquad \text{C}=\text{O} \\ \diagup \\ \text{TBA}^{+-}\text{O} \end{array}$$

(b)   TBA^{+-}O
$$\begin{array}{c} \text{TBA}^{+-}\text{O} \\ \diagdown \\ \qquad \text{C}=\text{O}\text{----H} \quad \text{H} \\ \diagup \qquad\qquad \diagdown / \\ \text{TBA}^{+-}\text{O} \qquad\quad \text{O} \end{array}$$

Figure 3.51 Schematic representation of the CO_3^{2-} anion in 0.1 M tetrabutylammonium tetrafluoroborate/CH_3CN.

The existence of the various species depicted in Figure 3.51 was proven in a series of experiments designed to allow the observation of the TBAC in acetonitrile of varying water content. In the presence of excess added water, stepping the potential down to successively lower values to induce dissolution of the CO_2 just lead to the gain of bands near $1387\,cm^{-1}$ and $1362\,cm^{-1}$ which the authors attributed to aqueous carbonate. The role of water in the solvation of the other forms of carbonate was shown by stepping the potential of a Pt electrode immersed in the electrolyte $+ 10\%$ by volume water and comparing the bands observed to those seen when repeating the experiment in the presence of 10% D_2O. Thus, the features near $1682\,cm^{-1}$, $1303\,cm^{-1}$ and $1274\,cm^{-1}$ are unaffected by the addition of D_2O, showing that these are due to the vibrations of carbonate/TBA$^+$ species solvated only by CH_3CN. In contrast, the bands near $1645\,cm^{-1}$, $1332\,cm^{-1}$, $976\,cm^{-1}$ and $683\,cm^{-1}$ are all shifted to lower frequencies, indicating the participation of water in the solvation.

The fact that the $1682\,cm^{-1}$ feature does not shift in D_2O while the $1645\,cm^{-1}$ band does, suggests that the latter feature is due to the C=O stretch of carbonate that is solvated in the manner shown in Figure 3.51(b). The reduction of the frequency of the C=O stretching mode from $1682\,cm^{-1}$ to $1645\,cm^{-1}$ is consistent with the general observation of reduction in v(C=O) on hydrogen bonding to the oxygen. Additional evidence for the involvement of water is shown by the presence of the broad feature extending from $2400\,cm^{-1}$ to $3400\,cm^{-1}$ in Figure 3.48 which underlies the C–H stretches of the TBA$^+$ ion. This structure is very similar to that found in the enol form of acetylacetone which has been ascribed to the presence of intramolecular hydrogen bonding.

The authors reported that in the absence of any water, i.e. after extensive drying of the acetonitrile, stepping the potential to successively lower potentials only gave rise to bands near $1682\,cm^{-1}$, $1303\,cm^{-1}$ and $1274\,cm^{-1}$. These were attributed by the authors to CO_3^{2-} ion-paired with two TBA$^+$ cations (see Figure 3.51(a)).

The results so obtained by Christensen and colleagues (1990) can be summarised as shown in Table 3.5.

The authors found no evidence for CO_2 reduction at Pt under these conditions at potentials $\geqslant -2.0\,V$ vs. SCE; in agreement with the general consensus of opinion, the CO_2 was simply dissolving to form CO_3^{2-}. At gold, in contrast, CO_2 was reduced at potentials $< -1.6\,V$, as shown by a strong band at $1606\,cm^{-1}$ (marked with * on the figure) assigned to formate and a feature near $2140\,cm^{-1}$ that was unequivocally assigned to CO (see Figures 3.52(a) and (b)).

The study by Christensen and colleagues (1990) provided important direct evidence for the existence of different forms of solvated carbonate and the dramatic effect on the IR absorption frequencies that this variation in solvation has, as well as clarifying the data of Desilvestro and Pons (1989). The work again proved the worth of a technique capable of providing a

Table 3.5 Assignments of the IR bands observed during the cathodic polarisation of a Pt electrode immersed in CO_2-saturated 0.1 M TBAF/CH_3CN at potentials between 0 V and -2.0 V vs. SCE (after Christensen et al., 1990).

Gain features	
Observed (cm^{-1})	Assignment
2971	
2929	TBA^+, CH_3CN C–H stretch
2880	
1682	$CO_3^{2-}[TBA^+]_2$
1645	$CO_3^{2-}[TBA^+]_2 \cdot H_2O$
1481	TBA^+ C–H def.
1387	
1362	$CO_3^{2-}(aq)$
1332	$CO_3^{2-}[TBA^+]_2 \cdot H_2O$
1303	$CO_3^{2-}[TBA^+]_2$
1274	$CO_3^{2-}[TBA^+]_2$
976	$CO_3^{2-}[TBA^+]_2 H_2O$
840	$CO_3^{2-}[TBA^+]_2 H_2O$
828	$CO_3^{2-}[TBA^+]_2 H_2O$
683	$CO_3^{2-}[TBA^+]_2 H_2O$

direct, and straightforward, means of identifying the products of an electrochemical reaction.

3.3.2   The catalysed reduction of CO_2

One possible strategy in the development of low-overpotential methods for the electroreduction of CO_2 is to employ a catalyst in solution in the electrochemical cell. A few systems are known that employ homogeneous catalysts and these are based primarily on transition metal complexes. A particularly efficient catalyst is (Bipy)Re[CO]$_3$Cl, where Bipy is 2, 2′ bipyridine, which was first reported as such by Hawecker et al. in 1983. In fact, this first report concerned the *photochemical* reduction of CO_2 to CO. However, they reasoned correctly that the complex should also be capable of catalysing the electrochemical reduction reaction. In 1984, the same authors reported that (Bipy)Re[CO]$_3$Cl catalysed the reduction of CO_2 to CO in DMF/water/ tetraalkylammonium chloride or perchlorate with an average current efficiency of $>90\%$ at -1.25 V vs. NHE (c. -1.5 V vs. SCE). The product analysis was performed by gas chromatography and ^{13}C nmr and showed no other products.

Figure 3.53 shows the volume of CO produced, and the turnover number of the rhenium catalyst, as a function of the charge passed. Over 14 hours of electrolysis the Re catalyst underwent 300 catalytic cycles without loss of activity and showed no degradation, as seen by isolation and characterisation

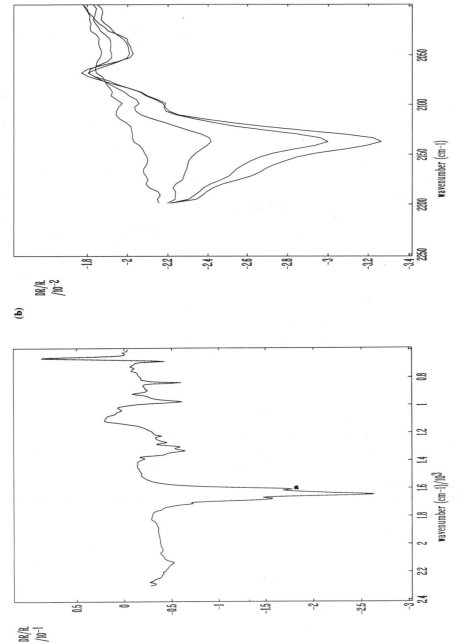

Figure 3.52 (a) Reflectance spectrum of a gold electrode immersed in CO_2-saturated 0.1 M tetrabutylammonium tetrafluoroborate/CH_3CN at -1.9 V and normalised to the reference collected at -0.8 V. The principal formate band is marked with an asterisk. (b) Reflectance spectra of the 2200–2000 cm^{-1} region collected off a gold electrode immersed in CO_2-saturated 0.1 M tetrabutylammonium tetrafluoroborate/CH_3CN at (top to bottom): -1.6 V, -1.8 V, -1.9 V and -2.0 V. From Christensen et al. (1990).

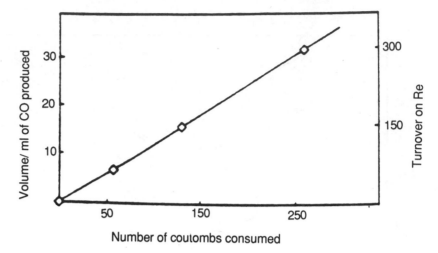

Figure 3.53   Volume of CO produced and turnover number with respect to catalyst as a function of coulombs consumed in the electroreduction of CO_2 to CO by Re(bipy)(CO)$_3$Cl. From Hawecker *et al.* (1983).

of the catalyst by thin layer chromatography and ^{13}C nmr at the end of the experiment.

The authors found that the water content of the system was important, as can be seen from Figure 3.54(a) which shows the volume of CO produced after 4 h of electrolysis as a function of the percentage of added water, the maximum yield being obtained for the electrolyte having 10% added water. In the absence of water, the reduction occurred much more slowly and an orange species accumulated in solution, disappearing slowly when the current was switched off.

In one of the experiments reported by the same authors in a later publication, the electrolysis at -1.25 V vs. NHE in 5% H_2O/DMF was interrupted after 2 h, when the solution was purged with N_2 (see Figure 3.54(b)). No CO was produced in the following 30 min electrolysis of the N_2-saturated solution but the orange species mentioned above was seen to accumulate again. On readmitting the CO_2, the orange colour disappeared instantaneously and CO was produced with the same efficiency as prior to the interruption. In the absence of CO_2 the reduction of the (Bipy)Re[CO]$_3$Cl at -1.25 V was quantitative (i.e. 1 e$^-$ passed per Re atom), with the orange species apparently the product.

In the presence of 10% H_2O but no CO_2, the same orange species accumulated in the solution and electrocatalysis was slow, with H_2 being generated with a current efficiency of *c.* 85% (only a tiny amount of H_2 was observed under the same conditions in the absence of the complex). In the presence of CO_2 the water reduction reaction was completely inhibited, showing that the orange species is less reactive towards water than CO_2 and hence is a highly specific catalyst for the conversion of CO_2 to CO.

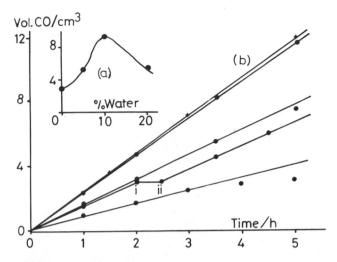

Figure 3.54 Volume of CO generated electrochemically as a function of the percentage of water added (a) and time (b); the supporting electrolyte was Et$_4$NCl (●) or Bu$_4$NClO$_4$ (+). The electrolysis cell was flushed and kept under N$_2$ during the 30 minutes represented by points (i) and (ii), after which the N$_2$ was replaced by CO$_2$. In (a) the volume of CO was measured after 4 hours electrolysis. From Hawecker *et al.* (1984).

Under the optimum conditions in the presence of CO$_2$, the orange species was observed close to the cathode but does not accumulate and was considered by the authors to be the active species, which was formulated as $\{(Bipy)Re[CO]_3Cl\}^-$.

When NH$_4$PF$_6$ was employed as the supporting electrolyte, no CO was produced but near-quantitative formation of H$_2$ was observed. During the electrolysis an air-stable, green, sparingly soluble material was produced, which was isolated and characterised as the dimer, $\{(Bipy)Re[CO]_3\}_2$. It was fairly reasonable to assume that this was formed via dimerisation of the radical (Bipy)Re[CO]$_3$. The authors postulated that more dimerisation occurs in the presence of a non-coordinating anion such as PF$_6^-$, rather than coordinating anions such as Cl$^-$ or ClO$_4^-$, due to the labilisation (and loss) of Cl$^-$ and thus the exposure of the sixth coordination site and subsequent dimerisation.

In 1985, Sullivan *et al.* reported a voltammetric study on (Bipy)Re[CO]$_3$Cl in acetonitrile with tetrabutylammonium hexafluorophosphate (TBAHFP) as the supporting electrolyte.

Figure 3.55(a) shows a cyclic voltammogram of a Pt electrode immersed in a solution of the complex in acetonitrile. The peak labelled A was found to be insensitive to replacing the Cl by other ligands, an observation that was taken as strong evidence that the first reduction process is localised on the bipyridine ligand. This was supported by the fact that the peak potential of the wave, -1.35 V vs. SCE, occurs relatively close to the first cathodic peak for (Bipy)$_3$Ru^{2+}, -1.29 V, which is known to involve the Bipy.

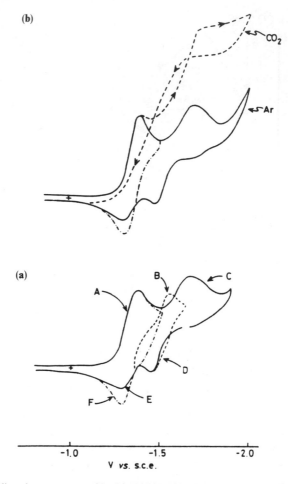

Figure 3.55   Cyclic voltammograms of Re(Bipy)(CO)$_3$Cl in CH$_3$CN/0.1 M tetrabutylammonium hexafluorophosphate as supporting electrolyte at a button Pt electrode, and with a sweep rate of 200 mV s^{-1}. (a) The switching potential characteristics of the coupled chemical reactions in the absence of CO$_2$. The lettered redox processes are discussed in the text. (b) The effect of saturating the solution with CO$_2$. From Sullivan *et al.* (1985).

The second, more irreversible, process was attributed to the reduction of the Re(I) to Re(0) because of the sensitivity of its cathodic peak potential to the identity of any ligand replacing the chloride.

Electrolysis of a solution of the catalyst at -1.4 V to -1.5 V vs. SCE in the absence of CO$_2$, giving 1 e$^-$/Re atom, gave the sparingly soluble green dimer (as characterised by UV-visible, IR, ^{1}H nmr and elemental analysis on the isolated material) in agreement with the work of Lehn and colleagues (Hawecker, 1983) and showing that the first reduction is coupled to the formation of the dimer if chloride loss is allowed to occur. If the cyclic voltammetric scan was reversed when the potential reached -1.5 V, at a scan rate of 200 mV/s (see Figure 3.55(a)), then the first reduction was found to be reversible, indicating that dimer formation is slow on the timescale of the experiment.

Electrolysis at -1.8 V, i.e. past the second metal-based reduction, consumed $2e^-$/Re atom and gave an intense red/purple solution, attributed by the authors to $\{(Bipy)Re[CO]_3\}^-$ on the basis of the known electrochemical behaviour of the 1, 10-phenanothroline analogue. The redox couple observed when cycling between -1.35 V and -1.6 V, peaks B and D in Figure 3.55(a), was attributed to the anion/dimer couple:

$$\{(Bipy)Re[CO]_3\}_2 + 2e^- \longleftrightarrow 2\{(Bipy)Re[CO]_3\}^- \tag{3.73}$$

Figure 3.55(b) shows the first sweep of the voltammogram obtained in the presence of CO_2 and also in Ar saturated solution for comparison. The authors concluded that the reductive scan showed that little or no catalytic enhancement occurs at the Bipy-based reduction potential and that the onset of the catalytic current occurs at -1.4 V continuing through the region characteristic of the metal-based reduction process (and the $2e^-$ reduction of the dimer) in direct contrast to the results obtained by the majority of other workers (see below).

Electrolysis of the $(Bipy)Re[CO]_3Cl$ solution at -1.5 V to -1.55 V gave sustained electrocatalysis, with $>98\%$ current efficiency, with only CO and CO_3^{2-} being detected as products. In contrast, electrolysis at -1.8 V gave only CO, with current efficiency $>85\%$.

On the basis of their results the authors postulated two distinct pathways for the electroreduction of CO_2 by the rhenium catalyst, with the intermediates being (a) $\{(Bipy)Re[CO]_3\}^-$ and (b) the radical $(Bipy)Re[CO]_3$ (see Scheme 3.1). The former is indicated by the fact that production of the green dimer is completely suppressed in CO_2-saturated electrolyte, as a result of the CO_2 intercepting the active species present.

$$(Bipy)Re[CO]_3Cl + e^- \longleftrightarrow (Bipy^-)Re[CO]_3Cl$$

then (A):
$$(Bipy^-)Re[CO]_3Cl + e^- \longrightarrow \{(Bipy)Re[CO]_3\}^- + Cl^-$$
$$\{(Bipy)Re[CO]_3\}^- + CO_2 \longrightarrow \{(Bipy)Re[CO]_3CO_2\}^-$$
$$\{(Bipy)Re[CO]_3CO_2\}^- + A + e^- \longrightarrow \{(Bipy)Re[CO]_3\}^- + AO^- + CO$$

(where A is an oxide ion acceptor)
and (B):
$$(Bipy^-)Re[CO]_3Cl \longrightarrow (Bipy)Re[CO]_3 + Cl^-$$
$$(Bipy)Re[CO]_3 + CO_2 \longrightarrow (Bipy)Re[CO]_3CO_2$$
$$(Bipy)Re[CO]_3CO_2 + CO_2 + 2e^- \longrightarrow (Bipy)Re[CO]_3 + CO + CO_3^{2-}$$

Scheme 3.1

The identity of the oxide ion acceptor species in Scheme 3.1 was unknown and the authors provided no *direct* evidence for the intermediates postulated in the scheme.

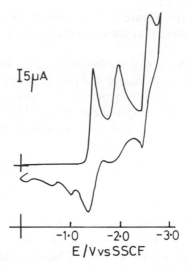

Figure 3.56 Initial negative-going sweep of the cyclic voltammogram of a Pt electrode in CH_3CN/0.1 M tetrabutylammonium perchlorate containing Re(dmbpy)(CO)$_3$Cl. (dmbpy) = 4, 4′-dimethyl-2, 2′-bipyridine. From Breikss and Abruna (1986).

In 1986, Breikss and Abruna reported electrochemical and mechanistic studies on a close analogue of the rhenium complex, (Dmbpy)Re[CO]$_3$Cl, where Dmbpy = 4,4′ dimethyl 2,2′ bipyridine. The cyclic voltammogram of the complex at platinum in CH_3CN/tetrabutylammonium perchlorate is shown in Figure 3.56; for simplicity we will consider only the electrochemistry taking place above c. -2.3 V vs. SCE.

The first cathodic wave was studied by cycling the potential across it at various scan rates and the peak potentials were found to increase as $v^{1/2}$, indicative of a reversible, diffusion-controlled system, with $E^0 = -1.43$ V vs. SCE. However, at sweep rates $\leqslant 20$ mV/s the peak anodic current is much smaller than expected which was interpreted by the authors as indicating that the reduced species undergoes a subsequent chemical reaction, i.e. an EC process.

The authors then examined the voltammograms obtained for two consecutive sweeps between 0 V and -2.1 V (see Figure 3.57(a)). The first complete scan shows a very similar voltammogram to that observed in Figure 3.56 down to -2.1 V, with cathodic waves at -1.47 V and -1.98 V. On the second sweep these two waves have decreased substantially and a new wave appears near -1.65 V and -0.23 V. Initiating the second sweep in the negative direction at -1.45 V (see Figure 3.57(b)) shows that the new cathodic wave is the counterpart to the anodic wave at -1.57 V. The authors attributed these features to reversible couples with E^0 at -1.62 V and -0.19 V.

The fact that the -1.47 V and -1.98 V cathodic peaks both have similar peak currents on the first sweep and both decrease on the second (and subsequent) sweeps was taken by the authors as evidence that the lower

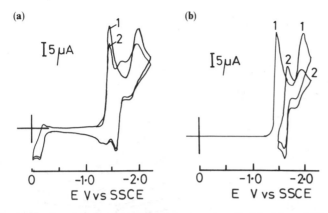

Figure 3.57 (a) Cyclic voltammograms for the system described in Figure 3.56 showing two complete, consecutive sweeps between 0 and -2.1 V vs. SSCE, in the absence of CO_2. (b) Same as in (a) except the second negative-going sweep was initiated at -1.45 V. From Breikss and Abruna (1986).

potential wave represents the further reduction of the $\{(Dmbpy)Re[CO]_3\}^{\cdot-}$ generated at -1.47 V. The difference between the first and subsequent sweeps was interpreted in terms of a slow chemical step in which reaction of the $\{(Dmbpy)Re[CO]_3Cl\}^{\cdot-}$ results in a product, or products, with cathodic waves at -0.23 V and -1.65 V. This chemical step was postulated as involving the loss of Cl^- followed by the formation of (a) the hydrido complex $\{(Dmbpy)Re\cdot[CO]_3H\}^{\cdot-}$, (b) the acetonitrile complex $\{(Dmbpy)Re[CO]_3\text{-}CH_3CN\}^\cdot$, or (c) the dimer $\{(Dmbpy)Re[CO]_3\}_2$.

To determine if Cl^- was indeed lost in the chemical step the potential was held at -1.7 V for 30 s, then a positive-going scan initiated up to 1.5 V. A number of anodic peaks were observed with the largest and most significant at 1.2 V. This was unequivocally attributed to the oxidation of Cl^- to Cl_2 on the basis of a second experiment in which tetrabutylammonium chloride was added to the base electrolyte and the potential regime repeated. Hence, the chemical step after the addition of the first electron involves the ejection of the chloride anion. The identity of the species formed subsequent to this process was determined thus: O'Toole *et al.* prepared and characterised the hydrido and acetonitrile complexes (as the bipyridine derivatives) and determined their E^0 values as -1.46 V and -1.25 V, respectively, far removed from the observed value -1.62 V; hence neither of these species were taken as being the product.

Further evidence came from *in situ* UV-visible studies by Breikss and Abruna (1986), apparently using external reflectance from a Pt electrode. Figure 3.58(A) shows the absorption spectrum for the rhenium complex in solution in the absence of CO_2 prior to electrolysis. The potential of the working electrode was then slowly ramped to more cathodic values until it reached -1.7 V, whereupon a reddish material formed with a spectrum labelled B in Figure 3.58, having a $\lambda_{max} = 512$ nm. The potential was then

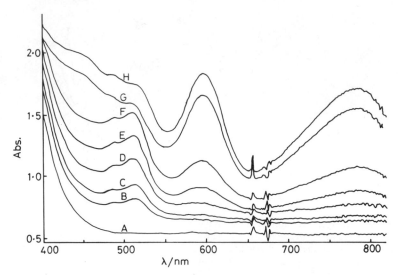

Figure 3.58 A–H Spectroelectrochemical studies of Re(dmbpy)(CO)$_3$Cl in CH$_3$CN/0.4 M tetra-butylammonium perchlorate. (A) Prior to electrolysis. (B) after applying a potential of -1.7 V vs. SSCE. (C–H) Sequential spectra after holding the potential at -1.7 V. From Breikss and Abruna (1986).

held at -1.7 V and a series of spectra collected as a function of time. As can be seen from spectra C–H, two additional features grow with time, near 598 nm and 792 nm, with the concomitant decrease in the intensity of the 512 nm band. Spectrum H was the limiting spectrum since it remained unchanged after a further 3 min. Again, $\{(Dmbpy)Re[CO]_3H\}^{\cdot-}$ and $\{(Dmbpy)Re[CO]_3CH_3CN\}^{\cdot}$ had been reported in the literature as having λ_{max} at 415 nm and 536 nm, respectively, so none of the species observed were these. Furthermore, the shift of λ_{max} to such a long wavelength is consistent with the formation of a metal-bonded dimer. Reduction of the parent complex in dry THF with sodium amalgam gave a green material having the same spectrum as H, and which was characterised by mass spectroscopy and elemental analysis as the dimer $\{(Dmbpy)Re[CO]_3\}_2$. On the basis of their results, they proposed Scheme 3.2.

$$(Dmbpy)Re[CO]_3Cl \longleftrightarrow \{(Dmbpy)Re[CO]_3Cl\}^{\cdot-} \qquad E^0 = -1.43\,V$$
$$\{(Dmbpy)Re[CO]_3Cl\}^{\cdot-} \longrightarrow \{(Dmbpy)Re[CO]_3Cl\}^{2-} \qquad E_{p,c} = -1.96\,V$$
$$\{(Dmbpy)Re[CO]_3Cl\}^{2-} \longrightarrow \{(Dmbpy)Re[CO]_3\}^- + Cl^- \qquad (fast)$$
$$\{(Dmbpy)Re[CO]_3\}^- + CH_3CN \longrightarrow \{(Dmbpy)Re[CO]_3CH_3CN\} + e^-$$
$$\{(Dmbpy)Re[CO]_3CH_3CN\} \longleftrightarrow \{(Dmbpy)Re[CO]_3CH_3CN\}^+ + e^-$$
$$E^0 = -1.34\,V$$

$$2(Dmbpy)Re[CO]_3Cl\}^{\cdot-} \longrightarrow \{(Dmbpy)Re[CO]_3\}_2 + 2Cl^-$$
$$\{(Dmbpy)Re[CO]_3\}_2 + e^- \longleftrightarrow \{(Dmbpy)Re[CO]_3\}_2^{\cdot-} \qquad E^0 = -1.62\,V$$
$$\{(Dmbpy)Re[CO]_3\}_2 \longleftrightarrow \{(Dmbpy)Re[CO]_3\}_2^{\cdot+} + e^- \qquad E^0 = -0.19\,V$$

Scheme 3.2

The authors also found that no dimer was formed in the presence of CO_2 and that the catalytically active species was the product of the first (Dmbpy-based) reduction.

Again, the conclusions of Breikss and Abruna (1986) were reached largely on the basis of indirect evidence and the investigation of the behaviour of (Bipy)Re[CO]$_3$Cl, or its Dmbpy analogue, up until 1992 lacked any molecular evidence for the various intermediates postulated in the presence and absence of CO_2. This was rectified by the work of Christensen *et al.* (1992) who employed *in situ* FTIR to study the reduction of (Dmbpy)Re[CO]$_3$Cl in the presence and absence of CO_2. The various carbonyl intermediates so observed were identified by comparison with CO absorption frequencies predicted on the basis of an energy-factored forced constant method. The method gives a root mean square (rms) error in the observed vs. calculated CO frequencies of $c. \pm 5\,cm^{-1}$ for uncharged compounds in hydrocarbon solvents. The method is less accurate for anionic or cationic complexes in polar solvents but the error is still only $c. \pm 8\,cm^{-1}$. The details of the predictive routine employed are beyond the scope of this book and the interested reader is referred to the paper itself and the references therein.

A cyclic voltammogram of the (Dmbpy)Re[CO]$_3$Cl in acetonitrile at a glassy carbon electrode is shown in Figure 3.59. The supporting electrolyte was the non-coordinating BF_4^-. On stepping down from $-1.0\,V$ down to $-1.4\,V$ vs. SCE (see Figure 3.60), large changes are seen in the CO stretching frequencies, with the parent molecule showing loss features near $2023\,cm^{-1}$, $1906\,cm^{-1}$ and $1893\,cm^{-1}$, all of which are shifted to *lower* values on reduction. This reduction in absorption frequency as the complex is reduced is a result of the metal becoming electron-rich which increases the amount of $d\pi(Re) \rightarrow p\pi(CO)$ back donation.

At $-1.3\,V$ small loss and gain features are observed that are very nearly symmetrical, with gains at $1993\,cm^{-1}$ and $1875\,cm^{-1}$. On the basis of the arguments of Cabrera and Abruna (1986), the authors reasoned that the parent compound is t_{2g}^6 low spin, hence addition of the e^- to the metal would give

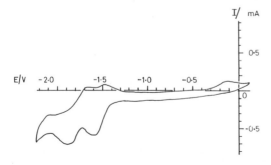

Figure 3.59 Cyclic voltammogram of a glassy carbon electrode immersed in N_2-saturated acetonitrile/0.2 M tetraethylammonium tetrafluoroborate containing 5×10^{-3} M Re(dmbpy)(CO)$_3$Cl. The scan rate is $100\,mV\,s^{-1}$. From Christensen *et al.* (1992).

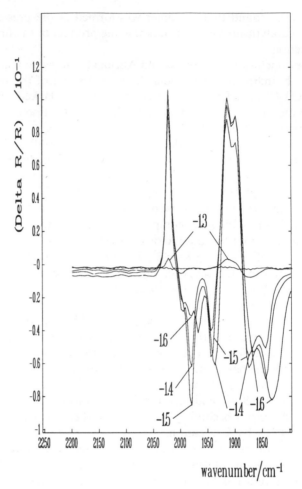

Figure 3.60 FTIR spectra collected from a glassy carbon electrode immersed in the solution of Figure 3.59. Spectra were collected at $-1.2\,V$ to $-1.6\,V$ vs. SCE in 100 mV steps and normalised to the reference spectrum taken at $-1.0\,V$. From Christensen *et al.* (1992).

a $19\,e^-$ complex, with the extra electron in a very high lying orbital, and such a compound would not be expected to be stable. Thus, the first reduction must be ligand-based, in agreement with the work of O'Toole *et al.* (1989), Breikss and Abruna (1986), Cabrera and Abruna (1986), etc., and the bands near 1993 cm^{-1} and 1875 cm^{-1} were attributed to (Dmpby$^{\bullet-}$)Re[CO]$_3$Cl.

At $-1.4\,V$ new features are observed near 1979 cm^{-1}, 1943 cm^{-1}, 1843 cm^{-1} and 1876 cm^{-1}, while at $-1.5\,V$ the only additional changes are a further increase in peak intensities, with the loss of the peak near 1993 cm^{-1} associated with the (Dmbpy$^{\bullet-}$)Re[CO]$_3$Cl, and the gain of a feature near 1828 cm^{-1}. As was discussed above, Breikss and Abruna postulated that the (Dmbpy$^{\bullet-}$)Re[CO]$_3$Cl underwent slow loss of chloride to give

(Dmbpy)Re[CO]$_3$ which then dimerises. The calculated v_{CO} for the 5-coordinate (Dmbpy)Re[CO]$_3$ species are 1984 cm^{-1}, 1862 cm^{-1} and 1849 cm^{-1}, in agreement with three of the four bands observed at -1.4 V. The peaks of the dimer have been reported to be at 1975 cm^{-1}, 1940 cm^{-1} and 1860 cm^{-1}, hence the authors found it not unreasonable to associate the 1943 cm^{-1} band in Figure 3.60 with 1940 cm^{-1} feature of the dimer on the basis that the 1943 cm^{-1} and 1975 cm^{-1} bands may well lie under the features attributed to the monomer.

At -1.6 V the spectrum changes considerably (see Figure 3.61) with the higher frequency bands being replaced by features at 1960 cm^{-1} and 1930 cm^{-1}, the lower frequency region dominated by the band at 1830 cm^{-1}, and the smaller feature at 1865 cm^{-1}. At very low potentials, the dominant bands are 1943 cm^{-1} and a broad feature near 1828 cm^{-1}.

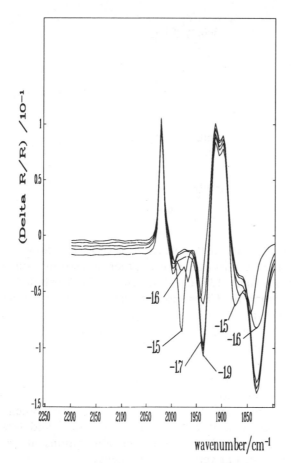

Figure 3.61 FTIR spectra collected at -1.5 V to -1.9 V in 100 mV steps and normalised to the reference spectrum taken at -1.0 V for the system in Figure 3.60. From Christensen *et al.* (1992).

At the most negative potentials the gains at $1943\,cm^{-1}$ and $1828\,cm^{-1}$ were attributed to the formation of $(Dmbpy^-)Re[CO]_3$, for which the calculated frequencies are $1967\,cm^{-1}$, $1836\,cm^{-1}$ and $1821\,cm^{-1}$; the two lowest frequencies were taken as corresponding to the broad feature near $1828\,cm^{-1}$. The large difference between the observed and calculated values for the remaining peak were attributed to the difficulty in calculating the effect of the $Dmbpy^-$.

The species with frequencies at $1960\,cm^{-1}$, $1930\,cm^{-1}$, $1830\,cm^{-1}$ and $1865\,cm^{-1}$ was tentatively assigned to $(Dmbpy)[CO]_3Re-Re[CO]_3(Dmbpy^-)$. Thus, in the absence of CO_2, the electrochemical behaviour of the parent complex can be described by Scheme 3.3.

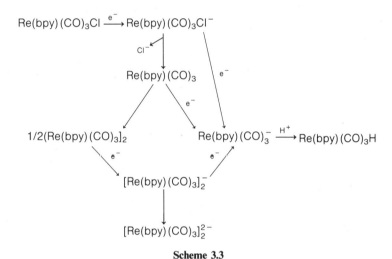

Scheme 3.3

The spectra obtained in the presence of CO_2 are shown in Figure 3.62 and 3.63. As can be seen from Figure 3.62 there is a strong gain in absorptions at $1640\,cm^{-1}$ and near $1350\,cm^{-1}$ which can be attributed to CO_3^{2-} (see discussion on work by Christensen et al. (1990) in previous section). In addition, there is the gain of a weak absorption near $2150\,cm^{-1}$ due to solution CO; with the loss of the CO_2 feature near $2340\,cm^{-1}$, this shows that the CO_2 is being reduced, as CO is not produced at these potentials on glassy carbon.

The carbonyl region at potentials down to $-1.5\,V$ is shown in more detail in Figure 3.64 which shows formation of a species which has features near $2010\,cm^{-1}$, $1902\,cm^{-1}$ and $1878\,cm^{-1}$. A more negative potentials a second species is observed with absorptions near $1997\,cm^{-1}$ and $1860\,cm^{-1}$, with small amounts of a third species, having absorptions at $1930\,cm^{-1}$ and $1828\,cm^{-1}$, being seen at very negative potentials.

None of the above features can be assigned to $(Dmbpy^{\cdot-})Re[CO]_3Cl$, $(Dmbpy)Re[CO]_3$ or the dimer, strongly supporting the suggestion by Lehn

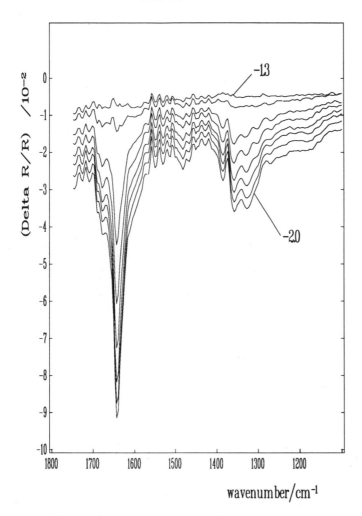

Figure 3.62 FTIR spectra collected from the glassy carbon electrode immersed in CO_2-saturated acetonitrile/0.2 M tetraethylammonium tetrafluoroborate containing 5×10^{-3} M Re(dmbpy)(CO)$_3$Cl. The spectra were collected at -1.1 V, -1.3 V, -1.4 V $\rightarrow -1.7$ V and -2.0 V, and normalised to the spectrum taken at -1.0 V. 1100 cm^{-1}–1750 cm^{-1} region. From Christensen *et al.* (1992).

and colleagues that the dimer does not form under CO_2. Both the CO and CO_3^{2-} bands in the figures are weaker than would be expected for complete conversion of the CO_2 to these species, suggesting that most of the CO_2 is held in the complexes present. The mechanisms that have been proposed for CO_2 reduction are summarised in Scheme 3.4.

In the absence of a proton source, i.e. water, the third mechanism can be rejected, particularly since it cannot account for the formation of CO_3^{2-}. There is no evidence for $\{(Dmbpy)Re[CO]_3\}^-$ at the relatively positive

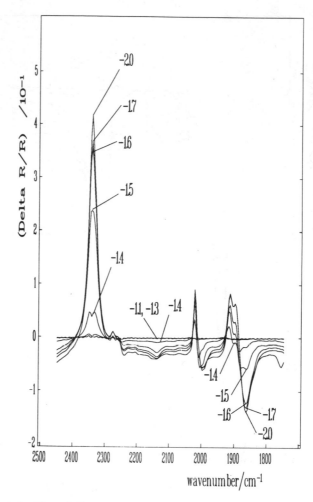

Figure 3.63   As for Figure 3.62 except over an extended spectral range. From Christensen *et al.* (1992).

potentials at which CO_2 is first reduced, which rules out the second mechanism. Consequently, the first mechanism seems most likely, particularly in view of the fact that $(Dmbpy^{\cdot-})Re[CO]_3Cl$ is not observed, suggesting that direct attack of the CO_2 on this species is most logical. Reduction is thought to proceed via a ReCOOH intermediate. The predicted frequencies for $(Dmbpy)Re(I)[CO]_3(COOH)$ are $2012\,cm^{-1}$, $1905\,cm^{-1}$ and $1876\,cm^{-1}$. The authors assigned the bands due to the species formed at potentials down to $-1.5\,V$ to the protonated form of the carboxylate. The calculated frequencies for the mono- and di-reduced forms of this species, $(Dmbpy^-)Re(I)[CO]_3(COOH)$ and $\{(Dmbpy^-)Re(I)[CO]_3(COOH)^-$, are $1992\,cm^{-1}$, $1867\,cm^{-1}$ and $1866\,cm^{-1}$, and $1936\,cm^{-1}$, $1816\,cm^{-1}$ and $1815\,cm^{-1}$,

(1)

$$(Dmbpy)Re[CO]_3Cl + e^- \longrightarrow (Dmbpy)Re[CO]_3 + Cl^-$$
$$(Dmbpy)Re[CO]_3 + CO_2 \longrightarrow (Dmbpy)Re[CO]_3CO_2$$
$$2(Dmbpy)Re[CO]_3CO_2 + 2L \longrightarrow 2\{(Dmbpy)Re[CO]_3CO_2L\}^{n+} + CO + CO_3^{2-}$$

where $L = Cl^-$ or CH_3CN and $n = 0$ or 1, respectively; this is a one electron pathway.

(2)

$$(Dmbpy)Re[CO]_3Cl + e^- \longrightarrow (Dmbpy)Re[CO]_3 + Cl^-$$
$$(Dmbpy)Re[CO]_3 + e^- \longrightarrow \{(Dmbpy)Re[CO]_3\}^-$$
$$\{(Dmbpy)Re[CO]_3\}^- + CO_2 \longrightarrow (Dmbpy)Re[CO]_3(CO_2)^-$$
$$(Dmbpy)Re[CO]_3(CO_2)^- + CO_2 + L \longrightarrow \{(Dmbpy)Re[CO]_3(L)\}^{n+} + CO + CO_3^{2-}$$

Again, $n = 0$ if L is Cl^-, or 1 if $L = CH_3CN$. Here, another molecule of CO_2 is the oxide ion acceptor.

(3)

$$(Dmbpy)Re[CO]_3Cl + e^- \longrightarrow \{(Dmbpy)Re[CO]_2Cl\}^- + CO$$
$$\{(Dmbpy)Re[CO]_2Cl\}^- + CO_2 + 2H^+ \longrightarrow \{(Dmbpy)Re[CO]_3Cl\}^+ + H_2O$$
$$\{(Dmbpy)Re[CO]_3Cl\}^+ + e^- \longrightarrow (Dmbpy)Re[CO]_3Cl$$

Scheme 3.4

(i) $(Dmbpy)Re[CO]_3Cl + e^- \longrightarrow (Dmbpy^-)Re[CO]_3Cl$

(ii) $(Dmbpy^-)Re[CO]_3Cl + CO_2 \longrightarrow (Dmbpy)Re[CO]_3CO_2 + Cl^-$

(iii) $(Dmbpy)Re[CO]_3CO_2 + e^- + H_2O \longrightarrow (Dmbpy)Re[CO]_3COOH + OH^-$

and

$$OH^- + CO_2 \longrightarrow CO_3^{2-} + H^+$$

or:

(iiia) $2(Dmbpy)Re[CO]_3CO_2 + 2Cl^- \longrightarrow 2(Dmbpy)Re[CO]_3Cl + CO + CO_3^{2+}$

(iv) $(Dmbpy)Re[CO]_3COOH + e^- \longrightarrow (Dmbpy^-)Re[CO]_3COOH$

or:

(iva) $(Dmbpy)Re[CO]_3COOH + e^- \longrightarrow (Dmbpy)Re[CO]_3COO^- + 1/2H^+$

Scheme 3.5

respectively, in good agreement with the observed values. The mechanism postulated by the authors on the basis of these results is given in Scheme 3.5.

Reaction (ii) is fast, ensuring that the stationary concentration of $(Dmbpy^-)Re[CO]_3Cl$ is low. Reaction (iii) must also be fast, given that the radical was not detected, and the main evidence for (iiia) is the appearance of the weak CO band.

The above experiment was then repeated under the conditions previously reported by Lehn and colleagues (Hawecker *et al.*, 1983, 1984, 1990) to be ideal, namely using 10% added water and tetrabutylammonium chloride as the supporting electrolyte (see Figure 3.65). Again, there are gains of absorbance to lower wavenumber of the parent complex, the most prominent of which are at $2010\,cm^{-1}$, $1935\,cm^{-1}$ and $1873\,cm^{-1}$. However, the most notable feature of the spectra in Figure 3.65 is the appearance of a gain to

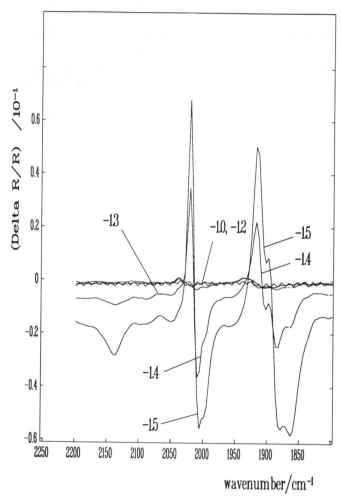

Figure 3.64   As for Figure 3.62 except over the spectral range $2050\,cm^{-1}$–$2250\,cm^{-1}$ and the spectra were normalised to the spectrum taken at $-0.6\,V$. Spectra were taken at the potentials shown. From Christensen *et al.* (1992).

higher wavenumber of the parent complex at $2040\,cm^{-1}$. In addition, the amount of CO produced is greatly increased compared to the dry electrolyte containing the non-coordinating BF_4^- and *no* CO_3^{2-} was observed.

The spectrum in Figure 3.65 was found to closely resemble that of a cationic complex observed by the authors in experiments investigating the anodic behaviour of the $(Dmbpy)Re[CO]_3Cl$. The complex, $\{(Dmbpy)Re[CO]_3 \cdot (CH_3CN)\}^+$, showed absorptions at $2039\,cm^{-1}$, $1948\,cm^{-1}$ (as a poorly-resolved shoulder) and $1935\,cm^{-1}$. The predicted frequencies were $2043\,cm^{-1}$, $1959\,cm^{-1}$ and $1944\,cm^{-1}$. In both cases, considerable structure was observed in the $C\equiv N$ absorption region between $2250\,cm^{-1}$ and $2300\,cm^{-1}$ (not

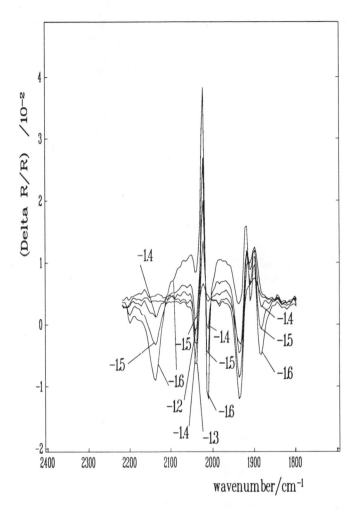

Figure 3.65 FTIR spectra collected from the glassy carbon electrode immersed in CO_2-saturated acetonitrile-water (9:1)/0.2 M tetraethylammonium tetrafluoroborate containing 5×10^{-3} M (Dmbpy)Re(CO)$_3$Cl. The spectra were collected at potentials between -1.2 V and -1.6 V in 100 mV steps and are normalised to the spectrum taken at -1.0 V. From Christensen *et al.* (1992).

observed in the absence of added water), suggesting the coordination of acetonitrile. Hence the authors postulated that one of the species indicated by the gain features in Figure 3.65 was $\{(Dmbpy)Re[CO]_3(CH_3CN)\}^+$. From the spectra it can be seen that the maximum concentration of this complex is observed at -1.3 V; below this potential, it rapidly disappears as further reduction takes place. From this observation, the authors concluded that sustained turnover of CO_2 can only take place at potentials of -1.4 V or below since it is only at these potentials that a complete cycle can be obtained.

As was stated above, no features that can be attributed to CO_3^{2-} were observed. However, peaks at $1690\,cm^{-1}$ and $1720\,cm^{-1}$ were present in the spectra. The lower of these was attributed to the carboxylate complex.

On the basis of their results, the authors postulated the mechanism depicted in Scheme 3.6.

(i) $(Dmbpy)Re[CO]_3Cl + e^- \longrightarrow (Dmbpy^-)Re[CO]_3Cl$

(ii) $(Dmbpy^-)Re[CO]_3Cl + CO_2 \longrightarrow (Dmbpy)Re[CO]_3CO_2 + Cl^-$

(iii) $(Dmbpy)Re[CO]_3CO_2 + e^- + H^+ \longrightarrow (Dmbpy)Re[CO]_3COOH$

(iv) $(Dmbpy)Re[CO]_3COOH + CH_3CN + H^+ \longrightarrow$
$\{(Dmbpy)Re[CO]_3CH_3CN\}^+ + CO + H_2O$

(v) $\{(Dmbpy)Re[CO]_3CH_3CN\}^+ + 2e^- \longrightarrow (Dmbpy^-)Re[CO]_3 + CH_3CN$

Scheme 3.6

The authors concluded that steps (ii) and (iii) are certainly fast but that step (iv) must be sufficiently slow for the carboxylato complex to diffuse away from the electrode before CO formation takes place.

The authors were able to follow the details of the complicated reduction of CO_2 by $(Dmbpy)Re[CO]_3Cl$ and to identify most of the intermediates present at each stage of the process.

3.4 Reactive film formation on electrodes

3.4.1   Oxide formation on metals: the nature of the passive film on iron

The corrosion of iron represents an electrochemical reaction of huge economic significance, accounting literally for billions of dollars of waste every year. The phenomenon has been investigated since the time of Faraday and still presents many controversial and puzzling aspects which only the arrival of *in situ* spectroscopic techniques has begun to clarify.

Corrosion is a mixed-electrode process in which parts of the surface act as *cathodes*, reducing oxygen to water, and other parts act as *anodes*, with metal dissolution the main reaction. As is well known, iron and ferrous alloys do not dissolve readily even though thermodynamically they would be expected to. The reason is that in the range of mixed potentials normally encountered, iron in neutral or slightly acidic or basic solutions *passivates*, that is it forms a layer of oxide or oxyhydroxide that inhibits further corrosion.

The process of passivation is shown in Figure 3.66(a). In the potential region just positive of the E^0 value for the process

$$M \longrightarrow M^{z+} + z\,e^- \tag{3.74}$$

the metal dissolves. As the potential is increased the iron dissolves with increasing rate, reaching a plateau in acid solution above $0\,V$ vs. NHE. Above

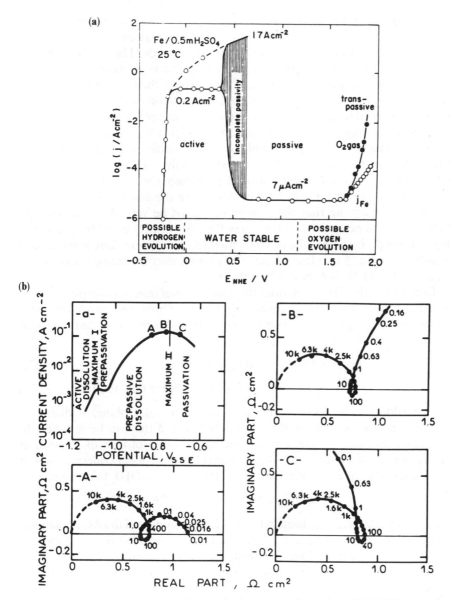

Figure 3.66 (a) Potentiostatic anodic polarisation curve of iron in $0.5\,M$ H_2SO_4, showing characteristics of anodic passivation. J_{Fe} is the iron dissolution current density. From N. Sato and G. Okamoto in *Comprehensive Treatise of Electrochemistry*, Vol. 4, eds. J.O'M. Bockris, B.E. Conway, E. Yeager and R.E. White, Chapter 4, Plenum Press, New York, 1981. (b) Polarisation curves of Fe disc in 1 M Na_2SO_4 acidified by addition of 1 M H_2SO_4, buffered by 4 mM CH_3COONa, $25 \pm 0.2°C$ deoxygenated by Ar bubbling. Rotation rate 1600 rev min^{-1}. (a) Steady state polarisation curve. The impedance diagrams A, B and C plotted at the corresponding points on (a). The numbers on the curves refer to the frequency of measurement in hertz. From I. Epelboin, C. Gabrielli, M. Keddam and H. Takenouti in *Comprehensive Treatise of Electrochemistry*, Vol. 4, eds. J.O'M. Bockris, B.E. Conway, E. Yeager and R.E. White, Chapter 3, Plenum Press, New York, 1981.

c. 0.5 V, however, the current suddenly drops to a very low value, a process apparently associated with the further partial oxidation of Fe^{2+} to Fe^{3+} and the precipitation of a mixed valence oxyhydroxide. The precise potential at which this passivation process takes place depends very much on the experimental conditions. However, independent of the formation conditions for the passive film, once it *has* formed, interruption of the electrochemical conditions by placing the passivated electrode at open circuit will result in its potential changing to a fixed value. This value is the Flade potential, Φ_F, at which reduction of the passive film back to soluble Fe(II) species can take place. Above the Flade potential, therefore, only a small current (the rest current, I_R) flows until, at very positive potentials, the current rises again in the 'transpassive' region. On iron in this region the main process is oxygen evolution, suggesting that the passive film must be very thin. For iron, in fact, the Flade potential has the value 0.58–(0.059.pH) vs NHE and if any iron electrode is cycled above this potential the current falls from several hundred mA cm^{-2} to a few μA cm^{-2}. Again on iron the transpassive region begins in 0.5 M H_2SO_4 at c. 1700 mV vs. NHE.

Even before the Flade potential the dissolution of iron is a complex process which has been thoroughly investigated by Epelboin and coworkers with AC impedance. The starting point is the same basic scheme as that proposed for chlorine evolution:

$$Fe + H_2O \longrightarrow FeOH_{ads} + H^+ + e^- \tag{3.75}$$

$$FeOH_{ads} \longrightarrow FeOH^+(aq.) + e^- \rightleftharpoons Fe^{2+}(aq.) \tag{3.76}$$

This would give rise to two semicircles in the complex plane, as discussed above, together, possibly, with a Warburg region if this can be resolved. In fact this behaviour is indeed seen close to the onset of iron dissolution in acidic media but at slightly higher potentials. As the local concentration of Fe(II) rises near the electrode, formation of insoluble $Fe(OH)_2$ takes place as:

$$FeOH_{ads} \rightleftharpoons [Fe(OH)_2]_{ads} + e^- \tag{3.77}$$

leading to a characteristic loop in the low-frequency part of the AC impedance response, as shown in Figure 3.66(b). In fact, the detailed interpretation of the features shown in Figure 3.66(b) requires the introduction of other adsorbed species, the final reaction scheme showing three dissolution pathways (scheme 3.7).

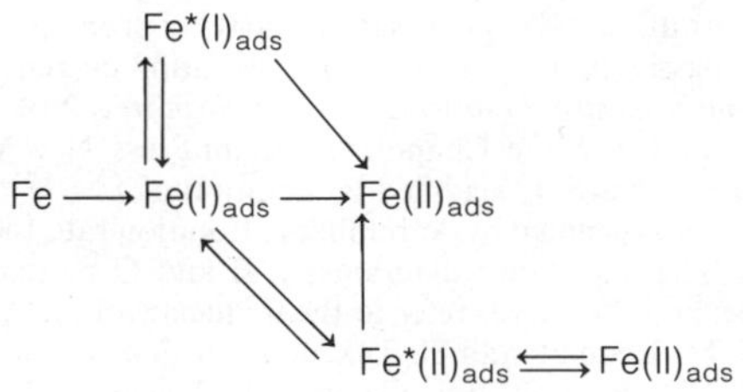

Scheme 3.7

At even higher potentials oxidation to Fe(III) becomes possible and further complexity is encountered.

Interpretation of the AC data for iron passivation in $1\,M\,H_2SO_4$ can be carried out with the basic model shown in Scheme 3.8.

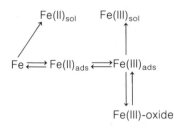

Scheme 3.8

The passive film itself could not be treated as purely insulating since the known thickness of the film should give rise to a capacitive effect that would easily be seen in the impedance diagram. In fact, the film is apparently relatively conducting, probably through proton migration.

The nature of the passive film has been the object of innumerable further studies with arguments over its thickness, composition, structure and electronic properties raging. What has become very clear is that that removal from solution and drying of the passivated electrode alters the film profoundly and any studies of the film *ex situ* must be treated with considerable caution. For that reason we will concentrate here primarily on *in situ* studies or at the least those studies carried out on passive films that retain their hydration.

The thickness of the passive layer has been determined by ellipsometry starting from an iron surface that was prepared entirely free of oxide in a UHV chamber and then passivated in nitrite solution in the presence of oxygen. Such films grow logarithmically with time reaching a thickness of 40 Å after 100 min and undoubtedly take the form of a phase oxide or oxyhydroxide rather than a two-dimensional film. Even if the detail of the interpretation of the ellipsometric data is challenged, as indeed it has been by virtue of the assumption of a single composition for the film, the ellipsometric data show that for long enough times, or at high enough potentials, the surface film substantially exceeds one monolayer in thickness.

The most controversial aspect of the film is its composition, for which there are three aspects:

1. To what extent do Fe(II) and Fe(III) co-exist within the film?
2. To what extent are protons and water molecules incorporated within the film?
3. Is the film uniform in composition, or is there more than one layer?

The third of these questions was first posed on thermodynamic grounds: it is easy to show that a pure Fe(III) oxide or hydroxide will *not* be thermodynamically stable in contact with metallic iron and a thin layer of a mixed oxide phase, such as Fe_3O_4, is expected next to the electrode surface. Ellipsometrically, this presents serious problems since the existence of two separate films makes solution of the ellipsometric equations extremely difficult. It may also be the case that composition of the film shows a quasi-continuous variation. Cahan and Chen (1982) have suggested that the film might resemble a conducting polymer within which a single structural framework can accommodate either a mixed valence oxide or a pure Fe(III) species with charge compensation being by protons. It has been suggested that these protons might occupy Fe sites but it would seem more likely that the protons are present as hydroxy ions.

The evidence for two reasonably well defined layers comes primarily from electrochemical studies, ellipsometry and Raman spectroscopy. The ellipsometric studies of Ord and DeSmet (1976) show good evidence of an abrupt

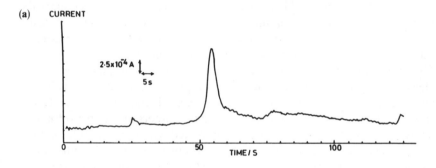

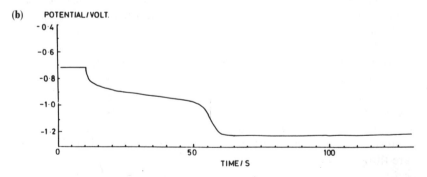

Figure 3.67 Typical disc potential and ring current behaviour during galvanostatic reduction of the passive film on pure iron (area $0.5\,cm^2$) at a rotation speed of 25 Hz. The Pt ring potential is maintained at 0.2 V to oxidise all Fe(II) to Fe(III), and the collection efficiency is 0.28. Note that the residual current detected on the ring beyond 70 s corresponds to re-oxidation of hydrogen generated galvanostatically on the disc. Reprinted from *Corrosion Science*, **28**, P. Southworth, A.M. Hamnett, A.M. Riley and J.M. Sykes, 'An Ellipsometric and RRDE Study of Iron Passivation and Depassivation in Carbonate Buffer', pp. 1139–1161 (1988), with kind permission from Pergamon Press Ltd., Headington Hill Hall, Oxford OX3 0BW, UK.

change in optical constants during film reduction. Electrochemically, the most compelling evidence is provided by the rotating ring-disc studies of Southworth *et al.* (1988) which show that if the passive film on the disc is reduced galvanostatically, and the ring monitored for re-oxidation of solution Fe(II), the results of Figure 3.67 are obtained. The Fe(II) current at the ring consists of two portions: there is a small but continuous current passed at all times during the reduction of the flm, accompanied by a sudden burst in current at the end-point. If the charge under this final burst is plotted against the formation potential of the passive film and the total reduction charge of the film, the resultant graph strongly suggests that the passive film has a bilayer structure with an inner, rather thin, layer of mixed-valence hydrous oxide and an outer layer of a hydrous Fe(III) oxide whose thickness varies approximately linearly with formation potential. Raman spectroscopy has been carried out in the presence of stabilising agents such as molybdate (see the work of Hugot-Le Goff and colleagues, 1990). In this case, the resultant spectrum shows an inner Fe_3O_4 layer and an outer layer composed of a mixture of γ-FeOOH and δ-FeOOH, the former being stabilised by Cl^- ions and the latter always detected in iron passivity.

The verification of the presence of hydrogen in the film has proved more controversial, primarily because many of the structural investigations have been carried out after the film has been dried *in vacuo*. An example of the problems here is the fact that electron diffraction, which has to be carried out *in vacuo*, reveals a relatively well-crystallised spinel lattice whose origin may be the comparatively high sample heating encountered in the electron beam. Moreover, the use of *in situ* techniques, such as Mössbauer and X-ray absorption spectroscopy, clearly reveals marked differences between the spectra of the films *in situ* and the spectra of the same films *ex situ* as well as the spectra of γ-Fe_2O_3 and γ-FeOOH standards. These differences are most naturally ascribed to hydration of the spinel forms.

The evidence for the spinel form is otherwise indirect. EXAFS studies are consistent with a *local* spinel structure, though the derived bond lengths are different from those found in either anhydrous Fe_3O_4 or γ-Fe_2O_3, and the width of the EXAFS peaks is also significantly wider in the passive film, a consequence of the poorer crystallinity of the film compared to the oxide standards. Incorporation of hydrogen in the film thus leads to a more 'glassy' matrix with the increase in average Fe–O bond length being most naturally associated with Fe–O–Fe bridges changing to Fe–OH–Fe bridges.

The composition of the film changes with potential and with the incorporation of both anions and cations. EXAFS data on the passive films grown on stainless steel or normal steel with CrO_4^{2-} show considerable incorporation of Cr into the film with further alterations in bond lengths and covalency parameters. Indeed, as the amount of Cr incorporated increases, so does the flexibility of the structure. It is well known, by contrast, that Cl^- incorporation leads to poorer quality films and to enhanced rates of corrosion.

Increasing the potential leads to dehydration of the film, and increasing

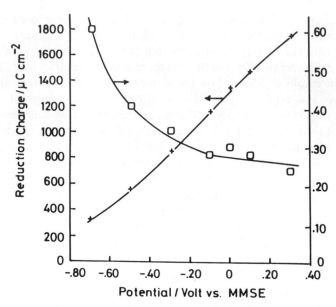

Figure 3.68 Total reduction charge (+) for the outer film and water content (□) of the outer film (as estimated from the effective medium model and the data at 633 nm) (see text for details). Reprinted from *Corrosion Science*, **28**, P. Southworth, A. Hamnett, A.M. Riley and J.M. Sykes, 'An Ellipsometric and RRDE Study of Iron Passivation and Depassivation in Carbonate Buffer', pp. 1139–1161 (1988), with kind permission from Pergamon Press Ltd., Headington Hill Hall, Oxford OX3 0BW, UK.

crystallinity. This is shown for the passive film grown on iron in carbonate buffer in Figure 3.68 in which the ellipsometric data have been interpreted within an effective medium model to yield the water content, which is plotted as x in $Fe_2O_3 \cdot xH_2O$. This, in turn, can be seen to decrease monotonically with increasing growth potential. An important consequence of this is seen in photocurrent measurements on passive films. Such photocurrents are associated with the generation of electron-hole pairs in the oxide: provided the film is sufficiently crystalline and continuous these holes and electrons will be mobile and can be separated by the internal electric field present in the film before recombination can take place. This carrier separation can, in turn, lead to the possibility of a nett photocurrent being observed and this is indeed found, *provided the potential is high enough*. Crucially, the photocurrent then persists even at lower potentials since the field-induced crystallisation process is relatively irreversible.

3.4.2 Conducting polymers

The discovery of the metallic properties of polymeric $(SN)_x$ by Walatka *et al.* in 1973 brought to notice a new class of materials: electronically conducting polymers. The importance of this discovery was enhanced greatly by the demonstration by MacDiarmid and colleagues in 1977 (Shirakawa *et al.*, 1978) of the semiconducting and metallic properties of the chemically synthesised organic polyacetylene. In the intervening years a wide range of polymeric organic species, including heterocyclic materials such as polythiophene, polypyrrole, etc., has been prepared as stable, adherent films on inert electrodes via both chemical oxidation and electropolymerisation from solution.

The most unusual and interesting feature of these polymers is their capacity to switch between insulating and conducting (or semiconducting) states. All other materials, with the only additional exception of some intercalation compounds, are normally found only as conductors *or* semiconductors *or* insulators, without the facility to switch between these states.

This section is concerned with polypyrrole, one of the earliest conducting polymers to be identified, and, more particularly, describes the increasing understanding of the means by which these materials conduct electricity that has arisen as a direct result of the application of (mainly) *in-situ* electrochemical techniques. Polypyrrole forms as a (semi)conducting film on the working electrode when this is made the anode in a cell containing an aqueous or non-aqueous solution of the monomer. The as-grown film (i.e. the freshly grown film prior to any electrochemical cycling) is in the oxidised form and is formed according to Scheme 3.9.

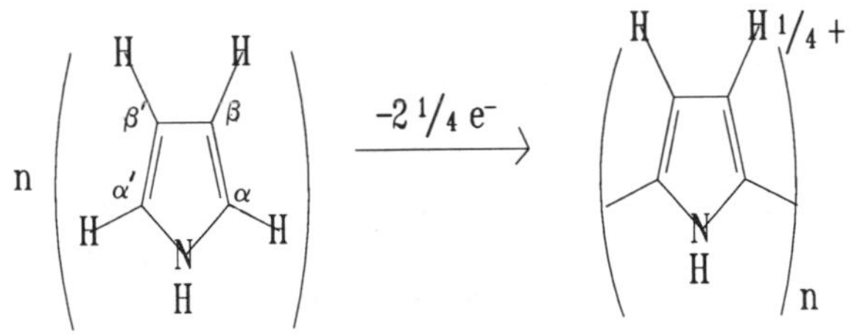

Scheme 9

The film can then be reduced to the non-conducting, insulating form by stepping the potential to more negative values, according to Scheme 3.10.

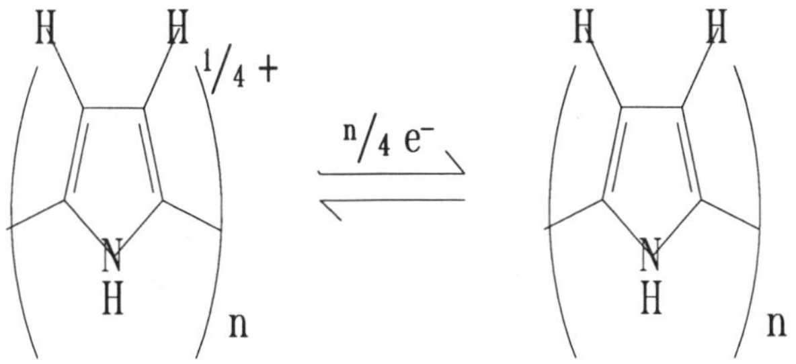

Scheme 10

The potential cycling of the film between the insulating and (semi)conducting forms can be repeated many times without the loss of the electroactivity of the film.

The neutral form of polypyrrole is weakly coloured while the oxidised form is a deep blue/black so that switching the state of the film not only changes its conductivity but is also accompanied by a marked colour change, a phenomenon termed *electrochromism*.

Materials such as polypyrrole are exciting in terms of their future techno-logical impact, not just because of the obvious applications of such a simple, cheap electrochromic but because it may be possible to develop them sufficiently to replace the more expensive, and often toxic, metallic conductors commonly employed in the electronics industry. This may not be such a distant dream since it has been calculated that the intrinsic conductivity of these materials, i.e. without the defects that are currently defeating attempts to increase their conductivity of $c. <1000\,\Omega^{-1}\,cm^{-1}$, may be many times that of copper.

3.3.2.1 Structure.

In 1968 Dall'Olio and colleagues prepared black films of an oxypyrrole on platinum by the electrochemical polymerisation of pyrrole from its solution in sulphuric acid. A decade or so later Diaz and co-workers (1979) employed a modification of the Dall'Olio approach to polymerise pyrrole onto Pt in acetonitrile, demonstrating that polymerising the pyrrole in this way led to a black, adherent film. XPS and elemental analysis of the as-grown film showed that the pyrrole unit is retained in the polymer. Elemental analysis also showed that the polymer contained $c.$ one BF_4^- anion for every 3 to 4 pyrrole units, suggesting that the polypyrrole is generated in its oxidised form, according to Scheme 3.9, with a charge of $1/3+$ per monomer unit being the maximum sustainable by the film.

This conclusion was supported by coulometry measurements that showed that the charge passed during the growth was typically $c.$ 9 or 10 times that under the anodic wave of the voltammogram up to the potential corresponding to the maximum doping level (cf. Schemes 3.9 and 3.10).

It was reasoned that the polymer must consist of α, α'-coupled pyrrole units, with β coupling less important, because of the fact that α-substituted pyrroles do not polymerise whereas β-substituted species do, and on the basis of magic angle spinning ^{13}C nmr and IR techniques. The method of labelling the monomer ring positions is also shown in Scheme 3.9.

In general it has been found that polypyrrole is extremely poorly crystalline, limiting the information that can be obtained from direct structural techniques such as X-ray crystallography, and hence much of our knowledge has been obtained from indirect measurements and/or experiments on model com-pounds (e.g. X-ray studies on pyrrole trimers and dimers). It is now generally accepted that the 'ideal' structure of the polymer is a planar $(\alpha-\alpha')$-bonded chain in which the orientation of the pyrrole molecules alternates.

3.3.2.2 Static measurements.   Diaz and colleagues (1972) showed that fully oxidised polypyrrole films could be grown thick enough (i.e. up to 50 μm) to allow the film to be peeled off the electrode, allowing standard four-probe conductivity measurements to be performed. The films grown from acetonitrile gave room temperature conductivities typically in the range $10–100\,\Omega^{-1}\,cm^{-1}$ (cf. copper, $10^6\,\Omega^{-1}\,cm^{-1}$).

The early work in the field of conducting polymers was performed by physicists and hence the terminology employed by them has found its way into electrochemistry. Thus, the films conduct only when they are oxidised, suggesting that conduction is via positive carriers (i.e. holes) and the polymer is termed *p-(for positive)-doped*, which is merely a descriptive term for the conduction. However, it was quickly realised that if they are to fulfil their full technological potential, a full understanding of the conduction process is essential and this includes determining the identity of the carriers.

Diaz and colleagues (1979), on being faced with a new conducting material, had a choice of two probable conduction mechanisms, differentiated only in terms of the parameter τ, the residence time. The residence time is the time that a carrier spends on a particular site in the lattice. Thus:

1. If τ is significantly shorter than the time over which the atoms in the lattice vibrate (phonon vibrations), then the carrier appears to move on an essentially rigid background and is termed 'free'. This is the form of conduction found in most simple metals.

2. In contrast, if τ is much longer than the time taken for lattice vibrations to occur, then the atoms of a particular site can relax around the occupying carrier. The carrier then cannot move as easily and must wait until the vibrations of the lattice provide a favourable pathway for it to escape to the next site. The carriers in this case are polarons (POLAR phonONs) or bipolarons and can be described in terms of Figures 3.69(a) and (b). Thus the removal of an electron from the chain leaves behind a localised charged entity having a finite conjugation length. Removal of a single charge generates a cation radical (a polaron) and removal of a second charge results in a spinless dication (bipolaron). Both the polaron and bipolaron have associated with them considerable perturbation of the chain. In addition, those ring vibrations that are able to couple with the charge movement give rise to selectively enhanced absorptions in vibrational spectra, infrared active vibrations (IRAVs).

Diaz and colleagues (1979) sought to identify the nature of the conduction mechanism by performing a range of *ex situ* experiments on the free-standing, as-grown film. They measured the conductivity, σ, of the film as a function of temperature and found that it varies as:

$$\sigma \propto \exp[-a/T^{1/4}], \tag{3.78}$$

where a is a constant. Equation (3.78) suggests a three-dimensional variable-range hopping mechanism, that is most certainly *not* metallic. In addition,

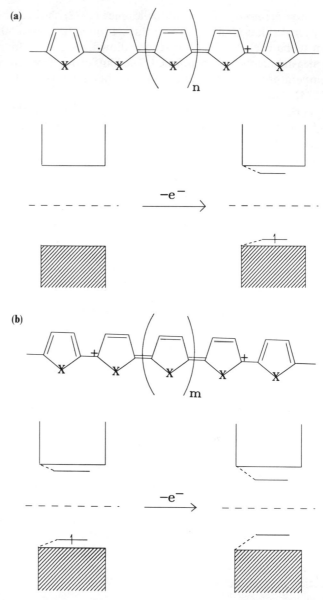

Figure 3.69 Schematic representations of the nature and energetics of (a) polaron and (b) bipolaron formation (see text for details).

the Hall coefficient was anomalously small and negative indicating n-type conduction involving localised states. However, the thermopower measurements, i.e. the voltage difference across the film generated by a temperature gradient, showed a value 2 or 3 orders of magnitude smaller than that expected for a semiconductor and typical of metallic behaviour. Hence the authors concluded that oxidised polypyrrole was metallic but this conclusion was almost immediately questioned.

UV-visible spectroscopy was one of the earliest techniques to be applied to try and resolve the question over the identity of the carriers in polypyrrole and was employed *ex situ* on the free-standing as-grown film. Two important experimental approaches were employed:

1. The as-grown film was fully reduced prior to its removal from the electrode. The film was then removed, placed in a UV-visible cell and exposed to an oxidising gas (i.e. O_2 or I_2), with spectra being collected at various times thereafter, each spectrum corresponding to a different degree of doping.
2. The film is reduced to the desired doping level prior to its removal and transfer to the UV-vis cell and the UV-visible spectra at different doping levels are obtained on different films.

Both approaches suffer from the high susceptibility of polypyrrole to oxidation by even trace amounts of oxygen. The latter method has the disadvantage that the comparison of the absorbances at different doping levels has to

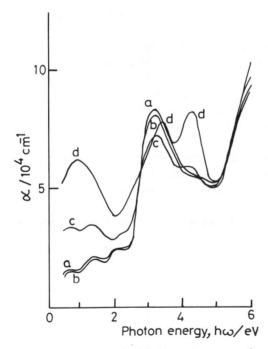

Figure 3.70 Room temperature optical absorption spectra of a 4500 Å-thick film of neutral polypyrrole doped with I_2 at 0.03 torr. (a) before exposure to I_2, conductivity $< 10^{-6} \Omega^{-1} cm^{-1}$; (b) after 2 minutes I_2 exposure, conductivity $4.8 \Omega^{-1} cm^{-1}$; (c) after 7 minutes I_2 exposure, conductivity $6.7 \Omega^{-1} cm^{-1}$; (d) after 22 minutes I_2 exposure, conductivity $32 \Omega^{-1} cm^{-1}$. The three structures seen on the low-energy side of (a)–(c) are possibly artifacts due to interference effects in the films. From Pfluger *et al.* (1983).

allow for different film thicknesses. However, it also has the great advantage that the redox state of the film is being set by electrochemical, rather than chemical, means. Later work consistently showed that the materials such as polypyrrole are irreversibly oxidised by O_2 and the optical spectra obtained by the two methods are somewhat different, as will be seen below.

Figure 3.70 shows optical absorption spectra of polypyrrole obtained via method 1: (a) the neutral polymer and (b)–(d), the film at various stages after exposure to I_2 vapour, from the paper of Pfluger et al. (1983). The conductivity was measured concomitantly with the optical spectra. The 1 eV (8000 cm^{-1}) absorption peak that grows with increasing doping was originally thought to be related to the free-carrier absorption. However, this was later discounted, as the intensity of the band can be increased by further exposure to the oxidant without causing a corresponding rise in the conductivity. This can clearly be observed in the figure: after 2 min exposure to I_2, the electrical conductivity of the film increases by $> 10^6$, whereas the major growth in the intensity of the 1 eV band only becomes apparent after the films have reached a conductivity within an order of magnitude of their final value.

In 1983, Yakushi et al. obtained UV-visible spectra of polypyrrole at various doping levels via the second approach described above and the results are shown in Figure 3.71. The authors employed a two-electrode cell and hence the potentials quoted are the potentials between the working and counter electrodes prior to removing the film.

The main features of the spectra are, in the completely neutral form, (i) a band at 3.2 eV, with a shoulder near 4.5 eV and (ii) the as-grown fully oxidised film has bands near 1.0 eV, 2.7 eV and a shoulder near 3.6 eV. On reduction the intensities of the 1.0 eV and 2.7 eV features decrease and shift to lower energy, while the shoulder near 3.6 eV increases in intensity and also shifts to lower energy.

An important point concerning the spectra in Figure 3.71 is that at intermediate doping levels three principal absorption bands can be seen, at c. 1.0 eV, 2.7 eV and 3.6 eV, that are not simply the superposition of the as-grown and neutral polymer absorptions. The authors interpreted this observation in terms of the homogeneous doping of the film throughout its bulk, not just the oxidation of the surface layer or the layer next to the electrode.

The authors attributed the 3.2 eV band in the neutral film to the $\pi \rightarrow \pi^*$ transition of π electrons in the HOMO and the 2.7 eV band was assigned to this transition in the fully oxidised form based on, for example, the systematic shift of this absorption to lower energy as the conjugation length shortens in pyrrole oligomers.

The authors postulated that since it was obvious from X-ray data that polypyrrole is a poorly crystalline material, for which there is no evidence for long conjugation lengths, it does not seem unreasonable to suppose that there is a wide distribution of conjugation lengths, a hypothesis supported by other workers. The cations of shorter conjugated units will be reduced

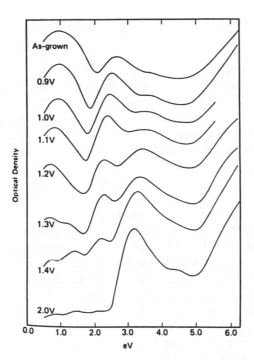

Figure 3.71 Dependence of the absorption spectrum of polypyrrole perchlorate on the reduction potentials. The voltages written in the figure are the applied voltages between working and counter electrodes. Each spectrum is obtained by using a different free-standing film. From Yakushi *et al.* (1983).

at higher potentials than those of longer, hence the red shift of the 2.7 eV band for the more reduced films.

As was mentioned above, the 1.0 eV band was originally thought to be associated with the free carrier absorption. However, it is possible to calculate the optical conductivity spectrum of the oxidised polypyrrole from its optical absorption spectrum. For a metal having a half-filled band the optical conductivity should decrease monotonically with increasing frequency of the absorbed light, v, from a maximum at $hv = 0$. This was not found to be the case and the authors, reflecting the general consensus of opinion, turned against a metallic conduction mechanism in polypyrrole and favoured the involvement of polarons and/or bipolarons as the carriers. Thus, they postulated that the 1 eV absorption was due to a transition between conjugated segments on the same chain, or between chains.

Polarons are epr-active by virtue of their unpaired spin. Consequently, another technique to be applied relatively soon after the discovery of conducting polymers was epr. Again the early studies were carried out *ex situ* on free-standing films, usually by monitoring the epr response (and the conductivity) of the completely reduced film on various exposures to oxidising gases such as I_2 or O_2 (cf. the UV-visible experiments discussed above).

Pfluger *et al.* (1983) studied the epr activity and conductivity of polypyrrole by epr using O_2 as the oxidant. Doping in this manner lead to an initial large increase in the conductivity associated with a corresponding increase in the intensity of the epr signal. When the electrical conductivity reached a constant value, the epr intensity was observed to pass through a maximum before decreasing. The authors thus concluded that the early stages of oxidation led to the generation of polarons. However, in a later paper, Scott and colleagues (1983) reported a more comprehensive study, also *ex situ*, on electrochemically cycled films which remained highly conducting but showed little or no detectable epr absorption. Hence they concluded that the observed epr signal from polypyrrole arose from neutral π-radical defects.

This paper is also of note as a result of the fact that the authors produced one of the first (qualitative) descriptions of the electronic structure of polypyrrole in terms of band theory. However, this was later modified considerably by Bredas *et al.* (1984) to produce the now widely accepted description. Bredas and colleagues (1984), on the basis of the data of Yakushi *et al.* (1983), performed tight-binding band structure calculations on a deformable polypyrrole chain as a function of doping. As a result, they were able to interpret the development of the spectra in Figure 3.71 in terms of the energetics depicted in Figure 3.72. At low levels of oxidation (see Figure 3.72(a)) the strong absorption maximum at 3.2 eV is associated with the interband $\pi \rightarrow \pi^*$ transition. On introducing a single positive charge onto a chain, the authors calculated that a polaron is obtained, extending over approximately four monomer units with a binding energy of *c*. 0.12 eV. The presence of the polaron results in an unoccupied localised (antibonding) level being brought down 0.53 eV from the conduction band into the gap and a singly occupied localised (bonding) level correspondingly being raised 0.49 eV out of the valence band. These two states in the gap account for the three transitions

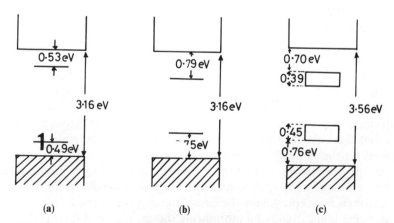

(a)   (b)   (c)

Figure 3.72 Electronic structure diagrams for a polypyrrole chain containing (a) a polaron and (b) a bipolaron. (c) The band structure obtained for highly oxidised (33 mol. %) polypyrrole showing the presence of two broad bipolaron bands in the gap. After Brédas *et al.* (1984).

observed within the gap at low doping in Figure 3.71, at 0.7 eV (valence band →
bonding polaron level, ω_1), 1.4 eV (bonding polaron level → antibonding
polaron level, ω_3) and 2.1 eV (valence bond → antibonding polaron level, ω_2).

At higher doping levels the polaron states interact; the authors calculated
that a bipolaron level is 0.49 eV more stable than two polaron levels so
unfavourable polaron interactions are avoided via the formation of a more
stable bipolaron, in agreement with the epr data of Scott and colleagues
(1983).

A bipolaron introduces two states in the gap, both now *empty* (see Figure
3.72(b)), 0.75 eV above the valence band and 0.79 eV below the conduction
band. As a result of the bonding state being empty, only two transitions
within the gap are now possible, hence the loss of the middle 1.4 eV absorption
peak in Figure 3.71.

The authors also calculated the band structure expected for the fully
oxidised form, taken as 33% doping or 2 charges per 6 rings, and the result
is depicted in Figure 3.72(c). Continued removal of the states from the valence
and conduction bands widens the gap to 3.56 eV, with the two intense
absorptions in the gap observed in the optical spectra now accounted for
by the presence of wide bipolaron bands. The authors stated that, on the
basis of other workers' calculations, the lowest energy absorption should have
the most intense oscillator strength, as is indeed observed.

Hence the authors concluded that bipolarons are the major current carriers
on the basis of the calculation showing them to be 0.49 eV more stable than
two polarons, with polarons combining where possible to give bipolarons.
This paper dominated the interpretation of experiments on polypyrrole for
several years until *in situ* techniques became routinely applied.

3.3.2.3 Dynamic properties. On the basis of cyclic voltammetry, Diaz *et al.*
(1981) showed that thin films of polypyrrole on an electrode immersed in
acetonitrile could be repeatedly driven between the conducting and insulating
states, as shown by the stability of the cyclic voltammograms of the films
(see Figure 3.73).

From the figure it can be seen that there is a pronounced anodic peak
near 0.15 V vs. SCE that has also been observed in aqueous electrolyte. The
current at this peak scales linearly with the scan rate, as expected for a
surface-bound redox material (see section on cyclic voltammetry of adsorbed
species).

The cyclic voltammograms in Figure 3.73 show an interesting characteristic
in that, at potentials greater than the anodic peak potential, the current tends
towards a constant value. The charge passed, Q, in charging a capacitor C
to a potential V is (see section 2.1.1.):

$$Q = CV \tag{3.79}$$

and:

$$dQ/dt = I = C \, dV/dt = Cv \tag{3.80}$$

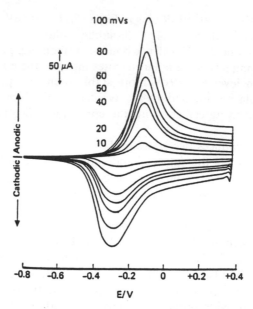

Figure 3.73 Cyclic voltammograms of a 20 nm-thick polypyrrole on Pt in tetraethylammonium tetrafluoroborate/CH_3CN solution. The scan rates are as shown. From Diaz *et al.* (1981).

where v is the scan rate. Since v is a constant, then it is clear from equation (3.80) that I will be independent of V if a capacitive current flows. In addition, if the charge under the cyclic voltammogram is plotted against potential, then the plot will be linear if the current is capacitive. The latter phenomenon is indeed observed at potentials greater than the anodic peak potential.

On the basis of observations such as that above, Feldberg proposed a theory whereby the cyclic voltammogram of a film such as polypyrrole was the result of two processes. On increasing the potential from that at which the film is in the neutral form, the charge carriers are generated via a single, reversible electron transfer followed by a capacitive charging of the very high surface area internal polymer/electrolyte interface. The current flowing in the latter region is only due to the carriers moving to the interface, not to the generation of further carriers. As will be seen below, the theory is not sustained by experiment.

The broad nature of the current peaks in the voltammogram of conducting polymers such as polypyrrole has been interpreted in a number of ways, one of which was to attribute it to the movement of anions across the polymer/electrolyte interface, a vital process if the overall charge neutrality of the film is to be maintained. The participation of the electrolyte in the electrochemistry of the polymer film is easily seen by comparing the response of polypyrrole in a variety of different electrolytes (see Figure 3.74).

In 1985, Feldman *et al.* reported *in situ* conductivity measurements on polypyrrole. These were obtained using a twin electrode, thin-layer cell

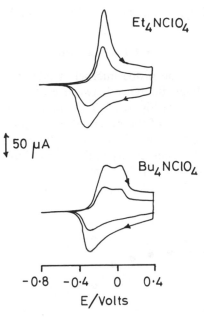

Figure 3.74 Cyclic voltammograms of a 20 nm-thick polypyrrole film on Pt in CH_3CN containing different electrolytes. Two sweep rates are shown, the voltammograms showing the lower currents were taken at $50 \, mV \, s^{-1}$, and the larger currents were obtained at $100 \, mV \, s^{-1}$. From Diaz *et al.* (1981).

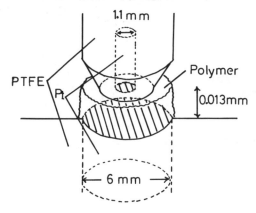

Figure 3.75 Schematic illustration of a twin electrode thin layer cell used for 'static' conductivity measurements. From Feldman *et al.* (1985). Copyright 1985, American Chemical Society.

depicted schematically in Figure 3.75. The film was grown onto Pt from a solution of the monomer in acetonitrile. The uncoated microelectrode probe was brought into contact with the polymer film after the latter had reached equilibrium at the desired potential. The potential of the probe was then stepped to a value $10 \rightarrow 100 \, mV$ different from that of the polymer and the current, I, measured. The static (DC) conductivity of the film, σ, is then given by:

$$\sigma = I\,d/AV\,\Omega^{-1}\,cm^{-1}, \tag{3.81}$$

where A is the area of the probe microelectrode, d is the thickness of the film and V is the voltage difference between probe and coated electrode.

The results are shown in Figures 3.76(a) and (b). Figure 3.76(a) shows the conductivity as a function of the potential of the polymer-coated electrode and Figure 3.76(b) shows the conductivity as a function of the charge injected per monomer ring. The conductivity of the fully oxidised polymer was found to be significantly lower than that obtained by other workers via *ex situ* measurements, $10^{-2}\,\Omega^{-1}\,cm^{-1}$ compared with $10^2\,\Omega^{-1}\,cm^{-1}$. The difference

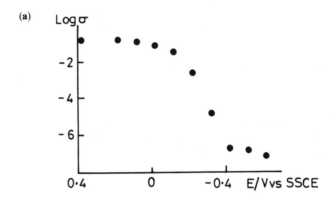

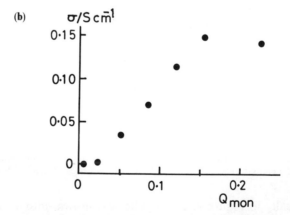

Figure 3.76   (a) Plot of the log of the polypyrrole film conductivity (σ) vs. the polymer potential, E, for a 10 mV potential difference applied across a 13.9 μm-thick film in 0.1 M tetraethylammonium tetrafluoroborate/CH_3CN. (b) Plot of polypyrrole conductivity vs. fractional charge per pyrrole monomer subunit (q_{mon}) in 0.1 M tetraethylammonium tetrafluoroborate/CH_3CN. From Feldman *et al.* (1985). Copyright 1985 American Chemical Society.

was not attributed to the presence of the electrolyte, however, since measurement of the conductivity of the dried film, followed by the addition of solvent, gave the same conductivity to within a factor of 2.

The oxidation of the neutral film causes a 10^6 increase in the magnitude of σ up to $c.$ 0 V vs. SSCE (saturated sodium chloride electrode). At potentials >0 V, however, the conductivity becomes effectively potential independent.

In terms of the charge passed during the oxidation at very low doping, the films are poor conductors and between 0 and 0.15 charges/ring the conductivity is proportional to the injected charge. Finally, at higher doping, a limiting conductivity is attained.

Feldman and colleagues (1985), therefore, concluded that the electrical conductivity of polypyrrole depends on the state of the polymer charge only at potentials <0.1 V, corresponding to a doping level of $c.$ 1 charge per 6 rings. They postulated that the non-linear relationship between σ and Q suggests that σ is determined by different limiting factors depending on the film oxidation state, i.e. a balance between factors such as the population of the chains by bipolarons and the rate of *inter*chain hopping, etc.

The critical role of the latter process was clearly shown in the extremely elegant work of Wegner and Rühe (1989) who measured the temperature dependence of the DC conductivity of a range of polythiophene and polypyrrole derivatives as a function of the interchain separation. The derivatives

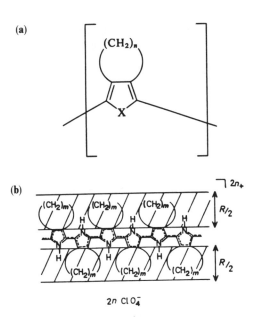

Figure 3.77 (a) The cycloaliphatic monomer subunits employed by Wegner and Rühe (1989). n varied between 3 and 10. (b) Schematic description of the polypyrrole backbone wrapped in a layer of poorly conducting methylene moieties of the alkylene chains fused to the pyrrole units in the 3,4 positions. The minimum separation distance, R, between adjacent chains can be estimated from molecular models. From Wegner and Rühe (1989).

were of the general formula shown in Figure 3.77(a) in which a cycloaliphatic ring was fused to the constitutional unit of the polymer backbone. This results in each polypyrrole chain being cylindrically surrounded by a layer of methylene groups, the thickness of which (and hence the interchain separation) is determined by the size of the alkyl ring fused to the pyrrole units, as shown in Figure 3.77(b). The interchain distance, given by the parameter R in the figure, was determined using X-ray crystallography and varied with the number of methylene units, n, as indicated in Table 3.6.

Table 3.6 Characteristic interchain distance, R, of polypyrrole with cycloalkyl rings fused to the pyrrole moiety

n	R (nm)
3	0.76
4	0.79
5	0.97
10	1.38

The alkyl substitution forced adjacent pyrrole units on the same chain out-of-plane by at least 40°. A consequence of this was that the authors could only evaluate their data at higher temperatures (including room temperature), in terms of the assumption that *inter*chain electronic hopping was the dominant factor in determining the macroscopically measureable electronic conductivity, rather than *intra*chain hopping. Under these conditions, the

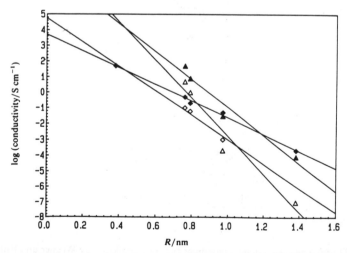

Figure 3.78 Dependence of conductivity on the parameter R, defined by Figure 3.77(b), for: (◆) poly(3,4-cycloalkylpyrrole) perchlorates; (◇) poly(3,4-cycloalkylpyrrole) hexafluorophosphates; (△) poly(3,4-cycloalkylthiophene) tetrachloroferrates; (▲) poly(3,4-cycloalkylthiophene) hexafluorophosphates; all data at 24°C. From Wegner and Rühe (1989).

conductivity should be described by:

$$\sigma = \sigma_0 \exp(-2\alpha R)\exp(-E_a/kT),$$

where α and σ_0 are constants and E_a is the activation energy barrier between the initial and final states. At constant temperature therefore, a plot of $\log \sigma$ vs. R should be linear and this was indeed found to be the case, as can be seen in Figure 3.78. Moreover, extrapolating the plots in Figure 3.78 back towards $R = 0$ gives σ values comparable to those found in practice for the parent polymers.

The results of Wegner and Rühe (1989) clearly show that electronic conduction in conducting polymers such as polythiophene and polypyrrole occurs via a hopping mechanism that is dominated by *inter*chain rather than *intra*chain hopping.

Returning to the debate over the identity of the carriers in polypyrrole, another major contribution was made by the epr studies of Genoud and colleagues (1985; Devreux et al., 1987; Nechtschein et al., 1986) who, in contrast to Scott and colleagues (1983), performed the measurements *in situ*.

The authors found the neutral form of the polymer to be epr inactive. With increasing potential the number of spins detected increased from zero, reached a maximum and then decreased, as shown in Figures 3.79(a) and (b). The data were interpreted by the authors in terms of the initial formation of polarons followed by their subsequent recombination into bipolarons.

As can be seen from Figure 3.79(a), the ratio of spins (i.e. polarons) to charge injected is 1:1 up to *c.* 1 charge per 30 monomer units. Above this doping level, the ratio declines, indicating that both polarons *and* bipolarons are being generated. The maximum number of spins (1 *spin* per *c.* 11 monomer rings) is observed at -0.2 V vs. Ag/Ag$^+$ (see Figure 3.79(b)) or *c.* 1 *charge* per 6 monomer units, and was interpreted by the authors in terms of the recombination of the polarons to form bipolarons.

On the basis of their results the authors concluded that the crossover from the neutral to oxidised form of polypyrrole, and vice versa, requires the number of spins to pass through a maximum, i.e. the creation and anihilation of bipolarons involves the passage through the polaron state. They expressed this as a two-step redox reaction:

$$PP^0 \longleftrightarrow PP^+ + e^- \quad E_1 \tag{3.82}$$

$$PP^+ \longleftrightarrow PP^{2+} + e^- \quad E_2 \tag{3.83}$$

where E_1 and E_2 denote the quasistandard potentials for the formation of polarons and bipolarons, respectively, and the equilibrium potential E is given by:

$$E = E_1 + (RT/F)\ln(x/[1-x-y]) \tag{3.84}$$

$$E = E_2 + (RT/F)\ln(y/x) \tag{3.85}$$

where x is the concentration of polarons, y the concentration of bipolarons

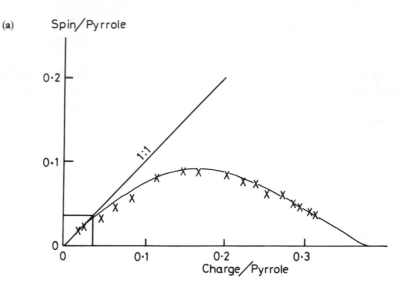

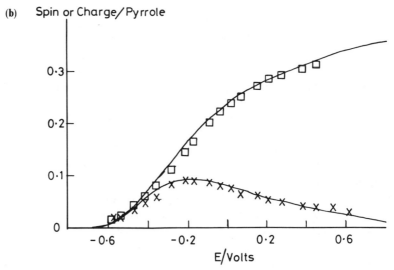

Figure 3.79 (a) Spin vs. charge concentration in polypyrrole. (b) Charge ($\square$) and spin ($\times$) concentration vs. potential in polypyrrole. From Devreux *et al.* (1987).

and $(1 - x - y)$ the concentration of neutral sites. $\Delta E = E_2 - E_1$ gives the bipolaron potential with respect to that of the polarons, i.e. if ΔE is negative, bipolaron formation is favoured and polarons will attract each other to form bipolarons.

Using the fact that the charge passed is related to x and y by:

$$Q/V = x + 2y \qquad (3.86)$$

where V is the volume of the polymer, and equations (3.84) and (3.85), it is possible to obtain x and thus the number of spins as a function of Q, with ΔE as a 'fitting' parameter. The authors found that the best fit to the data in Figure 3.79(a) was obtained with $\Delta E \approx 0.04\,V$, i.e. with the polarons and bipolarons being nearly degenerate, such that they can coexist, in complete contrast to the conclusions of Bredas and colleagues (1984). This conclusion was supported by the work of Conwell (1985) who calculated, on the basis of statistical arguments, that if a bipolaron is favoured over two polarons by $0.45\,eV$, the polaron concentration would be negligible at room temperature. Hence, the observed generation of bipolarons via an appreciable number of polarons cannot be explained in terms of bipolarons being notably more stable than polarons.

This means that for a given polypyrrole chain if a 'unit cell' (chosen as 6 monomer units on the basis of the observed maximum doping level of 2 charges per 6 monomer units) already has a polaron on it and a second charge is injected onto the same chain, it is more favourable to form a second polaron on another vacant unit cell than to force spin pairing to give a bipolaron. If, however, all the other unit cells are already occupied, then a bipolaron will be formed in the unit cell, rather than two polarons.

The model initially proposed by Genoud and coworkers (1985) was very simple, e.g. it did not take into account interparticle interactions. In their later papers, the authors did take into account such interactions but the conclusions remained essentially the same: bipolarons and polarons are almost degenerate and can coexist. At low doping, only polarons are produced and then both polarons and bipolarons until the latter dominate at very high doping levels. They also found that their data were best understood in terms of bipolarons that are 5 monomer units long and polarons $c.$ 4.5 units long.

The data of Genoud and colleagues were strongly supported by the work of Waller and Compton (1989) who also employed *in situ* epr to study the electrochemical cycling of polypyrrole as well as AC impedance to obtain the conductivity of the film as a function of doping. In agreement with the work of Feldman and colleagues (1985) discussed above, the authors found that the conductivity of the film increased with the charge injection until $c.$ 1 charge/6 monomer units, after which the conductivity reaches a plateau and further oxidation leads to no corresponding increase in conductivity. In addition it was found that the generation of polarons also declined at doping levels greater than 1 charge/6 monomer units, indicating that polarons were recombining to give bipolarons.

The authors postulated that the conductivity is proportional to the number of carriers in the film and that the mobility of polarons is equal to that of bipolarons, hence the conductivity is independent of carrier type. Thus, the conductivity increases steadily as polarons, and then both polarons and bipolarons, are generated but should attain a steady value when polaron recombination to give half as many bipolarons becomes important.

Clearly, the *in situ* studies such as those discussed above resulted in a cohesive picture emerging of the electronic conduction in polypyrrole. Thus, the conduction is not metallic in nature but involves definite species moving along, and between, the polymer chains. These species, polarons and bipolarons, are almost degenerate and so coexist, with both contributing to the conductivity. At very low doping levels only polarons are generated. As the doping level increases, both carriers are produced until polaron recombination results in bipolarons being the dominant species in the highly oxidised polymer. The doping levels at which bipolarons start to be generated, and then polaron recombination commences, will depend on the exact relationship between the relative numbers of the two carriers at low and medium doping levels. This, in turn, will clearly be dependent on the average chain length of the polymer or, more precisely, the average *conjugation* length between defect or β-coupled monomer units, the distribution of conjugation lengths throughout the polymer, and the size of the polarons and bipolarons. Based on these considerations Hamnett (1989) stressed the diverse nature of growth conditions as a possible explanation for the spread of results commonly obtained from the same conducting polymer grown in different ways in different laboratories.

However, while the evidence for the existence of polarons was extremely convincing, that for bipolarons was rather more problematical in that it was largely effectively *negative* in nature: the *absence* of an absorption peak in the optical spectrum, the *absence* of a signal in epr studies on the decline of the observed signal. In essence, bipolarons had not been actually *observed*. This fact was remedied by the work of Christensen and Hamnett (1991) who employed ellipsometry and FTIR to study the growth and electrochemical cycling of polypyrrole *in situ* in aqueous solution.

It should be emphasised here that very different growth mechanisms are found according to the conditions of the experiment and that the study of the growth of conducting polymers such as polypyrrole represents a major area of investigation in its own right. The work of Christensen and Hamnett (1991) on the growth of polypyrrole should not, therefore, be taken as representative, it is specific to the conditions they used.

Figures 3.80(a) and (b) show the behaviour of n, k and the thickness L of a polypyrrole film during its growth on a Pt electrode in aqueous perchlorate solution obtained via *in situ* ellipsometry. A definite lag can be seen between the change in L and the changes in n and k. Up to $c.$ 1.3 s the thickness of the film increases linearly with time while n and k remain constant. Between 1.3 and 2 s the growth almost tails off before increasing slowly up to 4 s and then more quickly until attaining a steady rate of increase at $t > 4$ s. Similarly at $1.3 \, \text{s} < t < 4 \, \text{s}$ both n and k increase, with n attaining a maximum value at 4 s before decreasing, while k increases at a lower rate at $t > 4$ s than at $t < 4$ s.

The data in Figures 3.80(a) and (b) were interpreted in terms of the precipitation of oligomers onto active sites on the surface. These oligomers then grow homogeneously with time in the first 1.3 s, during which the film consists largely of electrolyte, and hence n and k do not change, the chains attaining an average length of $c.$ 34 monomer units (i.e. the value of L prior to the

(a)

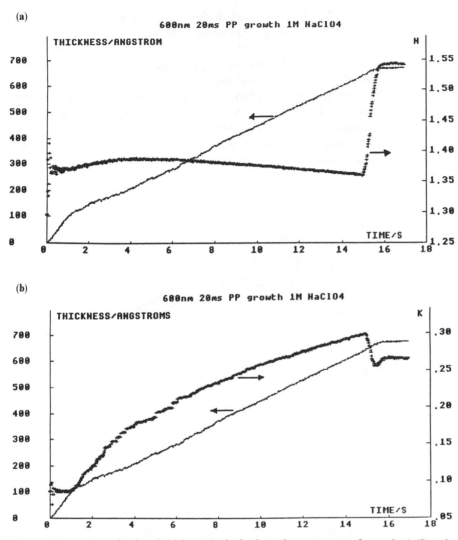

Figure 3.80 Values of n, k and thickness L obtained *via* three parameter fits to the Δ, Ψ and intensity data obtained during the growth of a polypyrrole film on a sputtered Pt electrode in N_2-saturated 1 M $NaClO_4$/0.1 M pyrrole. The potential was stepped from 0 V to 0.8 V vs. SCE for 15 s, and readings taken every 20 ms. Reprinted from *Electrochimica Acta*, **36**, P.A. Christensen and A. Hamnett, '*In situ* Spectroscopic Investigations of the Growth, Electrochemical Cycling and Overoxidation of Polypyrrole in Aqueous Solution', pp. 1263–1286 (1991), with kind permission from Pergamon Press Ltd., Headington Hill Hall, Oxford OX3 0BW, UK.

temporary tail-off in the growth). At $1.3\,\mathrm{s} < t < 2\,\mathrm{s}$ the chains are long enough such that the potential drop along their length precludes further growth and the polymer 'fills in' with nucleation and growth occurring on the remaining exposed Pt surface. Eventually growth recommences over the full surface since the filled in film is more conducting and the *IR* drop consequently falls.

On the basis of the charge passed during growth (and assuming 100% current efficiency), the authors calculated that the film thickness was 420 Å. From Figures 3.80(a) and (b), it can be seen that the 'best fit' to the data was obtained with a film thickness of 669 Å, suggesting that the as-grown film is 63% pyrrole, 37% solvent and electrolyte.

Typical voltammograms of a polypyrrole film obtained by the authors are shown in Figure 3.81. If the film was held at $-0.6\,V$ vs. SCE prior to cycling, the cyclic voltammogram showed a definite peak, near $c. -0.15\,V$, as was observed by other workers and is discussed above. The potential at which this peak occurs was found to mark a definite transition point in the behaviour

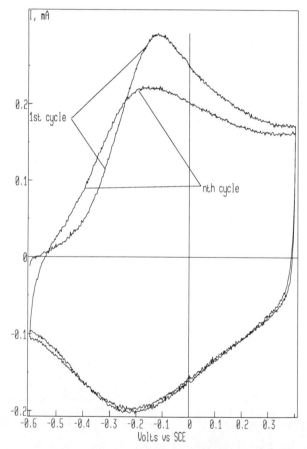

Figure 3.81 Typical cyclic voltammograms of a polypyrrole film on Pt in N_2-saturated 1 M $NaClO_4$. The voltammograms were collected immediately after holding the film at $-0.6\,V$ vs. SCE for 5 min and after cycling for 5 min. The scan rate was $100\,mV\,s^{-1}$ and the film thickness 84 nm. Reprinted from *Electrochimica Acta*, **36**, P.A. Christensen and A. Hamnett, '*In situ* Spectroscopic Investigations of the Growth, Electrochemical Cycling and Overoxidation of Polypyrrole in Aqueous Solution', pp. 1263–1286 (1991), with kind permission from Pergamon Press Ltd., Headington Hill Hall, Oxford OX3 0BW, UK.

of the film, e.g. it is the potential at which the charge vs potential plot obtained from the cyclic voltammogram becomes linear.

The behaviour of n, k and L as a function of potential during electrochemical cycling is shown in Figures 3.82(a)–(d). Taking the thickness first, it can be seen that L shows little change on oxidation until -0.2 V. Between -0.2 V and 0.2 V, however, the film contracts by a surprising 30%, associated with the expulsion of solvent. At potentials >0.2 V it expands slightly, $c.$ 7 Å. The dramatic, and reversible, collapse of the film was entirely unexpected, primarily because all previous theories involved *ingress* of anions and associated solvent, and this could only be tested with ellipsometry.

The authors postulated that the dramatic contraction of the film was due to one, or both, of the following:

1. An electrostrictive process associated with the generation of a particular carrier.
2. The expulsion of cations and/or protons with their associated large solvation sheath.

The absorption, k, of the film at the wavelength of the probing light, 600 nm, was observed to increase monotonically with potential. Concomitant with k, the real part of the refractive index shows a steady *decrease* in value as soon as the potential is increased beyond -0.5 V. In addition, n can be seen to pass through an inflection at $c. -0.2$ V. On the basis of optical considerations, it is possible to infer from the behaviour of n that:

1. Some absorption is occurring in the near-IR even at comparatively low potentials, i.e. long before changes are observed in the optical spectra.
2. Some particular change is taking place in the nature of the near-IR absorption at -0.2 V.

The authors also observed considerable hysteresis in the variation of L with potential between the first and subsequent scans (see Figure 3.82(d)) which they linked to the asymmetry of the cyclic voltammograms in Figure 3.81.

To explain these observations the authors drew on the theory of Heinze *et al.* (1987) who postulated that a film such as polypyrrole can exist in two conformations, A and B. The A form is the stable, twisted conformation of the neutral polymer but is metastable when the polymer becomes conducting. The authors proposed that during the oxidation the initially formed charge carrier undergoes a geometrical relaxation from the twisted conformation favoured by the neutral form to a planar conformation, B. This stabilises the charged system such that the redox energies of the charge carriers are lowered and, postulated Christensen and Hamnett (1991), will also have major implications with respect to the overall film thickness. In the reduction step the geometric distortion exists up to the end of the discharge then the chain relaxes relatively quickly to the twisted conformation (see Scheme 3.11). As the co-planar structure, B, of the charged polymer is energetically stabilised,

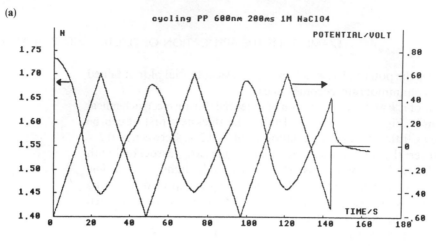

(a)

cycling PP 600nm 200ms 1M NaClO4

(b)

cycling PP 600nm 200ms 1M NaClO4

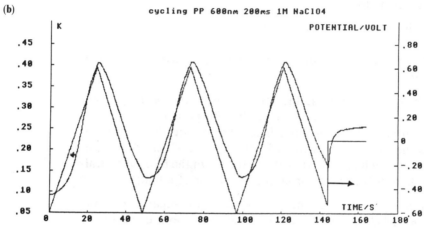

(c)

cycling PP 600nm 200ms 1M NaClO4

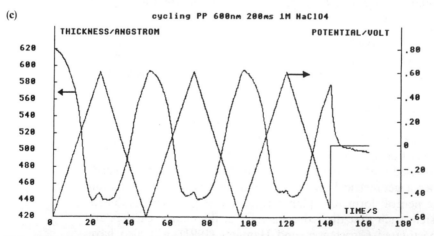

Figure 3.82 (a–d) Values of *n*, *k* and thickness *L* obtained *via* three parameter fits to the Δ, Ψ and intensity data obtained during the electrochemical cycling of the film in Figure 3.80. In N_2-saturated 1 M NaClO$_4$. The film was cycled between $-0.6\,V$ and $+0.6\,V$ at $50\,mV\,s^{-1}$, and readings taken every 20 ms. In (d) the variation in the thickness is plotted as a function of potential. Reprinted from *Electrochimica Acta*, **36**, P.A. Christensen and A. Hamnett, 'In situ Spectroscopic Investigations of the Growth, Electrochemical Cycling and Overoxidation of Polypyrrole in Aqueous Solution', pp. 1263–1286 (1991), with kind permission from Pergamon Press Ltd., Headington Hill Hall, Oxford OX3 0BW, UK.

(d)

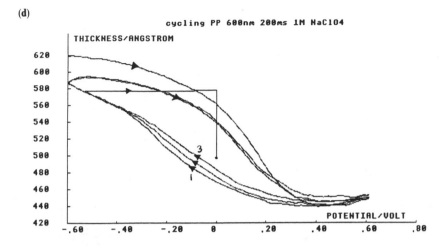

the redox energies for the discharge steps are now lower than for the formation of polarons and hence the asymmetry in the cyclic voltammograms.

$$A \longleftrightarrow B$$
$$\uparrow \qquad\quad \downarrow$$
$$A^{n+} \longleftrightarrow B^{n+}$$

Scheme 3.11

The authors showed this effect in the cyclic voltammograms in Figure 3.81. If the geometric relaxation $B \longrightarrow A$ is slow and the polymer is fully reduced before the first cycle, then the second cyclic voltammogram should give an oxidation current at a lower potential than the first and the voltammogram itself should be more symmetrical due to the fact that parts of the distorted chain will still exist. This is indeed the case as can be seen from the figure. The ellipsometric data support these results as can be seen from Figure 3.82(d) which clearly shows that the behaviour of the film during the first potential cycle after full reduction is markedly different than subsequent cycles, reflecting the continuous presence of some of the B conformation in the second, and subsequent, cycles. Heinze and colleagues suggested that a critical number of charge carriers was required before the chain could undergo the $A \longrightarrow B$ relaxation and, as was noted above, no change of thickness occurs until $-0.2\,\text{V}$ after which dramatic change does take place.

Figures 3.83(a) and (b) show *in situ* FTIR spectra taken of a polypyrrole film on Pt, also from the work of Christensen and Hamnett (1991). The spectra were collected at successively higher potentials after the reference spectrum taken of the fully reduced film at $-0.6\,\text{V}$. As can be seen from the spectra, an intense absorption extending out into the near-IR is steadily

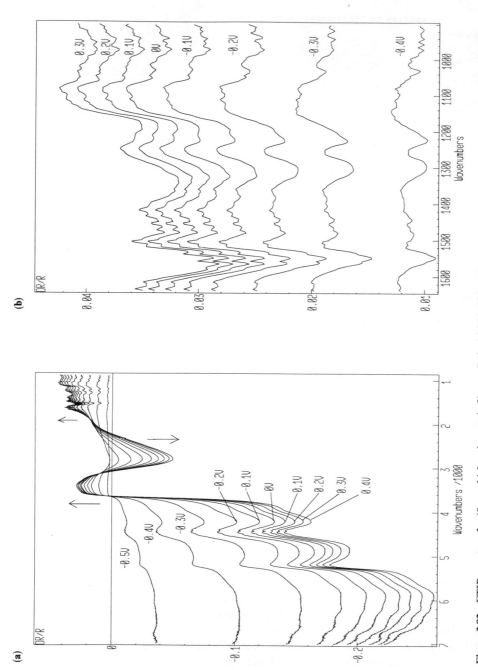

Figure 3.83 FTIR spectra of a 40 nm-thick polypyrrole film on Pt in 1 M NaClO₄. The spectra were collected successively at the potentials shown, and normalised to the spectrum, taken first, at − 0.6 V. (a) Full spectral range, (b) the IRAV region. In (b) the spectra collected at 0.4 V and − 0.5 V are omitted for clarity. Reprinted from *Electrochimica Acta*, **36**, P.A. Christensen and A. Hamnett, '*In situ* Spectroscopic Investigations of the Growth, Electrochemical Cycling and Overoxidation of Polypyrrole in Aqueous Solution', pp. 1263–1286 (1991), with kind permission from Pergamon Press Ltd., Headington Hill Hall, Oxford OX3 0BW, UK.

gained at all potentials $> -0.6\,V$, as expected from the low n values derived from the ellipsometric data. The electronic absorption increases in intensity dramatically with increasing potential up to $-0.2\,V$ after which it increases more slowly. Superimposed on this gain feature are some broad loss features near $5200\,cm^{-1}$ and $4400\,cm^{-1}$ that are artifacts. Below $1600\,cm^{-1}$ are several IRAV bands.

The electronic absorption can be attributed to the lowest polaron and/or bipolaron states in the gap (see Figure 3.72). It is clear from the above discussion that some form of transition in the carrier behaviour occurs near $c. -0.2\,V$; this is reinforced by a consideration of the IRAV absorptions in Figure 3.83(b). The IRAV bands are reasonably sharp up until $-0.2\,V$ after which they broaden and appear less well-defined.

Figures 3.84(a) and (b) show the same spectra as those in Figures 3.83(a) and (b), except only up to $-0.2\,V$, and Figure 3.84(c) and (d) show the spectra in Figures 3.83(a) and (b) from $-0.1\,V$ to $0.4\,V$ but normalised to that taken at $-0.2\,V$.

Clearly there are two sets of absorptions, as shown in Table 3.7, which were attributed to polarons and bipolarons by the authors. All the IRAV bands are relatively narrow and so suggest that the carriers occupy a well-defined number of monomer units. The highest frequency IRAV has been attributed to a combination of the intra-ring C=C vibration and the inter-ring C–C stretch which increases in frequency as the conjugation length decreases, consistent with the bipolaron being somewhat shorter than the polaron.

The initially formed carriers are the polarons, which increase in intensity up to $-0.2\,V$ where their intensity attains a constant value independent of any further doping. This can be deduced from the fact that no loss features attributable to polarons can be seen in Figures 3.84(c) or (d). In addition, from $c. -0.4\,V$ bipolarons are also generated and it is as a result of the growing-in of these absorptions that the IRAV bands appear to broaden out but they only become important at potentials $> -0.2\,V$.

A plot of the polaron and bipolaron electronic band intensities as a function of charge injected (as electrons removed per monomer ring) is shown in Figures 3.85(a) and (b). The point at which the polaron intensity attains a plateau and the bipolaron intensity starts to dominate, at the anodic peak

Table 3.7 Polaron and bipolaron absorptions in polypyrrole (after Christensen and Hamnett, 1991)

Polaron bands (cm^{-1})	Bipolaron bands (cm^{-1})
1196	1165
1227	1258
1308	1358
1558	1588
c. 6000	c. 5000

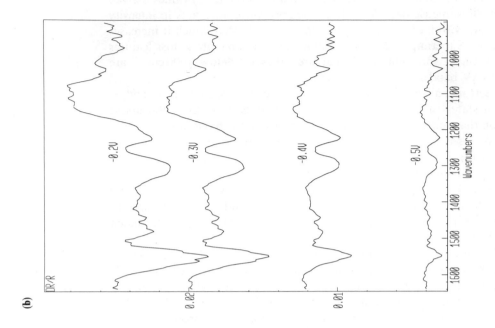

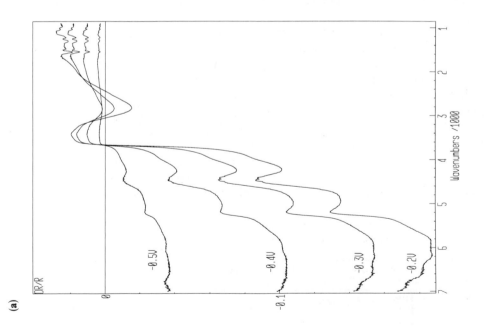

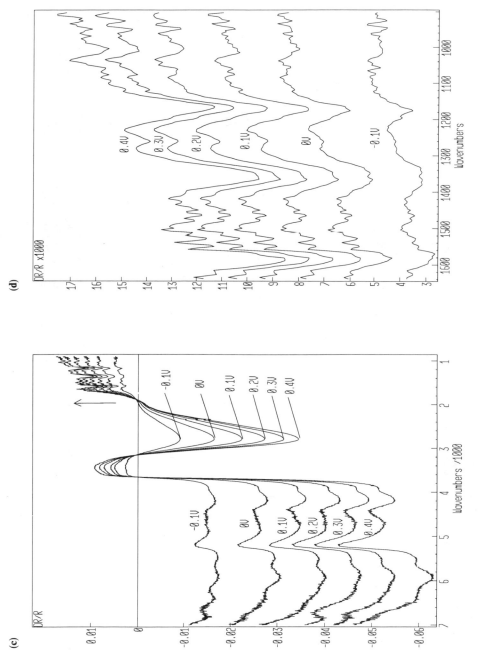

Figure 3.84 (a and b) The spectra in Figure 3.83 except only up to $-0.2\,V$. (a) Full spectral range, (b) the **IRAV** region. (c and d) The spectra in Figure 3.83 collected at potentials $> -0.2\,V$ normalised to that taken at $-0.2\,V$. (c) Full spectral range, (d) the **IRAV** region. Reprinted from *Electrochimica Acta*, **36**, P.A. Christensen and A. Hamnett, '*In situ* Spectroscopic Investigations of the Growth, Electrochemical Cycling and Overoxidation of Polypyrrole in Aqueous Solution', pp. 1263–1286 (1991), with kind permission from Pergamon Press Ltd., Headington Hill Hall, Oxford OX3 0BW, UK.

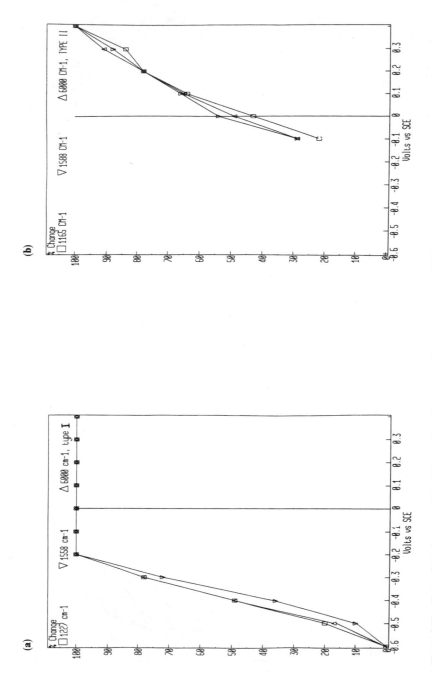

Figure 3.85 Plots of the integrated band intensities of representative polaron (a) and bipolaron (b), features from Figure 3.84(a)–(d). The intensities are normalised to their maximum value. Reprinted from *Electrochimica Acta*, **36**, P.A. Christensen and A. Hamnett, '*In situ* Spectroscopic Investigations of the Growth, Electrochemical Cycling and Overoxidation of Polypyrrole in Aqueous Solution', pp. 1263–1286 (1991), with kind permission from Pergamon Press Ltd, Headington Hill Hall, Oxford OX3 0BW, UK.

potential, corresponds to 1–2 charges per average chain length. On the basis of this observation, and the conclusions of Genoud and colleagues (1985), the authors attempted to model the data in Figure 3.85. The basic assumptions of their model were:

1. Polarons are longer than bipolarons (see above).
2. A Gaussian distribution of chain lengths about the mean of 34 monomer units (from the ellipsometric results on the growth).
3. A single charge is placed on each chain before a second charge is introduced onto a chain.
4. A chain can accommodate a second charge to form a second polaron, providing that the chain length is longer than two polarons. If the chain is shorter than two polaron lengths, a bipolaron is formed.

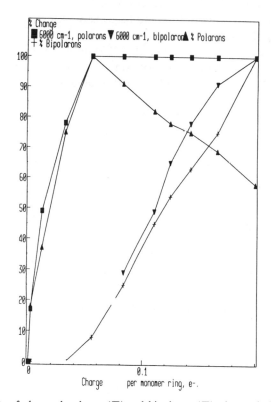

Figure 3.86 Plots of observed polaron (■) and bipolaron (▼), electronic band intensities at $6000\,cm^{-1}$ and the intensities predicted by the model of Christensen and Hamnett (% polarons ▲, % bipolarons +), against the injected charge (in e^- removed per monomer ring). The intensities are normalised to their maximum value. The polaron absorption intensities are taken from the spectra in Figure 3.84(a) and the bipolarons from Figure 3.84(c). Reprinted from *Electrochimica Acta*, **36**, P.A. Christensen and A. Hamnett, '*In situ* Spectroscopic Investigations of the Growth, Electrochemical Cycling and Overoxidation of Polypyrrole in Aqueous Solution', pp. 1263–1286 (1991), with kind permission from Pergamon Press Ltd., Headington Hill Hall, Oxford OX3 0BW, UK.

The results obtained on the basis of the model are also shown in Figure 3.86. The fit is reasonably good to the author's data and would seem to be a very good qualitative model of the epr data of Genoud and colleagues (1985), and Waller and Compton (1989). The best fit, as shown, was found for a polaron length of 12 monomer units and a bipolaron length of 3; the latter in agreement with the conclusions of Albery and co-workers (1989).

Thus, it appears that the 'transition' represented by the anodic peak in the cyclic voltammogram of polypyrrole is due to a changeover in the dominant carrier type and is accompanied by a dramatic contraction of the film. The authors strongly suspected that this contraction was due to electrostriction associated with bipolaron formation. As a further test they also carried out experiments intended to test if proton expulsion from the film occurred on oxidation. They found that it did indeed occur but monotonically at *all potentials* $> -0.6\,V$, in agreement with the extremely elegant work of Tsai *et al.* (1987), and so could not be responsible for the relatively sudden contraction at potentials $> -0.2\,V$.

Rajeshwar and colleagues (Tsai *et al.*, 1987) grew a polypyrrole film on a gold minigrid OTE in a spectroelectrochemical UV-visible transmittance cell. The film was then electrochemically cycled in aqueous $LiClO_4$ containing a pH-sensitive dye, the colour changes of which were employed as a means of monitoring the pH near the film. This, in turn, enabled the authors to show that protons were expelled from the polypyrrole throughout the oxidation process and re-incorporated on reduction.

Thus, Christensen and Hamnett (1991) concluded that polypyrrole grows on Pt in aqueous perchlorate via the deposition of oligomers of an average length of 34 monomer units. Oxidation of the neutral film generates polarons of about 12 units in length. At potentials $> -0.5\,V$, bipolarons are also generated but these do not become important until $\geqslant -0.2\,V$, represented by the anodic peak potential. Beyond this potential polarons cease to be formed but do not appear to actually decrease in number. The formation of large numbers of bipolarons is accompanied by large electrostriction effects and the ejection of electrolyte and solvent from the film.

The work of Christensen and Hamnett (1991) provided the first 'positive' evidence for bipolarons and showed, for the first time, that the oxidation of polypyrrole was accompanied by a dramatic *decrease* in the film thickness, linking this with the generation of these carriers. Taken in addition to all the work discussed above their work provided some of the final pieces in a workable theory of the conduction mechanism in polypyrrole.

**3.5
Adsorbed films for the promotion of enzyme electrochemistry**

In nature there are many redox enzymes which interact with biologically important molecules and in doing so eject or accept electrons. Consequently, it has long been realised that if such redox enzymes could be immobilised at an electrode surface, and rapid electron transfer between the two facilitated, the technological implications would be enormous as this would open up a whole realm of possible biosensors. However, for many years it was believed that direct enzyme electrochemistry would be unobservable because: (i) the slow diffusion rate of large enzymes would lead to very small currents, and (ii) the prosthetic, i.e. redox active group, is buried within the enzyme, hence interaction with the electrode would be prohibitive for all but a few favourable orientations of the enzyme at the surface. In practice, while enzymes such as cytochrome *c* do indeed adsorb on clean electrode surfaces, they do so with the loss of their redox activity and any electron transfer that is observed is too slow to be of any practical use.

Up until the pioneering work of Yeh and Kuwuna (1977), and Eddowes and Hill (1977), the only successful way of promoting rapid electron transfer between a redox enzyme and an electrode in aqueous solution (buffered at pH 7 to mimic the body pH) was to employ a species having a reversible redox couple with an E^0 value close to that of the enzyme, a *mediator*, which would then mediate between the enzyme and the electrode as shown in Figure 3.87. However, Yeh and Kuwuna (1977) showed that cytochrome *c* exhibited reversible electrochemistry at an indium-doped tin oxide electrode and Eddowes and Hill (1977) showed that quasi-reversible electron transfer to ferricytochrome *c* could be induced at a gold electrode in aqueous solution using 4,4′ bipyridine as a 'promoter'. Unlike previous mediators based on redox couples in solution, the 4,4′ bipyridine acted via adsorption onto the gold surface and yet was electro-inactive over the potential region of interest, suggesting an entirely novel mechanism of *promoting* the enzyme electro-chemistry. The adsorption of the 4,4′ bipyridine was clearly reversible as the continued electroactivity of the cytochrome *c* could only be obtained if the promoter was present in solution.

Another major discovery in this area came with the work of Taniguchi *et al.* in 1982. Up until this point attempts to implement the next logical step after the work of Eddowes and Hill (1977), i.e. to generate a surface-modified

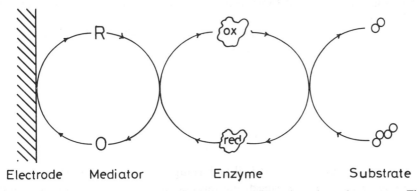

Figure 3.87 The *mediated* electron transfer between an electrode and a redox enzyme. The mediator facilitates the charge transfer between enzyme and electrode by cycling between its oxidised and reduced forms.

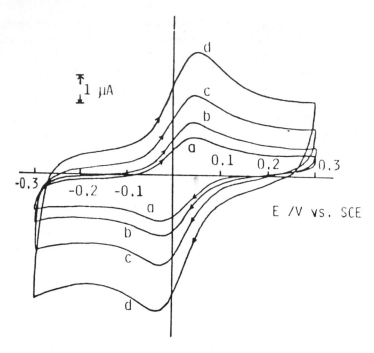

Figure 3.88 Cyclic voltammograms of 0.38 mM cytochrome c, in the presence of 0.47 mM SSBipy, in 0.02 M KH_2PO_4 + 0.026 M K_2HPO_4 + 0.1 M $NaClO_4$, (pH 6.7). The scan rate is: (a) 30; (b) 50; (c) 100; (d) 200 mV s^{-1}. From Taniguchi *et al.* (1982).

electrode capable of promoting reversible electron transfer to a redox enzyme in the absence of a solution promoter, had failed. Taniguchi and colleagues (1982) reported that a well-defined redox wave of cytochrome c could be observed in pH 7 buffer (see Figure 3.88) at a gold electrode that had previously been exposed to an aqueous solution of bis(4-pyridyl) bisulphide (SSBipy, **1**) for 5 minutes. No promoter was present in the solution and the modified electrode was reported to still function well after several weeks. At a bare gold electrode immersed in the same aqueous cytochrome c solution, no electrochemistry was observed.

$$:N\underset{}{\bigcirc}-S-S-\underset{}{\bigcirc}N:$$

(1)

The work of Eddowes and Hill and Taniguchi and co-workers opened up an extremely exciting prospect, the more as it appeared that the technique of surface modification could be extended to enzymes other than cytochrome c. However, to exploit this technology fully the mode of action of these 'surface mediators' or 'promoters' would have to be better understood, if only to enable the identification of other such species capable of promoting the electrochemistry of as wide a range of redox enzymes as possible. One of the first attempts to achieve such an understanding came with a detailed study by Alberry, Hill and colleagues (1981) who employed the RDE to study the promotion of cytochrome c electrochemistry by 4,4′ bipyridine. On the basis of their results the authors concluded that a well-defined sequence of events must take place if the promoter is to fulfil its function, i.e.:

1. Diffusion of the enzyme to the electrode.
2. Association of the enzyme with the adsorbed promoter to give the transient promoter: enzyme complex in the correct orientation for electron transfer.
3. Electron transfer.
4. Dissociation of enzyme.
5. Diffusion of enzyme away from electrode.

In the original paper, Eddowes and Hill (1977) had reported that, in addition to 4,4′ bipyridine, 1,2 bis(4-pyridyl) ethylene (**2**) was a more effective promoter of cytochrome c electrochemistry than the 4,4′ bipyridine and SSBipy was better than both. This lead Allen and colleagues (1984) to postulate that the prerequirement for successful promoter activity was bifunctionality of the form:

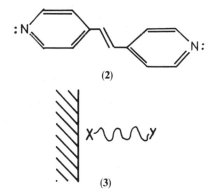

(2)

(3)

where X is a surface active group that binds to the electrode surface and Y is a functional group required to interact with the enzyme. Allen, Hill and Walton (1984) then tested 54 bifunctional organic compounds for their ability to promote the electrochemistry of cytochrome c via adsorption at a gold electrode. These compounds had X=HS–, RS–, RSS–, 3-thiophenyl, PR_3, 4-pyridyl or R_2N–, and Y=4-pyridyl, $-NH_2$, $-NR_2$, –COOH, $-SO_3H$, $-PO_3H$ or –OH. The central linking groups were aliphatic, aromatic or absent (i.e. the two groups were connected directly).

They found that they could assign the compounds tested to four main classes depending on the observed electrochemistry of cytochrome c (see Figure 3.89).

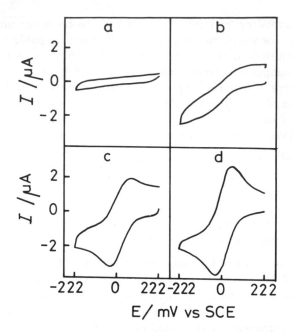

Figure 3.89 Cyclic voltammograms of 500 μm cytochrome *c* at a gold electrode modified by (a) 2 mercaptopyridine, (b) 2-mercaptosuccinic acid, (c) 4,4'-dithiobis(butanoic acid), (d) 4-mercaptoaniline. pH 7.0 phosphate buffer + 0.1 M NaClO$_4$. Scan rate 50 mV s^{-1}. From Allen *et al.* (1984).

Type I (see Figure 3.89(a)) showed no promoter activity.

Type II (see Figure 3.89(b)) yielded extremely poor voltammetric responses ($k^0 \leqslant 0.1 \times 10^{-3}$ cm s^{-1}) with the activity quickly declining to zero.

Type III (see Figure 3.89(c)) promoted quasi-reversible electrochemistry (0.1×10^{-3} cm s$^{-1} < k^0 < 1 \times 10^{-3}$ cm s^{-1}) with no loss of electroactivity.

Type IV (see Figure 3.89(d)) promoted quasi-reversible, stable electrochemistry with $k^0 \geqslant 1 \times 10^{-3}$ cm s^{-1}.

The compounds of type IV are depicted in Figure 3.90 along with their surface conformations as postulated by the authors.

The fundamental working hypothesis regarding the function of these promoters employed by the authors was that the surface of the electrode is modified by an adsorbed layer of the promoter, which then:

1. provides functional groups to which the protein can bind transiently;
2. orients the protein on binding such that the redox-active site (the prosthetic group) is disposed towards the electrode;
3. facilitates exchange with solution enzyme, because the binding is reversible, and hence continuous turnover of the enzyme can occur.

X-ray crystallography has shown that the haem prosthetic group is located in the protein with one edge of the haem ring exposed at the surface

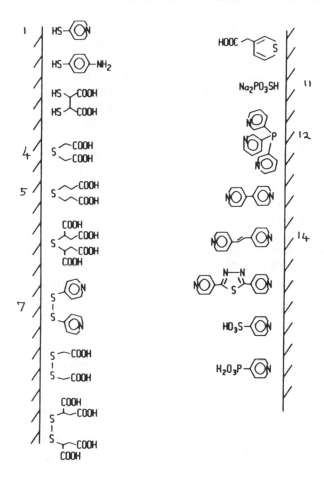

Figure 3.90   Structures and proposed surface conformations of the class IV modifiers (for details see text). From Allen *et al.* (1984).

in a cleft. This exposed edge occupies $<0.06\%$ of the total protein surface area. The cleft is surrounded by a number of positively charged lysine ($-CH_2CH_2CH_2CH_2NH_3^+$) residues, the excess of which gives the protein a $+9$ charge at pH 7. The authors remarked that in solution electron transfer between cytochrome *c* and its substrates (the biological species with which the enzyme reacts, e.g. cytochrome *c* oxidase, cytochrome *bc*, etc.) is facilitated via the formation of a transitory enzyme: substrate complex. This complexation occurs via the lysine residues on the cytochrome *c* and the complementary negatively charged domains on the substrate ($-CH_2CH_2COO^-$, $-CH_2COO^-$). Cytochrome *c* is also known to bind to phospholipid membranes. The importance of the lysine residues on the cytochrome *c* was later confirmed by Hill and colleagues (1985). They chemically modified the enzyme such that the majority of the lysine residues were carboxydinitro-phenylated (i.e. converted to negatively charged residues). The modified enzyme showed on electrochemistry at an SSbipy/Au electrode.

The above reasoning explained the activity of those promoters having carboxylate or phosphate groups as the Y functionality. The authors interpreted the activity of the pyridyl and amino compounds in terms of hydrogen bonding between these species and the lysine residues. This hypothesis is supported by the observation that the electroactivity of cytochrome c at an SSBipy-coated electrode decreases to zero as the pH is lowered, the pK_a of the pyridyl nitrogen on the promoter being 5.1. The hydrogen binding between the lysine residues and the pyridyl N on the thiopyridine fragment stabilises the transient protein: promoter complex and holds it in an orientation that allows rapid electron transfer to the haem group.

Hill and colleagues (1987) noted that the results obtained for SSbipy and 4 mercaptopyridine (PySH) (compound 1 in Figure 3.90) were indistinguishable and thus postulated that the SSBipy chemisorbed via S–S bond cleavage to give the same adsorbed layer as that from the chemisorption of PySH, i.e. a layer of PyS_{ads}.

The results clearly showed the importance of directionality. 4-thiopyridine is an excellent promoter, while the 2 isomer shows no activity. After adsorption the 4 isomer has the pyridine nitrogen direction out into solution while the 2 isomer points the N back towards the electrode where it is available for adsorption to the gold rather than the cytochrome c.

The length of the promoter seems to be of little importance with respect to activity. $HSPO_3^{2-}$, thiopyridine and 1,2 bis(4-pyridyl)ethene (compounds 11, 1 and 14 in Figure 3.90) all show roughly equivalent activities although they are markedly different in length (2.9 Å, 4.6 Å and 9.2 Å respectively).

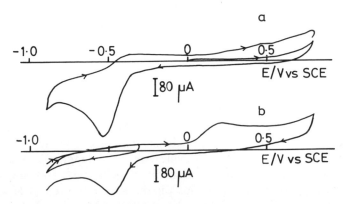

Figure 3.91 Cyclic voltammograms of 1 mM (a) SSBipy and (b) PySH in a phosphate buffer solution with 0.1 M $NaClO_4$ (pH 7.0) at a SERS-activated gold electrode. Scan rate 50 mV s^{-1}. From Taniguchi *et al.* (1982).

Similarly, there is no difference in activity between thiodiethanoic acid and 3,3′ thiobis(propanoic acid) (compounds 4 and 5 in Figure 3.90).

Although identifying the major criteria for a successful promoter, the work of Hill and colleagues (1987) did raise a fundamental question concerning the mode of adsorption of SSBipy. This was addressed directly by Taniguchi *et al.* (1985) who employed SERS to study the adsorption of both SSBipy and PySH at a gold electrode in aqueous buffer at pH 7.

Figures 3.91(a) and (b) show cyclic voltammograms of the SERS electrode in aqueous solutions of the SSBipy and PySH. In the potential range $-0.3\,V$ to $0.3\,V$ vs. SCE, which is the range of interest for the reversible reduction of cytochrome c, no notable faradaic currents were observed for either of these species. However, at potentials $< -0.4\,V$ SSBipy is reduced to PySH and the PySH so formed is re-oxidised to SSBipy at potentials $>0.1\,V$. Similarly, PySH is oxidised to SSBipy at potentials $>0.1\,V$ and this product re-reduced at potentials $< -0.4\,V$.

Figure 3.92 shows SERS spectra of adsorbed SSBipy and PySH at $0\,V$ in the absence of the solution species, together with the Raman spectra of PySH in solution and crystalline SSBipy. The activities of the modified electrodes were first confirmed in solution containing cytochrome c.

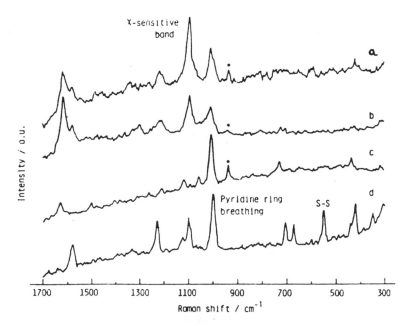

Figure 3.92   SERS spectra of (a) SSBipy-Au and (b) PySH-Au electrodes at $0\,V$ vs. SCE in pH 7.0 phosphate buffer/0.1 M NaClO$_4$, together with Raman spectra of (c) the buffer employed in (a) and (b) saturated with PySH ($c.$ 50 mM). (d) SSBipy in the solid state as a powder. A He/Ne laser (632.8 nm, 30 mW) was used. The signal labelled with an asterisk is due to aqueous ClO$_4^-$. From Taniguchi *et al.* (1982).

Several points are worthy of note:

1. The $550\,cm^{-1}$ absorption present in the spectrum of the solid SSBipy, due to the S–S stretch, is absent in both the SERS spectra.
2. The SERS spectra of the SSBipy and PySH are almost identical.
3. The band at $1120\,cm^{-1}$ in the PySH solution spectrum is the substituent-sensitive band which shifts to $1100\,cm^{-1}$ in the SERS spectra of the two adsorbed species and increases in intensity.

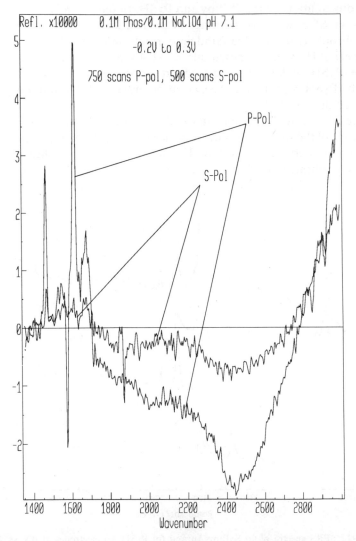

Figure 3.93 Reflectance spectra ($8\,cm^{-1}$ resolution, 750 scans) of SSBipy adsorbed at an Au electrode immersed in N_2-saturated 0.1 M NaH_2PO_4 + 0.1 M $NaClO_4$ + NaOH at pH 7.1. The spectra were collected at $+0.3$ V vs. SCE and normalised to the reference taken at -0.2 V using P- and S-polarised light as indicated on the figure. From Christensen *et al.* (1991).

These observations clearly indicate that both the SSBipy and PySH adsorb to give the same species on the electrode, adsorbed PyS·, via S–S bond cleavage. In addition, the pyridyl ring breathing mode of the two adsorbed species, and that of the solution PySH, appear at the same position ($1010\,\mathrm{cm}^{-1}$) and hence there can be no adsorption through the pyridine ring. This was supported by the fact that no signals attributable to the pyridine ring out-of-plane vibrations were observed in the SERS spectra which would be expected to be enhanced if the ring were lying flat. Thus the authors concluded that the chemisorption of SSBipy and PySH resulted in a PyS fragment adsorbed such that the pyridine ring pointed out into solution.

The mode of adsorption of SSBipy was supported by the work of Christensen *et al.* (1991) using *in situ* FTIR. Figure 3.93 shows reflectance spectra, using both S- and P-polarised light, of a gold electrode modified with SSBipy via adsorption from a solution of the promoter in pH 7 aqueous buffer at 0.3 V vs. SCE, and Figure 3.94 shows a plot of the intensity of the $1610\,\mathrm{cm}^{-1}$ loss feature as a function of potential. The spectra were normalised to the respective reference spectra collected at $-0.2\,\mathrm{V}$ and only the spectral region above $1400\,\mathrm{cm}^{-1}$ is shown, as below this value the spectra are dominated by perchlorate features. It is clear from the absence of any absorptions in the spectrum obtained with S-polarised light, that the bands in Figure 3.93 are due to a surface adsorbed species changing its orientation with potential. By applying the surface selection rule the authors were able to show that for a vertical orientation of adsorbed fragment only vibrational modes 8a and 19a, corresponding to the absorptions near $1610\,\mathrm{cm}^{-1}$ and $1470\,\mathrm{cm}^{-1}$, would be observed in the spectral region under study. If the molecule then showed a potential-dependent tilt towards the electrode surface these absorptions would be lost, as observed. If the pyridine ring tilts towards the surface, such that the plane of the ring is parallel to the surface, then no gain features would be expected as the out-of-plane absorptions are all below $1400\,\mathrm{cm}^{-1}$. If, however, the pyridine ring tilts such that the ring remains perpendicular to the electrode surface:

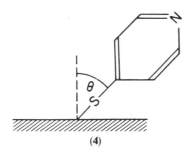

(4)

then we would expect to observe the appearance and subsequent growth in intensity of an absorption near $1580\,\mathrm{cm}^{-1}$ due to the 8b vibration, with a concomitant loss of the $1610\,\mathrm{cm}^{-1}$ and $1470\,\mathrm{cm}^{-1}$ absorptions. This is, of course, what is actually observed.

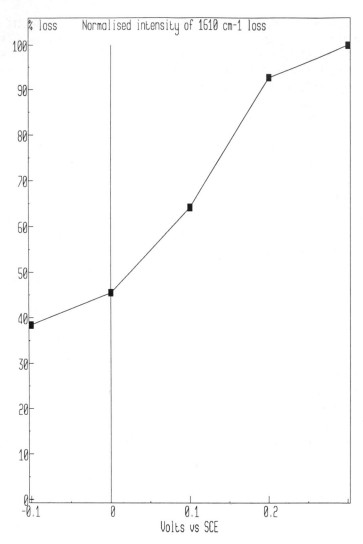

Figure 3.94 Plot of peak intensity vs. potential for the $1610\,cm^{-1}$ loss feature in Figure 3.93. The band intensity was normalised to that at $+0.3\,V$. From Christensen *et al.* (1991).

The next question to be addressed concerns the timescale over which chemisorption of the SSBipy occurred and the degree of adsorption that so results. Some indication of this could be gleaned from the experiments of Hill and colleagues (1987) discussed above wherein immersion of the electrode into SSBipy solution for $\leqslant 5$ minutes appeared to give saturation coverage. However, by the nature of the chemisorption process two PyS fragments result on the surface very close together. Does such chemisorption result in densely packed pairs of PyS fragments or do the SSBipy precursors space themselves out on the surface? The IR data of Christensen and colleagues suggested the latter since the PyS fragment is free to tilt with potential. Considerable light was thrown on these problems by the ellipsometric work of Elliott *et al.* in 1986.

The authors monitored a gold electrode immersed in aqueous buffer at

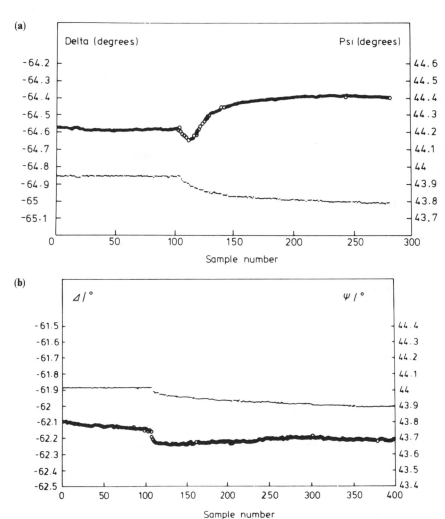

Figure 3.95   (a) The changes in Δ (top) and Ψ (bottom) at a wavelength of 633 nm, accompanying the formation of an adsorbed film of SSBipy on a gold electrode. One sample is equivalent to a time interval of 4 s and addition of the adsorbate occurred at sample 100. A 1 mM SSBipy solution was added from a burette to give a final concentration of 0.2 mM. The potential, initially at -0.2 V vs. MMSE (MMSE $= 0.36$ V vs. SCE), was stepped at sample 250 to 0.0 V. (b) As in (a) except 1 cm³ of 0.7 mM SSBipy was injected through a small plastic tube immediately adjacent to the electrode surface, resulting in an extremely dilute solution *c.* 0.5 μM. The potential, initially at -0.2 V, was stepped to 0.0 V at sample 300 (top, Ψ; bottom, Δ). From Elliott *et al.* (1986).

pH 7 during the addition of SSBipy solution in two experiments (see Figures 3.95(a) and (b)). In (a) 1 mM SSBipy solution was added to the solution in the cell to give a final concentration of 0.2 mM. In (b) 1 cm^3 of 0.7 mM SSBipy solution was injected into the electrolyte immediately adjacent to the electrode resulting, eventually, in an extremely dilute solution, c. 0.5 μM. In both cases a sharp decrease in Δ was observed on addition of the modifier followed by a gradual increase over the course of several minutes, much larger and more rapid at the higher concentration. In both cases, a monotonic decrease in Ψ was observed over the same period.

On the basis of a simple three-phase model (i.e. electrode, monolayer of modifier, solution), the authors' calculations suggested that the initially formed layer of SSBipy on the surface was formed without bond cleavage and with the pyridine rings essentially *flat* on the surface.

The subsequent evolution of the Δ and Ψ values with time was taken, in the light of the SERS results of Taniguchi and colleagues (1985), as being indicative of the chemical transformation of the initially formed physisorbed layer, via the slow dissociation of the S–S bond, to give two upright PyS fragments over a timescale of c. 400 s at the higher promoter concentration. The authors reported that such a transformation would result in a two-fold reduction in coverage leaving further sites available for adsorption. This explained the differences between the data in Figures 3.96(a) and (b). In (a) the high concentration of SSBipy allowed further adsorption on the sites revealed by the chemisorption process to take place rapidly. In contrast, the low concentration of the modifier in (b) ensured that the subsequent adsorption was mass-transport limited.

At this point it was clear how SSBipy was adsorbed on the electrode and the timescale over which this occurred, as well as the role of the concentration of the initial solution. However, the actual mode of action of the adsorbed species still remained somewhat obscure. An important insight into this was provided by the work of Hill *et al.* in 1987 who studied the effect of partial substitution of the layer of adsorbed promoter on the electrochemistry of cytochrome *c*.

The authors prepared a mixed monolayer of thiophenol and SSBipy on a gold electrode. The thiophenol adsorbs via the SH group and is more strongly bound than SSBipy, so irreversibly replaces the latter species on the surface. Adsorbed thiophenol does not have any exposed groups capable of binding to cytochrome *c* and so shows no promoter activity. Hence, the gold electrode was modified fully with SSBipy and the modified electrode was then rinsed and immersed in thiophenol solution for a given time *t*. Increasing the substitution by the thiophenol was achieved by increasing *t*. Figures 3.96(a) to (e) show the electrochemistry of cytochrome *c* at gold electrodes modified with different amounts of thiophenol/SSBipy. Figure 3.96(a) shows the electrochemistry at gold fully modified by SSBipy, i.e. *t* = 0. Figures 3.97(b) to (e) show cyclic voltammograms of cytochrome *c* obtained at gold electrodes which had been exposed to thiophenol solution for successively longer times.

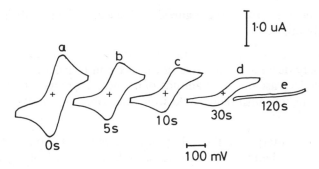

Figure 3.96 The effect of increasing time of exposure (as indicated) of a gold electrode once-modified with SSBipy to thiophenol on the cyclic voltammetry of horse heart cytochrome c (0.4 mM). 20 mM sodium phosphate/0.1 M $NaClO_4$ pH 7.0. Scan rate 20 mV s^{-1}. From Hill *et al.* (1987).

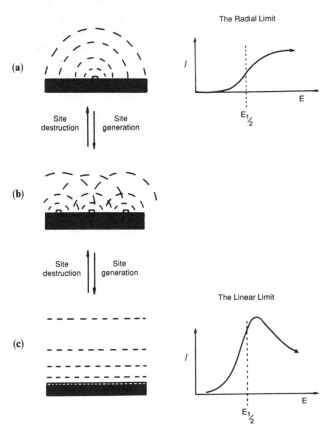

Figure 3.97 Schematic representation of the conversion of radial to linear diffusion, as the density of specific electroactive surface sites increases. From Guo and Hill, *Advances in Inorg. Chem.*, **36** (1991) p. 353.

As can be seen from the figures, two effects of the substitution are immediately apparent:

1. A steady increase in the separation of the current peaks corresponding to a steady decrease in the rate of the electron transfer at the electrode.
2. A steady decrease in the magnitude of the peak currents, with the voltammogram in (e) indistinguishable from that obtained at the unmodified electrode.

The authors interpreted these data in terms of a reduction in the number of active sites as the thiopyridine fragments were replaced *and* by a *decrease in their activity* in terms of a decrease in the heterogeneous rate constant. However, such an interpretation is not consistent with the change in shape of the cyclic voltammograms shown in Figures 3.97(a) to (e). The theoretical treatment originally employed by the authors was based on the assumption that the gold electrode was completely covered by a uniformly active layer of SSBipy, albeit each active site requiring several thiopyridine fragments. In this approach, mass transport to such a uniform 'macroscopic' electrode in linear and the two expected limits of behaviour, including the decreasing effectiveness of the active sites as they lose contributing thiopyridine fragments, are reversible and irreversible electron transfer. Thus (see chapter 2) the anodic and cathodic current peaks should not be lost, they should just move apart and decrease in magnitude. However, Figure 3.96(d) is definitely sigmoidal, clearly indicating that the mass transport is no longer through linear diffusion but via radial diffusion. Bond *et al.* (1990) proposed that the SSBipy adsorbs to give active sites in the form of microscopic islands. Under

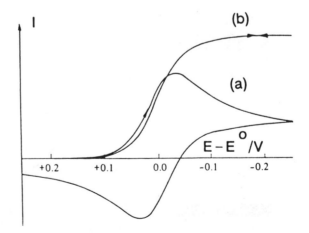

Figure 3.98 Comparison of a reversible conventional cyclic voltammogram (linear diffusion) and reversible steady-state voltammogram obtained at a single microelectrode disc where mass transport is solely by radial diffusion. Current axis not drawn to scale. From A.M. Bond and H.A.O. Hill, *Metal Ions in Biological Systems*, **27** (1991) 431. Reprinted by courtesy of Marcel Dekker Inc.

conditions of high SSbipy coverage these islands are close enough together such that their diffusion layers overlap (Figures 3.97(b) and (c)) and so lead to linear diffusion and the expected peaked cyclic voltammograms. However, as the thiopyridine fragments are replaced by the inactive thiophenol adsorbate, the effective separation of the remaining active islands increases until they effectively resemble an array of microelectrodes (see Figure 3.97(a)). Mass transport to a microelectrode of, for example, 10 μm diameter, is governed by radial diffusion and the current/voltage response is consequently sigmoidal (see Figures 3.98(a) and (b)).

Hence, the macroscopic model of Bond and Hill reinterpreted the data in Figure 3.96 in terms of a cross-over between the limiting forms of mass transport, i.e. linear to radial diffusion, and *not* in terms of a slowing down in the heterogeneous kinetics.

The advancement in our understanding of mediated enzyme electrochemistry since the pioneering work of Hill and colleagues can be easily appreciated when it is realised that a commercial blood glucose sensor, the size of a pen, became commercially available only about a decade later.

References

Albery, W.J., Eddowes, M.J., Hill, H.A.O. and Hillman, A.R. (1981) *J. Am. Chem. Soc.*, **103**, 3904.
Albery, W.J. *et al.* (1989) *Far. Diss.*, **88**, 247.
Allen, P.M., Hill, H.A.O. and Walton, N.J. (1984) *J. Electroanal. Chem.*, **178**, 69.
Angerstein-Kozlowska, H., Conway, B.E. and Sharp, W.B.A. (1973) *J. Electroanal. Chem.*, **43**, 9.
Armand, D. and Clavilier, J. (1987) *J. Electroanal. Chem.*, **255**, 205; (1987) **233**, 251.
Augustynski, J. and Balsenc, L. (1979) in *Modern Aspects of Electrochemistry*, Vol. 13, Bockris, J.O'M. (ed.), p. 313.
Aylmer-Kelly, A.W.B., Bewick, A., Cantrill, P.R. and Tuxford, A.M. (1973) *Far. Diss.*, **56**, 96.
Bagotzky, V.S. and Vassilyev, Yu.B. (1967) *Electrochim. Acta*, **12**, 1323.
Beden, B., Bewick, A., Kunimatsu, K. and Lamy, C. (1981) *J. Electroanal. Chem.*, **121**, 343.
Beigler, T. and Koch, D.F.A. (1967) *J. Electrochem. Soc.*, **114**, 904.
Beigler, T., Rand, D.A.J. and Woods, R. (1971) *J. Electroanal. Chem.*, **29**, 269.
Bewick, A. and Kunimatsu, K. (1980) *Surface Science*, **101**, 131.
Bewick, A. and Russell, J.W. (1982) *J. Electroanal. Chem.*, **132**, 329.
Bewick, A. and Tuxford, A.M. (1970) *Symp. Far. Soc.*, **4**, 114.
Bewick, A., Kunimatsu, K. and Pons, B.S. (1980) *Electrochim. Acta*, **25**, 465.
Bewick, A., Kunimatsu, K., Robinson, J. and Russell, J.W. (1981) *J. Electroanal. Chem.*, **119**, 175.
Bond, A.M., Hill, H.A.O., Page, D.J., Psalti, I.S.M. and Walton, N.J. (1990) *Eur. J. Biochem.*, **191**, 737.
Bowden, F.P. (1928) *Nature*, **122**, 647.
Bowden, F.P. (1929) *Proc. Roy. Soc. A.*, **125**, 446.
Bredas, J.L., Scott, J.C., Yakushi, K. and Street, G.B. (1984) *Phys. Rev. B.*, **30**, 1023.
Breikss, A.I. and Abruna, H.D. (1986) *J. Electroanal. Chem.*, **201**, 347.
Breiter, M.W. (1964) *J. Electroanal. Chem.*, **8**, 449.
Breiter, M.W., Kammermaier, H. and Knorr, C.A. (1956) *Z. Elektrochem.*, **60**, 37.
Cabrera, C.R. and Abruna, H.D. (1986) *J. Electroanal. Chem.*, **209**, 101.
Cahan, B.D. and Chen, C.-T. (1982) *J. Electrochem. Soc.*, **129**, 921.
Campbell, C.T., Ertl, G., Kuipers, H. and Segner, J. (1981) *Surface Sci.*, **107**, 207.
Christensen, P.A. and Hamnett, A. (1991) *Electrochim. Acta*, **36**, 1263.
Christensen, P.A., Hamnett, A., Muir, A.V.G. and Freeman, N.A. (1990) *J. Electroanal. Chem.*, **288**, 197.
Christensen, P.A., Hamnett, A. and Blackham, A. (1991) *J. Electroanal. Chem.*, **318**, 407.
Christensen, P.A., Hamnett, A., Muir, A.V.G. and Jimney, J.A. (1992) *J. Chem. Soc. Dalton Trans.*, **1992**, 1455.

Christensen, P.A., Hamnett, A., and Troughton, G.L. (1993) *J. Electroanal. Chem.*, in press.

Conway, B.E. and Gottesfeld, S. (1973) *J. Chem. Soc. Far. Trans. 1*, **69**, 1090.

Conwell, E. (1985) *Synth. Met.*, **11**, 21.

Dall'Olio, A., Dascola, Y., Varacco, V. and Bocchi, V. (1968) *C. R. Acad. Sci. Ser. C.*, **267**, 433.

Davisson and Germer (1927) *Phys. Rev.*, **30**, 705.

Desilvestro, J. and Pons, S. (1989) *J. Electroanal. Chem.*, **267**, 207.

Devreux, F., Genoud, F., Nechtschein, M. and Villeret, B. (1987) *Synth. Met.*, **18**, 89.

Diaz, A.F., Kazanawa, K.K. and Gardini, G.P. (1979) *J. Chem. Soc. Chem. Comm.*, 635.

Diaz, A.F., Castillo, J.I., Logan, J.A. and Lee, W.-Y. (1981) *J. Electroanal. Chem.*, **129**, 115.

Dickinson, T., Povey, A.F. and Sherwood, P.M.A. (1975) *J. Chem. Soc. Far. Trans. 1*, **71**, 298.

Eddowes, M.J. and Hill, H.A.O. (1977) *J. Chem. Soc. Chem. Comm.*, 771.

Elliot, D., Hamnett, A., Lettington, O.C., Hill, H.A.O. and Walton, N.J. (1986) *J. Electroanal. Chem.*, **202**, 303.

Eucken, A. and Weblus, B. (1951) *Z. Elektrochem.*, **55**, 114.

Feldberg, S.W. (1984) *J. Am. Chem. Soc.*, **106**, 4671.

Feldman, B.J., Burgamayer, P. and Murray, R.W. (1985) *J. Am. Chem. Soc.*, **107**, 872.

Fleischmann, M. and Mao, B.W. (1987) *J. Electroanal. Chem.*, **229**, 125.

Frumkin, A. and Slygin, A. (1934) *Dokl. Akad. Nauk.*, **2**, 173.

Genoud, F., Guglielmi, M., Nechtschein, M., Genies, E. and Salmon, M. (1985) *Phys. Rev. Lett.*, **55**, 118.

Greef, R. (1969) *J. Chem. Phys.*, **51**, 3148.

Gressin, J.C., Michelet, D., Nadjo, L. and Saveant, J.M. (1979) *Nouv. J. de Chim.*, **3**, 345.

Hamnett, A. (1989) *Far. Diss.*, **88**, 292.

Hawecker, J., Lehn, J.-M. and Zeissel, R. (1983) *J. Chem. Soc. Chem. Commun.*, 536.

Hawecker, J., Lehn, J.-M. and Zeissel, R. (1984) *J. Chem. Soc. Chem. Commun.*, 329.

Hawecker, J., Lehn, J.-M. and Zeissel, R. (1986) *Helv. Chim. Acta*, **69**, 1990.

Haynes, L.V. and Sawyer, D.T. (1967) *Anal. Chem.*, **39**, 332.

Heeger, A.J. (1989) *Far. Diss.*, **88**, 203.

Heinze, J., Storzbach, M. and Mortensen, J. (1987) *Ber. Bunsenges. Phys. Chem.*, **91**, 960.

Hill, H.A.O., Page, D.J., Walton, N.J. and Whitford, D. (1985) *J. Electroanal. Chem.*, **187**, 315.

Hill, H.A.O., Page, D.J. and Walton, N.J. (1987) *J. Electroanal. Chem.*, **217**, 141.

Hugot-Le, Goff, A., Flis, J., Boucherit, N., Joiret, S. and Willinski, J. (1990) *J. Electrochem. Soc.*, **137**, 2684.

Inada, R., Shimazu, K. and Kita, H. (1990) *J. Electroanal. Chem.*, **277**, 315.

Iwasita, T. and Nart, F.C. (1991) *J. Electroanal. Chem.*, **317**, 291.

Iwasita, T. and Vielstich, W. (1986) *J. Electroanal. Chem.*, **201**, 403.

Iwasita, T. and Vielstich, W. (1990) in *Advances in Electrochemical Science and Engineering*, Vol. 1, Gerischer, H. and Tobias, C.W. (eds.), VCH Publishers Inc., New York, Chapter 3.

Iwasita, T. Vielstich, W. and Santos, E. (1987) *J. Electroanal. Chem.*, **229**, 367.

Kazanawa, K.K., Diaz, A.F., Geiss, R.H., Gill, W.D., Kwak, J.F., Logan, J.A., Rabolt, J.F. and Street, G.B. (1979) *J. Chem. Soc. Chem. Comm.*, 854.

Kazanawa, K.K., Diaz, A.F., Gill, W.D., Grant, P.M., Street, G.B., Gardini, G.P. and Kwak, J.F. (1980) *Synth. Met.*, **1**, 329.

Kim, K.S., Wonograd, N. and Davies, R.E. (1971) *J. Am. Chem. Soc.*, **93**, 6296.

Kunimatsu, K. (1986) *J. Electroanal. Chem.*, **213**, 149.

Kunimatsu, K. and Kita, H. (1987) *J. Electroanal. Chem.*, **218**, 155.

Lamy, E., Nadjo, L. and Saveant, J.M. (1977) *J. Electroanal. Chem.*, **78**, 403.

Leung, L.-W.H. and Weaver, M.J. (1990) *Langmuir*, **6**, 323.

Lu, J. and Bewick, A. (1989) *J. Electroanal. Chem.*, **270**, 225.

McCabe, R.W. and Schmidt, L.D. (1977) *Surface Sci.*, **86**, 101.

Nechtschein, M., Devreux, F., Genoud, F., Vieil, E., Pernaut, J.-M. and Genies, E. (1986) *Synth. Met.*, **15**, 59.

Nichols, R.J. and Bewick, A. (1988) *J. Electroanal. Chem.*, **243**, 445.

Ord, J.L. and DeSmet, D.J. (1976) *J. Electrochem. Soc.*, **123**, 1976.

O'Toole, T.R., Sullivan, B.P., Bruce, M.R.-M., Margerum, L.D., Murray, R.W. and Meyer, T.J. (1989) *J. Electroanal. Chem.*, **259**, 217.

Parsons, R. and Ritzoulis, G. (1991) *J. Electroanal. Chem.*, **318**, 1.

Pfluger, P., Krounbi, M., Street, G.B. and Weiser, G. (1983) *J. Chem. Phys.*, **76**, 3212.

Pletcher, D. and Solis, V. (1982) *Electrochim. Acta*, **27**, 775.

Reddy, A.K.N., Genshaw, M.A. and Bockris, J.O'M. (1968) *J. Chem. Phys.*, **48**, 671.

Roth, J.D. and Weaver, M.W. (1991) *J. Electroanal. Chem.*, **307**, 119.

Russell, J.W., Overend, J., Scanlon, K., Severson, M. and Bewick, A. (1982) *J. Phys. Chem.*, **86**, 3066.

Schuldiner, S. (1959) *J. Electrochem. Soc.*, **106**, 891.

Scott, J.C., Pfluger, P., Krounbi, M.T. and Street, G.B. (1983) *Phys. Rev. B.*, **28**, 2140.

Shirakawa, H., Louis, E.J., MacDiarmid, A.G. and Chiang, C.K. (1978) *J. Chem. Soc. Chem. Comm.*, 578.

Southworth, P., Hamnett, A., Riley, A.M. and Sykes, J.M. (1988) *Corr. Sci.*, **28**, 1139.

Sullivan, B.P., Bollinger, C.M., Conrad, D., Vining, W.J. and Meyer, T.J. (1985) *J. Chem. Soc. Chem. Commun.*, 1414.

Taniguchi, I., Toyosawa, K., Yamaguchi, H. and Yasukuochi, K. (1982) *J. Electroanal. Chem.*, **140**, 187.

Taniguchi, I., Iseki, M., Yamaguchi, H. and Yasukuochi, K. (1985) *J. Electroanal. Chem.*, **186**, 299.

Tsai, E.W., Jang, G.W. and Rajeshwar, K. (1987) *J. Chem. Soc. Chem. Comm.*, 1776.

Vitus, C.M., Chang, S.-C., Scharot, B.C. and Weaver, M.J. (1991) *J. Phys. Chem.*, **95**, 7559.

Walatka, V.V., Labes, M.M. and Perlstein, J.H. (1973) *Phys. Rev. Lett.*, **31**, 1139.

Waller, A.M. and Compton, R.G. (1989) *J. Chem. Soc. Far. Trans. 1*, **85**, 977.

Wegner, G. and Rühe, J. (1989) *Far. Diss.*, **88**, 333.

Wick, E. and Weblus, B. (1952) *Z. Elektrochem.*, **56**, 169.

Wilhelm, S., Vielstich, W., Buschmann, H.W. and Iwasita, T. (1987a) *J. Electroanal. Chem.*, **229**, 377.

Wilhelm, S., Iwasita, T. and Vielstich, W. (1987b) *J. Electroanal. Chem.*, **238**, 383.

Yakushi, K., Lauchlan, L.J., Clarke, T.C. and Street, G.B. (1983) *J. Chem. Phys.*, **79**, 4774.

Yeh, P. and Kuwana, T. (1977) *Chem. Lett.*, 1145.

Further reading

Bond, A.M. and Hill, H.A.O. (1991) *Metal Ions in Biological Systems*, **27**, 431.

Chandler, G.K. and Pletcher, D. (1985) *Electrochemistry*, **10**, 117.

Collin, J.P. and Sauvage, J.P. (1989) *Coord. Chem. Rev.*, **93**, 245.

DiGleria, K. and Hill, H.A.O. (1992) *Advances in Biosensors*, **2**, 53.

Evans, G.P. (1990) in *Advances in Electrochemical Science and Engineering*, Vol. 1, Gerischer, H. and Tobias, C.W. (eds.), VCH Publishers Inc., New York, Chapter 3.

Feast, W.J. and Friend, R.H. (1990) *J. Mater. Sci.*, **25**, 3796.

Guo, L.-H. and Hill, H.A.O. (1991) *Adv. Inorg. Chem.*, **36**, 341.

Iwasita, T. and Vielstich, W. (1990) in *Advances in Electrochemical Science and Engineering*, Vol. 1, Gerischer, H. and Tobias, C.W. (eds.), VCH Publishers Inc., New York, Chapter 3.

Kruger, J. (1989) *Corr. Sci.*, **29**, 149.

Linford (ed.) (1987) *Electrochemical Science and Technology of Polymers*, Vols. 1 and 2, Elsevier, London.

O'Connell, C., Hommeltoft, S.I. and Eisenberg, R. (1987) in *Carbon Dioxide as a Source of Carbon*, Aresta, M. and Forti, G. (eds.), D. Reidel Publishing Company, Dordrecht, p. 33.

Parsons, R. and VanderNoot, T. (1988) *J. Electroanal. Chem.*, **257**, *Berichte der Bunsen-Gesellschaft fur Physikalische Chemie* (1990) **94**.

Simonet, J. and Rault-Berthelot, J. (1991) *Prog. Solid State Chem.*, **21**, 111.

Skotheim, T.A. (ed.) (1986) *Handbook of Conducting Polymers*, Vols. I and II, Marcel Dekker Inc., New York.

Skotheim, T.A. (1987, 1991) *Electroresponsive Molecular and Polymeric Systems*, Vols. 1 and 2, Marcel Dekker, New York.

Sun, S.G. and Clavilier, J. (1987) *J. Electroanal. Chem.*, **236**, 95, and references therein.

Taniguchi, I. (1989) in *Modern Aspects of Electrochemistry*, Vol. 20, Bockris, J.O'M., Conway, B.E. and White, R.E. (eds.), Plenum, New York, p. 327.

Ullman, M., Aurian-Blajeni, B. and Halmann, M. (1984) *Chemtech*, 235.

Charge conduction in polymeric systems. *Far. Diss.*, **88**.

Index